T0184868

Advanced Materials for Radiation Detection

Krzysztof (Kris) Iniewski
Editor

Advanced Materials
for Radiation Detection

 Springer

Editor
Krzysztof (Kris) Iniewski
Emerging Technologies CMOS Inc.
Port Moody, BC, Canada

ISBN 978-3-030-76463-0   ISBN 978-3-030-76461-6   (eBook)
https://doi.org/10.1007/978-3-030-76461-6

This Springer imprint is published by the registered company Springer Nature Switzerland AG
The registered company address is: Gewerbestrasse 11, 6330 Cham, Switzerland

Contents

About the Editor

Krzysztof (Kris) Iniewski is managing R&D development activities at Redlen Technologies Inc., a detector company based in British Columbia, Canada. During his 15 years at Redlen, he has managed the development of highly integrated CZT detector products in medical imaging and security applications. Prior to Redlen, Kris held various management and academic positions at PMC-Sierra, University of Alberta, SFU, UBC, and University of Toronto.Dr. Iniewski has published over 150+ research papers in international journals and conferences. He holds 25+ international patents granted in the USA, Canada, France, Germany, and Japan. He wrote and edited 75+ books for Wiley, Cambridge University Press, McGraw Hill, CRC Press, and Springer. He is a frequent invited speaker and has consulted for multiple organizations internationally.

Radiation Detection Materials Introduction

Paul Johns

Abstract Radiation detection science is still a field with room for technology growth. Now two decades into the twenty-first century, there is still no perfect sensor material that is practical for all missions. Because of this, research and development efforts on sensor materials and algorithms for radiation detection remains a thriving field. In this chapter we introduce the state of the art in radiation detection, present challenges that detection systems seek to overcome, and discuss the technologies behind the most prominent radiation detection methods. Focusing on scintillating and semiconducting compounds, reviews are presented on established and emerging materials for radiation detection. Discussion starts with scintillating materials, which are among the most common sensors in radiation detection instruments. These materials are optimized for high light output and rapid timing response to provide information that enables the quantification of radiation sources. Semiconducting materials, which currently provide the highest resolution sensors in deployed systems, are likewise reviewed. Strategies employed to produce large volume semiconductors that are stable at room temperature are presented. Physics, materials, and practical difficulties in these technologies each present a challenge that must be overcome in radiation detection missions. Despite this, recent advances in radiation detection science have led to systems with abilities to detect, identify, localize, quantify, and image radiation sources with higher fidelity than has ever before been seen. Continued research and development efforts, presented in this book, show the path forward to a future where radiation detection science becomes a field with a perfect sensor.

P. Johns (✉)
Pacific Northwest National Laboratory Applied Radiation Detection Group, Richland, WA, USA
e-mail: paul.johns@pnnl.gov

© The Author(s), under exclusive license to Springer Nature Switzerland AG 2022   1
K. Iniewski (ed.), *Advanced Materials for Radiation Detection*,
https://doi.org/10.1007/978-3-030-76461-6_1

1 Overview and Challenges in Radiation Detection

Radiation detectors are utilized in missions worldwide by government and commercial entities within areas of energy, research, medicine, military, and homeland security. There are a number of challenges imposed by physics, materials, and practical factors that these detectors need to overcome. To be useful in most missions, a radiation sensor must be able to detect and/or identify radioactive material in conditions where signatures are weak and confounded by interference sources. A detector also needs to avoid giving the "wrong answer," as false positives can have severe consequences in practice. Detectors that have success in their missions do so by utilizing algorithms that allow radioactive materials to be discriminated and quantified. And likewise, these algorithms have success by treating signals from sensor materials with the highest interaction efficiency, best resolution, and lowest noise.

Radiation sensors and detection systems have an impressive history of development over the twentieth century. Minimum detectable quantities of radiation have improved by orders of magnitude from the early 1900s when naked-eye ZnS scintillation panels were used, to the invention of the first photomultiplier tube (PMT)-coupled scintillators in the 1940s, to today where modern laboratory spectrometers can confirm the presence of sub-Bq quantities of radiation in a matter of seconds. Despite the improvements gained in radiation sensors over time, a number of standing challenges remain in the field. Many modern difficulties encountered by radiation detectors relate to the transition of technology from the laboratory to the field. In reality, size, weight, available infrastructure, and most importantly cost restrict what detection systems can be implemented in operational conditions. Because of this, no single sensor material or detection system currently exists that is applicable across all mission spaces. Focused research efforts in improving radiation detection capabilities can be separated into developing new sensor materials, designing smarter algorithms, or integrating better sensors and algorithms into detection systems that exceed the performance of legacy technology.

Discussion herein focuses primarily on the challenges associated with radiation sensor materials, particularly for gamma-ray detection and isotope identification. Research and development efforts in the field of radiation detection science seek to improve the current state of the art and enable better radiation detection, identification, quantification, localization, and imaging. A (non-comprehensive) list of some of the major standing problems in radiation detection science is presented in Table 1. These challenges are separated by the research or development category as driven by materials, algorithm, or overall detection system capability.

2 Design Factors for Radiation Sensor Materials

Most of the sensor materials widely integrated into radiation detectors today are not the compounds in existence with the best available detection efficiency, spectral resolution, or overall performance. Instead, materials are selected and utilized when

Table 1 Examples of standing challenges in the field of radiation detection science

Category	Challenge
Materials	Large-volume ($>10,000$ cm^3) radiation sensors that provide well-defined spectroscopic resolution
Materials	Sensors with moderate-to-high spectroscopic resolution that can be produced at $<\$1$/cm^3
Materials	Scintillator-photodetector combinations with quantum efficiencies that produce spectra at the intrinsic resolution of the scintillator
Materials	Moderate-volume (>25 cm^3) sensors that can operate at or near room temperature and provide sub-1% resolution
Materials	Spectroscopic sensors with material properties that remain constant over a range of realistic operating temperatures (i.e., $-20\,^\circ$C to $+50\,^\circ$C)
Materials	Ultrafast (sub-100 ps) sensors for medical imaging and high-energy particle experiments [1]
Materials	Robust sensor materials that do not degrade over >30-year life spans
Algorithms	Algorithms with step function receiver-operator characteristics (ROC curves), providing perfect classification of true positives and no detection or identification of false positives over relevant operating conditions
Algorithms	Calibration- and stabilization-agnostic identification algorithms
Algorithms	Fusion of radiation data with other telemetry that improves the probability of detection or identification over strictly radiation-based data
Detection systems	Imaging systems with fine pixel resolution ($<$ mm^2) and high contrast ($>$20:1) among material density increments of <0.1 g/cm^3
Detection systems	Sensors that provide adequate spatial resolution in sizes that enable fieldable micro-computed tomography (μCT) systems
Detection systems	Sensors that enable detection and identification at extremely long ($>$1 km) standoff distances
Detection systems	Highly efficient lightweight sensors for mobile and remote-operated missions
Detection systems	Gaseous sampling radiation detection systems with the ability to discriminate from single interactions and localize effluent origins
Detection systems	Systems immune to adversarial actions that could influence data quality
Detection systems	Passive imaging systems for nondestructive analysis (NDA) with fast (<seconds) response times

balance is achieved between practical factors and performance that fits mission needs. Three key indicators can determine whether a sensor material is a good solution for a radiation detector:

1. Is it cheap to produce?
2. Can it be produced in large volume?
3. Is the detector response adequate for the targeted application?

Sensor production cost is dictated primarily by the market price of precursor materials and the energy (and time) necessary to produce the material. Ideal sensor materials are comprised of elements that are abundant and easily accessible in large quantities at low cost. Even with stellar performance, it is difficult to imagine sensor

Table 2 Categories of radiation detection systems and sensor sizes representative of what are typically incorporated into the detection systems

Category	Description	Representative sensor size (cm^3)
Personal radiation detector	Detection systems worn on person that alert wearers to hazardous radiation environments	~5–10
Radioisotope identification devices	Handheld or portable detectors used primarily for isotope identification	~15–25
Backpack radiation detectors	Backpack-held radiation detectors used typically for area scanning, search, and identification	~50[a]
Mobile radiation detectors	Radiation detection systems used for detection and identification in vehicle or aerial platforms	~2000[a]
Radiation portal monitors	Passive detection systems positioned on both sides of a lane to scan pedestrians, vehicles, or cargo	~25,000[a]

[a]Full systems often incorporate multiple sensors

materials comprised of rare and expensive transition metals and lanthanide elements becoming prevalent in radiation detectors. High purity requirements for precursor materials and reagents often add cost to producing a sensor or introduce extended purification steps to the growth process. Low energy requirements for growth are often related to the melting temperature of the compound and how easily it can be grown into a single crystal. Compounds that melt congruently, or can be grown via solution, are more attractive than compounds that require high temperatures or pressures to induce crystallization. Often, other practical factors such as hygroscopicity, mechanical durability, and stability against environmental aging effects are desirable and can influence material production costs. Advanced engineering and design can also mitigate some material challenges. Consider that NaI(Tl), a highly hygroscopic material, is among the most widely produced and utilized sensor materials today.

Production of materials in "large volumes" is entirely dependent on the context of how a sensor will be used. A large-volume crystal for use in a handheld isotope identifier is a completely different scale from a large-volume sensor intended for portal monitoring. Table 2 provides a reference on the sensor sizes commonly employed in different types of detection systems, and how those instruments are used.

Particularly for scintillators, the maximum size in which a sensor can be reliably produced is an indicator of how ready a material is for application in commercial technologies. A key process in the development of any sensor material is improving yield to larger volumes in homogeneous single phases with low defect concentrations. Learning how to grow new compounds in large volumes is often a process of trial and error that can stall the advancement of emerging sensor materials with promising properties. Crystal growth is a challenging process where success is achieved by precisely managing many parameters that are difficult to control: temperature gradient across a boule, solidification speed, solid-liquid interface shape, stress and strain at the solid-liquid interface, prevention of secondary-phase nucleation, and many other factors [2]. Most radiation sensor materials are grown in furnaces through Bridgman or Czochralski techniques that control a solid-liquid growth interface to yield optically translucent or semiconducting single crystals. Due to these challenges, the largest sensors utilized in radiation detectors are not crystalline materials but amorphous plastics, and are typically produced by casting or extrusion methods [3].

The last, and most obvious, factor to consider when selecting a radiation sensor material is to ensure that the detector has a suitable response function for the application. A detector response function is the matrix that allows information from an incident gamma radiation source to be unfolded. Mathematically, an observed spectrum, $O(E)$, is the product between the detector response function, $R(E)$, and the incident gamma-ray energy spectrum, $I(E_0)$, expressed as [4]

$$O(E) = \int_0^\infty R(E, E_0) I(E_0) dE \qquad (1)$$

A sensor material with properties that allow an observed response to be well correlated to incident radiation is well suited for a radiation detector. Intrinsic material properties as well as external design factors such as the scattering environment and energy calibration make up the detector response function. Three important parameters of the detector response function are intrinsic to material properties of the sensor: the efficiency at which radiation interactions occur, the spectral resolution, and the manner in which energy carriers are extracted and translated into observable spectra. Optimizing these three properties in the design of a sensor material likewise optimizes radiation detection and identification capability.

The efficiency with which a material interacts with radiation, often simply called detection efficiency, is proportional to the effective atomic number (Z_{eff}) of a material [5]. Because interaction efficiency scales with Z, high-Z elements are the most well suited (but not mandatory) for radiation detection. More specifically, mass attenuation of gamma radiation is determined by the total photon interaction cross section for a given material, which increases with Z_{eff} and physical density. Good detection efficiency is also more nuanced than just prioritizing high interaction rates. Radiation also needs to interact "the right way" in a sensor to provide information useful for isotope identification. Among the interaction mechanisms that can occur between radiation and matter, the photoelectric effect is prioritized in gamma-ray sensors as it imparts the full energy of the gamma ray in a single interaction. Sensor materials designed for isotope identification from spectral features should therefore be based upon higher Z materials, in which case photopeaks will have more net interactions compared to lower Z alternates.

Spectral resolution is the key property that determines identification ability in a radiation sensor. "Better" resolution comes with lower percent values of the full-width at half-max (FWHM) in a gamma-ray spectrum photopeak, which is broadened by statistical, physical/material, and electronics factors. The absolute limit to the resolution is dictated by the statistical term, which is determined by the number of quanta generated in an interaction. The equations for the theoretical limit to resolution in semiconductors and scintillators are expressed by

$$R_{semiconductor} = 2.35\sqrt{\frac{F}{N_{e/h}}}, \tag{2}$$

and,

$$R_{scintillator} = 2.35\sqrt{\frac{F}{N_{ph}}}. \tag{3}$$

Equation (2) shows that the minimum resolution limit is determined by the number of electron and hole ($e-/h+$) pairs, $N_{e/h}$, generated in an interaction with a semiconductor. Likewise, the minimum achievable resolution in a scintillator, presented in Eq. (3), shows that the resolution limit is determined by the number of scintillation photons generated in an interaction, N_{ph}. The Fano factor, F, is a correction term that accounts for the finite number of energy states that orbital electrons can occupy, which slightly skews the quanta generation from what is expected by Poisson statistics. The relationship between minimum achievable detector resolution and number of information carriers generated is shown in Fig. 1, the key takeaway being that better resolution comes when more information-carrying quanta are generated in an interaction. In this regard, energy resolution is strongly coupled to the bandgap, E_g, of a sensor material. The bandgap is the separation energy between valence bands and conduction bands in the electronic structure of a

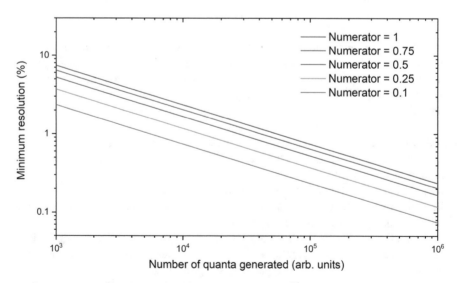

Fig. 1 Minimum photopeak resolution as a function of the number of quanta generated in a radiation interaction. More quanta generated allows for better resolution. Numerator terms refer to the numerator within the square root in Eqs. 2 and 3

material. Smaller E_g values allow for more quanta to be directly generated in semi-conductors, and tailored E_g values in scintillators enable high light yield from fluorescence activation states. Because of this relationship between bandgap and resolution, most semiconductors are tailored to have E_g values between roughly 1 eV and 3 eV, while most scintillators are tailored to have E_g values between approximately 3 and 9 eV [6].

The final material-based consideration in detector response is how efficiently a sensor is able to generate a signal once information carriers (ionized electron/hole pairs or scintillation light) are produced. Different mechanisms come into play during the processes where gamma-ray interactions are converted into interpretable signals in scintillators and semiconductors. In the next sections, we separately discuss these two predominant categories of radiation sensor materials and review the properties and considerations necessary to enable high performance as radiation sensors. Key material properties, development challenges, and promising emerging compounds that seek to displace the state of the art in detection and identification capability are discussed.

3 Scintillator Gamma-Ray Sensor Materials

Commercially available detection and identification technology, from smartphone-sized personal radiation detectors to lane-spanning portal monitors, predominantly consists of scintillator sensors. Scintillator materials fall into the classes of inorganic crystalline and ceramic compounds, amorphous glass compounds, and organic liquid, crystalline, or plastic materials. Applications where maximum detection efficiency is a requirement benefit more from larger sized scintillators than higher resolution semiconductor sensors. Scintillators provide the largest size scaling capability in radiation sensors because the mean free path of scintillation light in an optically clear medium greatly exceeds the mean drift length of charge carriers in a semiconductor. Because of this, scintillators will always be important radiation detection materials to fill the niche where extremely large-volume detection systems ($\geq 100,000$ cm^3) are required. Developing new scintillator materials that emit light on a very fast timescale (*i.e.*, ps to ns), exhibit no phosphorescence, and feature high scintillation yields (>100,000 ph./MeV) continues to make scintillator research a thriving field within radiation detection science.

Scintillators in general produce fewer quanta in radiation interactions than semiconductors and other direct energy conversion technologies. The amount of energy conversion mechanisms in place to generate a signal in detector introduces a number of stages where energy loss occurs. After energy from a gamma ray is transferred into a scintillator, at least five energy transfer or conversion mechanisms take place before a signal is produced at the output of the photodetector [7, 8]:

1. Conversion of primary ionized electron(s) energy into secondary ionized electron/hole pairs in the lattice
2. Efficiency of ionized electrons or holes entering scintillator centers (intrinsic or activator trap states) in the lattice

3. Quantum efficiency of the scintillation center undergoing a radiative transition and producing a scintillation photon
4. Collection efficiency of the scintillation photon being absorbed at the photodetector vs. being reabsorbed in the lattice
5. The quantum efficiency of which scintillation light is converted to a photoelectron at the photodetector
6. (For photomultiplier tubes) The efficiency of the photocathode ejecting a photoelectron that is collected on the first dynode amplification stage

Considering how many stages are present in this conversion process, the fact that advanced scintillator materials can produce spectra with resolution as low as 2% is quite remarkable. Emerging scintillating materials are continuously developed to provide higher light output, better timing response, better density, and emission photon wavelengths that overlap with the absorption spectra of photodetectors, so that minimum achievable resolution continues to improve. Many specific material properties are pursued for scintillators to have good utility as sensors in radiation detectors. First, a high light yield (photons/MeV) is necessary to drive a low spectral resolution limit. Luminosity can be intrinsic or, more commonly, enabled by activator dopants. Additionally, linear proportionality is highly desired between the light yield response and the incident gamma-ray photon energy. Scintillation spectra must also be emitted at wavelengths that highly overlap with the absorption spectra of photodetectors. Optical properties are then needed to provide complete transparency to light at the wavelength(s) at which scintillation photons are emitted. A large mean free path for emitted scintillation light is enabled by a lack of scintillation photon reabsorption and a refractive index that allows perfect reflection on non-photodetector surfaces. The luminescence of an intrinsic material is often significantly enhanced when optical center "activator" dopants are incorporated into the material. Lattice sites must therefore be available and suitable for the engineered incorporation of scintillation activators. The two most common activators, Eu^{2+} and Ce^{3+}, need lattice sites that can host the 3+ or 2+ valence state to enable scintillation [9]. Finally, a fast decay of trapped charges from optical centers in the bandgap is necessary to reduce dead time and pulse pileup. At a minimum, scintillators target completion of light response within 10μs for sensor applications.

Figure 2 presents an overview of the periodic table showing element selection considerations, based primarily upon the criteria presented by Cherepy et al. [10] and Derenzo et al. [11]. Elements in Fig. 2 that are unshaded are those that can be applied in scintillator design with the potential to grow useful compounds. Elements that are radioactive, have insufficiently low bandgaps, do not form into compounds, do not grow easily, or are optically absorptive are not strong contenders for scintillator materials [10]. Factors such as toxicity and price are also important, but do not rule out promising sensor materials.

Despite the stringent requirements that scintillating compounds must meet to serve as radiation sensors, there are many materials on the market that offer solutions for radiation sensors. In both detection and identification missions over both large and small sensors, perhaps the most ubiquitous of all radiation sensor materials is NaI, a halide scintillator typically doped with ppm levels of Tl to enhance its light

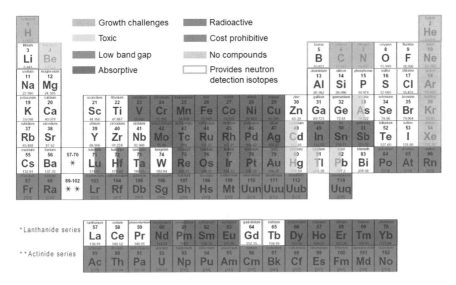

Fig. 2 Periodic table mapping of the elements investigated for scintillating materials, based on the criteria by Cherepy et al. [10] and Derenzo et al. [11]. Compounds that present challenges that inhibit use in scintillating materials are highlighted. Elements that, when enriched, are incorporated into scintillators to enable neutron detection are also highlighted

output. Since its first application as a particle counter in the 1940s, NaI(Tl) has become a prevalent sensor material in the nuclear industry [12, 13]. With a typical resolution of 7–10% at 662 keV, NaI(Tl) sensors provide the radioisotope detection capability in a majority of handheld detectors used in laboratory, environmental monitoring, and homeland security applications [5]. NaI(Tl) is neither impressive performance-wise, as dozens of other scintillators have better light emission and resolution, nor is it particularly robust, as it is notably hygroscopic and prone to fractures or cracking when mechanically or thermally shocked. Despite these drawbacks, NaI(Tl) is inexpensive, is relatively easy to grow in large volumes, and has a scintillation emission spectrum that couples well to bialkali PMTs and Si-based photodiodes. Because of these factors, NaI(Tl) is a workhorse sensor material in the nuclear industry, and the performance benchmark against which emerging sensor materials are most commonly compared. Other commercially available and deployed radiation detectors based on scintillation technology include $LaBr_3$, CsI (Tl), $Bi_4Ge_3O_{12}$ (BGO), and plastic scintillators such as polyvinyltoluene (PVT) and polystyrene (PS) [5, 14].

Comprehensive review articles on emerging scintillator materials over the last decade have been provided by Dorenbos [15], Nikl et al. [16], Martin et al. [17], Dujardin et al. [1], Zhou et al. [18], and Hajagos et al. [19], and the book by Korzhik [20]. An online database comprised of hundreds of scintillators and their properties has also been developed by Derenzo et al. and is hosted at scintillator.lbl.gov [21]. Some key scintillator materials drawn from the forementioned reviews and databases are reported with their key properties in Table 3. The presented

Table 3 Scintillator materials with light output greater than NaI(Tl) and those with historic significance. References to each material are provided along with reported luminosity, peak emission wavelength, and energy resolution at 662 keV

Material	Reference	Luminosity[a] ($\leq$photons/MeV)	Emission peak[b] (nm)	Energy resolution (% FWHM at 662 keV)
$C_{14}H_0$ (anthracene)	[22]	17,000	400–450	n.r.
$BaBr_{1.7}I_{0.3}(Eu)$	[23]	112,000	414	n.r.
$BaBr_2(Eu)$	[23, 24]	58,000	408–414	6.9
$BaBrCl(Eu)$	[23, 25]	59,500	405	4.3
$BaBrI(Eu)$	[23, 26–28]	97,000	413	3.4
$BaCl_2(Eu)$	[24, 29]	52,000	406	3.5
$BaFI(Eu)$	[23]	55,000	405	8.5
$Bi_4Ge_3O_{12}$	[8, 30, 31]	10,600	510	9.1
$CaF_2(Eu)$	[32–34]	25,000	435	6.1
$CaI_2(Eu)$	[10, 35]	110,000	470	5.2
$CeBr_3$	[36, 37]	68,000	371	3.6
$CsBa_2I_5(Eu)$	[26, 28, 38]	102,000	430	2.3
$Cs_4CaI_6(Eu)$	[39]	51,800	474	3.6
$Cs_3Cu_2I_5$	[40]	51,000	450	4.5
Cs_4EuBr_6	[41]	78,000	455	4.3
Cs_4EuI_6	[41]	53,000	465	5.0
Cs_2HfCl_6	[42, 43]	54,000	400	2.8
$Cs_2LiLaBr_6(Ce)$	[44]	60,000	410	2.9
$Cs_2LiLaCl_6(Ce)$	[44]	35,000	400	3.4
$Cs_2LiYCl_6(Ce)$	[44]	20,000	490	3.9
$Cs_2NaGdBr_6(Ce)$	[45]	48,000	418	3.3
$Cs_2NaLaBr_6(Ce)$	[46]	46,000	415	3.9
$Cs_2NaYBr_3I_3(Ce)$	[47]	43,000	350–465	3.3
$Cs_2NaLaBr_3I_3(Ce)$	[47]	58,000	350–460	2.9
$Cs_4SrI_6(Eu)$	[39]	62,300	474	3.3
$CsI(Tl)$	[8, 30]	61,000	540	5.6
$CsSrI_3(Eu)$	[48]	65,000	446–457	5.9
$GdI_3(Ce)$	[44, 49]	89,000	563	8.7
$K_2BaI_4(Eu)$	[50]	63,000	448	2.9
$K_2Ba_2I_5(Eu)$	[50]	90,000	444	2.4
$K_2LaCl_5(Ce)$	[51, 52]	30,000	340–380	5.0
$K_2LaI_5(Ce)$	[51]	55,000	400–440	4.5
$KCaI_3(Eu)$	[53]	72,000	465	3.0
$KSr_2Br_5(Eu)$	[54]	75,000	427	3.5
$KSr_2I_5(Eu)$	[55]	94,000	452	2.4
$LaBr3(Pr)$	[56]	74,500	490–730	3.2
$LaBr_3(Ce)$	[31, 57, 58]	63,000	380	2.6
$LaBr_3(Ce,Sr)$	[59]	78,000	n.r.	2.0
$LiCa_2I_5(Eu)$	[60]	90,000	472	5.6

(continued)

Table 3 (continued)

Material	Reference	Luminosity[a] ($\leq$photons/MeV)	Emission peak[b] (nm)	Energy resolution (% FWHM at 662 keV)
LiGdCl$_4$(Ce)	[61]	64,600	345–365	n.r.
LiSr$_2$I$_5$	[60]	60,000	417–493	3.5
LuI$_3$(Ce)	[62, 63]	115,000	522	3.3
NaI(Tl)	[5]	38,000	415	7.1
RbGd$_2$Br$_7$(Ce)	[64]	55,000	420	3.8
SrI$_2$(Eu)	[65, 66]	120,000	435	2.6
Tl$_2$GdCl$_5$(Ce)	[67]	53,000	389	5.0
TlGd$_2$Cl$_7$(Ce)	[68]	49,700	350–550	6.0
Tl$_2$LaCl$_5$(Ce)	[69, 70]	82,000	383	3.3
TlSr$_2$I$_5$(Eu)	[71]	70,000	463	4.2
Tl$_2$ZrCl$_6$	[72, 73]	50,800	450–470	4.3
TlSr$_2$I$_5$(Eu)	[71]	70,000	463	4.2
YI$_3$(Ce)	[44, 49]	98,600	549	9.3
ZnS(Ag)	[74]	50,000	450	n.r.
ZnSe(O)	[75]	71,500	595	7.4

[a]Peak reported emission, or integrated over all fluorescence components when applicable
[b]Average emission wavelength at room temperature. Emission wavelength can shift with varying activator dopant concentration and temperature

compounds have greater reported luminosity than NaI(Tl) or are historically significant in radiation detection. Luminosity, energy resolution at 662 keV, and emission wavelength are reported. It should be noted that timing response is an equally important property; however, due to wide variations seen across literature on emerging compounds, it is excluded from the table. It is also worth noting that light yield is often extracted by comparison to commercial scintillators such as BGO or NaI(Tl); these methods are prone to error based on the specific performance of the reference sensor.

Figure 3 presents light output (in ph./MeV) against the minimum resolution reported by scintillating compounds in the LBNL database [21]. As expected, the average resolution improves (lower % value) as light output increases up to the range of ~80,000 ph./MeV. Above this point, reported light output continues to improve; however, the minimum reported resolution does not trend as anticipated. These results suggest that there remains room for significant improvements in the minimum achievable resolution in scintillating radiation sensors. In his review, Dorenbos poses that a resolution of 1.5% is theoretically achievable in advanced scintillating materials with light output beyond 100,000 ph./MeV if high quantum efficiency overlap can be achieved between emission wavelength and photodetector absorption [15]. Figure 4 provides a comparison of spectra collected among the commercially available scintillator materials NaI(Tl), CsI(Tl), and LaBr$_3$(Ce) to emerging scintillator compounds with high resolution such as SrI$_2$(Eu), KBa$_2$I$_5$, and CsBa$_2$I$_5$(Eu).

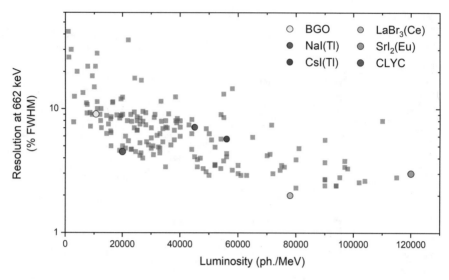

Fig. 3 Luminosity plotted against the reported resolution of compounds from Table 3. Data shows that the minimum reported resolution trends lower as the luminosity increases

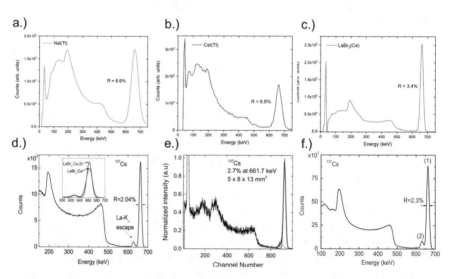

Fig. 4 Comparison of the spectral response to ^{137}Cs among commercially available (top row) and emerging high-performance scintillator materials (bottom row). Spectra from ^{137}Cs are presented as measured by (**a**) NaI(Tl) (commercial spectrometer, 2″ x 2″ sensor), (**b**) CsI(Tl) (commercial spectrometer, 0.5″ x 2″ sensor), (**c**) LaBr$_3$(Ce) (commercial spectrometer, 1.5″ x 1.5″ sensor), (**d**) SrI$_2$(Eu) (from Ref. [76]), (**e**) KBa$_2$I$_5$ (from Ref. [50]), and (**f**) CsBa$_2$I$_5$(Eu) (from Ref. [38])

Another important feature among scintillating materials is their ability to provide signals from the interaction of both gamma and neutron radiation. A number of solid-state neutron-detecting scintillators based on ^{6}Li have seen focused development over the past decade to reduce dependence on ^{3}He in times of market shortages.

Recently, scintillating compounds with the elpasolite crystal structure generated significant interest as dual-gamma and neutron detection sensors [77]. Isotopes with high neutron absorption cross sections, typically ^{6}Li or ^{157}Gd, are integrated into scintillators where pulse shape discrimination methods are then employed to determine if interaction origins come from gamma or neutron radiation [78]. The most prevalent inorganic scintillator with dual-gamma and neutron detection capabilities is the elpasolite material $Cs_2{}^{6}LiYCl_6$(Ce), known as CLYC. Personal radiation detectors comprised of small ~5 cm^3 CLYC crystals have recently entered the market [79]. Apart from CLYC, other dual-gamma and neutron-detecting scintillators include the Ce-doped elpasolites such as $Cs_2{}^{6}LiLaCl_6$ (CLLC), $Cs_2{}^{6}LiLaBr_6$ (CLLB), and $Cs_2{}^{6}LiYBr_6$ (CLYB). Commercially available neutron detectors have also recently emerged based on pulse shape discrimination in common halides such as NaI(Tl), doped with high neutron-interacting isotopes [80]. When NaI(Tl) is co-doped with ^{6}Li, "NaIL" sensors have been shown to have comparable neutron detection efficiency to CLYC without sacrificing much reduction in light yield or resolution [81, 82]. Another commercially available scintillating technology sensitive to neutron radiation is ^{6}Li-based glass fibers. A number of glass or glass ceramic compositions, typically doped with Ce and drawn into bundles of fibers, are applied in neutron detectors and imaging systems [83–85].

A final major class of scintillating materials are plastics and organic scintillators. Plastic compounds used for radiation detection are comprised of a polymer matrix base, typically polyvinyltoluene (PVT) or polystyrene (PS), in which soluble flour compounds are added to promote the generation of light upon radiation interaction. Due to the low wavelengths of emission typically seen in plastic scintillators, additional soluble wavelength-shifting compounds can be added to vary the emission wavelength (typically as a redshift) to wavelengths more easily absorbed by a photodetector [86]. The most widely implemented plastic scintillator is PVT, which serves as the sensor component of radiation portal monitoring (RPM) systems deployed at borders across the world. Economical and large-scale PVT formulations have relatively low light output of ~6500 ph./MeV and do not produce spectra with discernable photopeaks due to the low Z_{eff} of the compound [87]. Despite this, PVT can be produced in large volumes that make it particularly attractive for large-volume detection applications. For context, a typical RPM system can contain ~100,000 cm^3 of PVT sensor volume. Apart from low spectroscopic quality, a key drawback in PVT is the tendency for moisture-induced defects, known as "fogging," to reduce the opacity of the material over years of thermal cycling in operational environments [88, 89].

4   Semiconductor Gamma-Ray Sensor Materials

The second major class of gamma-ray sensors, semiconductor detectors, are direct-energy converters that innately produce higher resolution signals than scintillators. A large number of electron/hole pairs are generated when gamma rays ionize the atoms within semiconductors. The large number of quanta generated per interaction provides good resolution in photopeaks within the resulting gamma-ray spectra, and enables them to identify and characterize sources at farther standoff distances and shorter measurement times than equivalently sized scintillators. Additionally, unlike scintillators, there are not a large number of resolution-reducing conversion processes necessary to produce the signals that make up a spectrum. The resolution of a semiconductor instead depends on the Poisson statistics behind the generation of e^-/ h^+ pairs, material properties intrinsic to the sensor, and the manner in which electronics shape the signal [5].

One of the drawbacks in semiconducting sensor materials is that they are highly sensitive to atomic scale phenomena that quench charges generated in radiation interactions. Due to this, a good response to radiation depends on defect-tolerant crystal structures, thermodynamically stable lattices, chemical and structural stability, and, most importantly, electronic properties that enable stable charge generation and collection across operating temperatures. High material purity is more important in semiconductor sensors than in scintillators; impurity concentrations of just a few billion atoms per cubic centimeter can significantly impact the charge transport of electrons and holes. As ionized charge carriers drift through the lattice after a gamma-ray interaction, they can become trapped at electronic states within the bandgap that arise from the presence of defects or impurities. Optimization of material properties at the atomic scale is therefore crucial for most semiconductor sensors. For a semiconducting compound to perform well as a radiation sensor it must also exhibit electronic properties that enable the detection of ionization-induced charges on the order of fC to pC. The properties that enable this are determined by the electronic structure and the bandgap of the compound: high resistivity (at least 10^8 Ω-cm), electron/hole pair creation energies that approach the value of the bandgap, and charge carrier mobilities and lifetimes that enable optimum charge collection efficiency (mobility-lifetime product of at least 10^{-5} cm^2/V-s). Schlesinger et al. lay out at least six key properties for semiconductors to be capable radiation detectors [90]:

1. Adequate electron density to support high stopping power and interaction rates between sensor matter and gamma radiation
2. A bandgap in the "sweet spot" that is low enough to promote a high number of electron/hole ionization pairs upon interactions, but not also large enough to ensure high resistivity and low leakage current
3. High intrinsic mobility-lifetime product, such that the charge carrier drift length is far larger than the detector thickness
4. Single crystal phases with low impurity concentrations and few intrinsic lattice defects or trap states
5. Compatibility with electrode metals such that space charge buildup and polarization effects are eliminated
6. High surface resistivity to prevent shunt current across detector surfaces and ensuring that electric field lines are not perturbed at surfaces

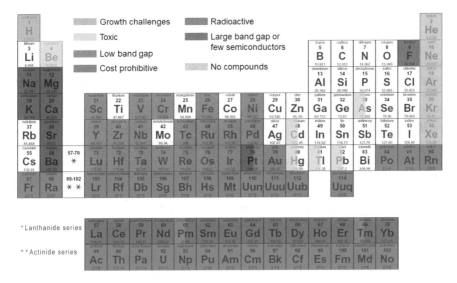

Fig. 5 Periodic table mapping the elements frequently incorporated into semiconductor materials. Compounds that present challenges that inhibit use in semiconducting materials are highlighted

Similar to scintillators, Se, Br, Te, and I are commonly selected as the high-Z anions and are paired with elements that produce semiconducting characteristics when bonded. Like Fig. 2, in Fig. 5 the elements commonly selected (and avoided) for semiconductor gamma-ray sensors are presented. The main considerations in semiconductor element selection are pairs of anions and cations that provide compounds with bandgaps suitable for sensor application, which is typically limited to between 1.5 and 2.5 eV. Many of the same criteria for scintillator element selection apply to semiconductors: elements that are radioactive, have insufficiently high or low bandgaps, do not form into compounds, or do not grow easily. Of note, few lanthanide elements have been shown to be incorporated into feasible semiconducting materials.

The most obvious contenders for semiconducting gamma-ray sensors are the most utilized semiconducting materials worldwide: Si and Ge. Si is one of the most ubiquitous and developed materials in the modern world and is found in nearly every electronic device worldwide. The attractiveness of using Si as a radiation sensor comes from its excellent electronic transport properties, such as its 3.62 eV pair creation energy, electron mobility of 1350 cm^2/V-s, and hole mobility of 480 cm^2/V-s at 300 K [5]. Unfortunately, Si is prevented from serving viably in most gamma radiation detectors due to its moderately low stopping power, and the high leakage currents that stem from its 1.12 eV bandgap at room temperature. Si therefore is most often used in X-ray detection systems, and typically operates at cryogenic temperatures to increase its bandgap and resistivity. Germanium, which has a greater atomic number and density, has instead arisen as one of the most widely used sensors in laboratory gamma-ray detection applications. Reducing impurity concentrations to the order of 10^{10} atoms/cm^3 and cooling the crystal with liquid nitrogen enable Ge detectors to serve as the world's highest resolution semiconductor sensors for gamma-ray spectrometers. Currently, high-purity germanium (HPGe) detectors are the industry performance benchmark for spectroscopy due to HPGe's ability to produce spectra with photopeak resolution below 0.8% over a large range of energies [91].

Due to their intrinsically low 0.67 eV room-temperature bandgap, the Ge crystals in HPGe spectrometers must be cooled to liquid nitrogen temperature (~80 K) in order to operate as a gamma-ray sensor. Without sufficient cooling, thermal phonons in the Ge create too high of an electron density in the conduction bands to distinguish signal from noise. Cooling HPGe spectrometers most often requires liquid nitrogen reservoirs or electromechanical coolers that add weight and power draw to detectors. Because of this, many semiconductors with larger bandgaps have come under investigation as candidate materials for room-temperature semiconductor detectors (RTSDs).

RTSDs remain a R&D focus in detection science because they can generate theoretically higher resolution spectra than comparably sized scintillators at ambient operating temperatures. Many semiconductors with sufficient bandgaps have come under investigation as candidate materials for radiation sensors. Despite the number of requirements for materials to serve effectively as RTSD sensors, new promising compounds are frequently discovered. Within the last decade, review articles on RTSD sensor materials have been written by McGregor et al. [92], Owens and Peacock [93], Luke and Amman [94], Sellin and Vaitkus [95], Zaletin and Varvaritsa [96], and Johns and Nino [97], as well as the books by Owens [98], Awadalla [99], and Iniewski [100]. A compilation of emerging RTSD compounds and their reported properties are presented in Table 4.

The highest technology-readiness-level RTSD compounds are currently CdZnTe (CZT), TlBr, and HgI$_2$. Among these, CdTe and CdZnTe (CZT) are the forefront RTSD sensor materials and currently are the only RTSD compounds found widely among commercial detection systems. CZT in particular is the performance benchmark RTSD compound against which emerging sensor materials are compared. CZT is CdTe alloyed with <15% Zn to increase the bandgap and prevent polarization effects from occurring after long periods under bias [173]. CZT that features a bandgap around 1.57 eV (for 10% Zn doped) is the industry standard RTSD material [174]. With modern charge sensing and depth of interaction correction techniques, it is possible for CZT to exhibit photopeak resolution well below 1% at 662 keV [175–177]. Figure 6 provides a comparison between HPGE and CZT spectra obtained from commercially available spectrometers.

TlBr and HgI$_2$ are other sensor materials widely seen in RTSDs. TlBr's density is among the highest of RTSD candidates at 7.56 g/cm^3, and its bandgap of 2.56 eV maximizes resistivity at the cost of a slightly higher pair creation energy. One of the historical drawbacks of TlBr is polarization that leads to the degeneration of electronic performance after extended periods under bias [178]; however, by engineering the crystals with Tl contacts, this issue has been solved [179]. Today, both pixelated [180] and capacitive Frisch grid detectors [181] have demonstrated a resolution <1%

Table 4 Electronic properties of candidate RTSD materials as reported in literature. Value references taken from Owens [93] along with sources given in column 2. The table is a reproduction of the property table provided in reference [97] with new data provided on compounds that have emerged since the original publication

Material	Refs.	Bandgap (eV)	Resistivity (Ω-cm)	$(\mu\tau)_e$ (cm^2/V)	Spectroscopic response?	Pair creation energy
CdTe	[101, 102]	1.44	1×10^9	1–2×10^{-2}	Yes	4.43
CdZnTe	[103, 104]	1.57	10^9–10^{10}	7.5×10^{-3}	Yes	4.64
HgI$_2$	[98]	2.15	1×10^{13}	3×10^{-4}	Yes	4.20
TlBr	[105, 106]	2.68	1×10^{10}	3×10^{-3}	Yes	6.50
CdMnTe	[107–110]	1.61	3×10^{10}	7×10^{-3}	Yes	2.12
CdSe	[111, 112]	1.73	1×10^{12}	6.3×10^{-5}	Yes	5.50
CdZnSe	[113]	2.00	2×10^{10}	1×10^{-4}	No	6.00
CdTeSe	[114, 115]	1.47	5×10^9	4×10^{-3}	Yes	(4.56)
GaAs	[116]	1.42	1×10^8	8×10^{-5}	Yes	4.20
GaSb	[117]	0.72	n.r.	3×10^{-5}	Yes	(2.51)
GaSe	[118–120]	2.03	10^8–10^{10}	3.7×10^{-5}	Yes	4.49
GaTe	[120]	1.66	1×10^9	n.r.	No	(5.08)
GaN	[96]	3.40	1×10^{11}	1×10^{-4}	No	10.2
GaP	[96]	2.26	1×10^9	1×10^{-5}	No	7.00
InP	[121]	1.34	1×10^9	5×10^{-6}	Yes	4.20
AlSb	[122]	1.65	1×10^8	1.2×10^{-4}	No	4.71
TlHgInS$_3$	[123]	1.74	4×10^9	3.6×10^{-4}	No	(5.30)
Cs$_2$Hg$_6$S$_7$	[124, 125]	1.63	6×10^7	1.7×10^{-3}	Yes	(4.99)
CsHgInS$_3$	[126]	2.30	9×10^{10}	3.6×10^{-5}	No	(6.83)
TlGaSe$_2$	[127]	1.93	1×10^9	6×10^{-5}	No	(5.82)
AgGaSe$_2$	[128]	1.70	1×10^{11}	6.0×10^{-6}	No	(5.19)
TlInSe$_2$	[129–131]	1.10	10^6–10^7	1×10^{-2}	No	3.60
LiInSe$_2$	[132, 133]	2.85	1×10^{11}	3.0×10^{-6}	No	(8.33)
LiGaSe$_2$	[134]	2.80	1×10^8	n.r.	No	(8.19)
CsCdInSe$_3$	[135]	2.40	4×10^9	1.2×10^{-5}	No	(7.10)
Tl$_3$AsSe$_3$	[136]	n.r.	10^6–10^7	n.r.	No	n.r.
Cs$_2$Hg$_3$Se$_4$	[137]	2.1	1.1×10^9	8.0×10^{-4}	No	(6.28)
Pb$_2$P$_2$Se$_6$	[138, 139]	1.88	5×10^{11}	3.1×10^{-4}	Yes	(5.68)
Cu$_2$I$_2$Se$_6$	[140]	1.95	10^{12}	n.r.	No	(5.87)
CsCdInTe$_3$	[135]	1.78	2×10^8	1.1×10^{-4}	No	(5.06)
Cs$_2$Cd$_3$Te$_4$	[137]	2.5	1×10^6	1.1×10^{-4}	No	(7.38)
PbI$_2$	[98]	2.32	1×10^{13}	1×10^{-5}	Yes	4.90
InI	[141, 142]	2.00	1×10^{11}	7×10^{-5}	Yes	(6.01)
SbI$_3$	[143]	2.20	1×10^{10}	n.r.	No	(6.56)
BiI$_3$	[144, 145]	1.67	10^8–10^{11}	1×10^{-4}	Yes	5.80
β-Hg$_3$S$_2$Cl$_2$	[146]	2.56	1×10^{10}	1.4×10^{-4}	No	(7.54)
CsPbBr$_3$	[147]	2.25	10^9–10^{11}	1.7×10^{-3}	Yes	(6.69)
Hg$_3$Se$_2$Br$_2$	[148]	2.22	1×10^{11}	1.4×10^{-4}	Yes	(6.56)

(continued)

Table 4 (continued)

Material	Refs.	Bandgap (eV)	Resistivity (Ω-cm)	$(\mu\tau)_e$ (cm^2/V)	Spectroscopic response?	Pair creation energy
SbSeI	[149]	1.70	1×10^8	4.4×10^{-4}	No	(5.19)
Tl$_4$CdI$_6$	[150]	2.80	2×10^{10}	6.1×10^{-4}	Yes	(8.19)
Tl$_6$SI$_4$	[151]	2.04	1×10^{10}	2.1×10^{-3}	Yes	(6.12)
Tl$_6$SeI$_4$	[152]	1.86	4×10^{12}	7.0×10^{-3}	Yes	(5.63)
Tl$_4$HgI$_6$	[136]	n.r.	10^{11}–10^{12}	8.0×10^{-4}	Yes	n.r.
TlSn$_2$I$_5$	[153]	2.14	4×10^{10}	1.1×10^{-3}	No	(6.39)
Hg$_3$S$_2$I$_2$	[154]	2.25	2×10^{11}	1.6×10^{-6}	No	(6.69)
Hg$_3$Se$_2$I$_2$	[154]	2.12	1.2×10^{12}	1.0×10^{-5}	Yes	(6.34)
Hg$_3$Te$_2$I$_2$	[154]	1.93	3.5×10^{12}	3.3×10^{-6}	No	(5.82)
Rb$_3$Bi$_2$I$_9$	[155]	1.93	3.2×10^{11}	1.7×10^{-6}	No	(5.82)
Rb$_3$Sb$_2$I$_9$	[155]	2.03	8.5×10^{10}	4.5×10^{-6}	No	(6.09)
Cs$_3$Bi$_2$I$_9$	[155]	2.06	9.4×10^{12}	5.4×10^{-5}	No	(6.17)
Cs$_3$Sb$_2$I$_9$	[155]	1.89	5.2×10^{11}	1.1×10^{-5}	No	(5.71)
CdGeAs$_2$	[156]	0.9–1.0	10^7–10^9	n.r.	No	(3.30)
LiZnAs	[157, 158]	1.51	10^6–10^{11}	n.r.	No	(4.67)
LiZnP	[157, 159]	2.04	10^6–10^{11}	n.r.	Yes	(6.12)
4H-SiC	[96, 160]	3.27	2×10^{12}	4×10^{-4}	Yes	7.78
Diamond	[161, 162]	5.47	5×10^{14}	2.7×10^{-7}	No	13.0
PbO	[163]	2.80	1×10^{13}	1×10^{-8}	No	(8.19)
ZnO	[164]	3.30	3×10^{13}	10^{-6}–10^{-4}	No	(8.37)
MAPbI$_3$	[165–167]	1.54	10^8–10^9	1.0×10^{-2}	Yes	(4.75)
FAPbI$_3$	[165, 168, 169]	1.43	10^6–10^9	1.8×10^{-2}	Yes	(4.40)
FACsPb (BrI)$_3$	[169]	1.52	10^8–10^9	1.2×10^{-1}	Yes	(4.69)
MAPb(BrI)$_3$	[165]	n.r.	n.r.	1.0×10^{-5}	Yes	n.r.
MAPbBr$_3$	[170, 171]	2.3	1.7×10^7	1.2×10^{-2}	No	(6.83)
MAPbBr$_3$: Cl	[172]	2.3	3.6×10^9	1.2×10^{-2}	Yes	(6.83)

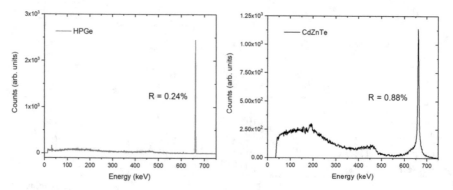

Fig. 6 Examples of sub-1% spectral resolution at 662 keV achieved from commercial HPGe and CZT sensors

at 662 keV. Likewise, HgI_2 has been honed into a quality RTSD gamma-ray sensor. HgI_2 originally proved to be challenging due to the hole ($\mu\tau$) product lagging orders of magnitude behind that of electrons. Modern techniques that allow spectra to be reconstructed from single charge carrier interactions have demonstrated remarkable resolution in HgI_2. Resolution of 1.55% at 662 keV has been demonstrated from collecting charge from pixelated anode electrodes and performing depth correction on interaction signals [182].

CZT, HgI_2, and TlBr each experiences a problem common to RTSD sensors: transport characteristics of one charge carrier (electron or hole) are dominant relative to the other. Equation (4) shows how induced charge, $Q_{induced}$, is generated on an electrode after radiation interaction in a semiconductor. Change in weighing potential, Ψ, experienced by electrons and holes with individual charge q, induces charge on the anode and cathode as carriers travel from the interaction position (x_1) to the anode (x_A) or cathode (x_C). When a single carrier is unable to transit effectively through the semiconductor, the induced charge on the electrode is reduced proportional to the weighing potential change not experienced by the electron or hole:

$$Q_{induced} = \underbrace{q[\psi(x_1) - \psi(x_C)]}_{holes} - \underbrace{\{-q[\psi(x_A - \psi(x_1)]\}_{eletrons}}. \tag{4}$$

To compensate for charge collection loss effects, single polarity charge sensing techniques are commonly used to improve the resolution in RTSD spectrometers. These techniques incorporate an electrode design that allows the majority charge carrier to experience a change in its weighing potential much greater than the minority charge carrier. The most common single polarity charge sensing techniques are pixel electrode grids and Frisch collars [183, 184] that enable the majority carrier to experience the most change in weighing potential. Application of these techniques creates weighing potential distributions where nearly all of the induced charge is due to the majority carrier, thereby improving overall spectral resolution.

5   Current State of the Art in Radiation Detection

The performance of a sensor is judged based on the specific function or role it is used for. Basic functions of radiation sensors fall into two categories: detection and identification. It is important to understand the distinction between detection and identification and how different responses in different missions shape the state of the art. One would not consider a moderate-volume ($\sim$50 cm^3) HPGe sensor to be outstanding in detection missions. Likewise, one would judge PVT as a terrible sensor choice for identification missions. Yet HPGe is among the best available sensor options for identification missions, and PVT the most suitable for detection. The simple definition of a detection mission is using a sensor to alert the presence of radiation levels above background. Likewise, an identification mission consists of using a sensor to identify any isotope(s) present in a radioactive material. The derivation of other useful information, such as source localization, quantification, and imaging, usually complements the detection and identification functions. Unique challenges are faced in these two missions that currently prevent any single radiation sensor material from being used generically in all mission spaces.

In identification missions, current industry standard detection systems based on NaI(Tl) can identify isotopes down to gamma-ray fluence rates between 1 and 0.1 γ/cm^2-s in a span of 1 or 2 min of measurement time [97]. Higher resolution sensors based on Ge semiconductors can identify sources in the same configurations in just a few seconds of measurement time. While Ge-based sensors are powerful identification tools, their cryogenic cooling requirements have driven the development of alternate sensors that can produce high-resolution responses at or near room temperature. While NaI(Tl) and Ge are widely available and found in a variety of detection systems, a number of other materials exist with good resolution, large volume growth capability, and abilities to detect both gamma and neutron radiation. Beyond NaI(Tl), the most common sensor materials found in commercially available identification systems are CsI(Tl)-, LaBr$_3$(Ce)-, CeBr$_3$-, Cs$_2$^{6}LiYCl$_6$(Ce) (CLYC)-, Bi$_4$Ge$_3$O$_{12}$ (BGO)-, Si-, Ge-, and CdTe-based compounds [6]. New compounds with increased performance potential such as TlBr and SrI$_2$(Eu) grow in their technology readiness by the year.

For decades, the highest fidelity isotope identification method has often been a trained spectroscopist's review of a spectrum. Consider the resolution of the various commercially available radiation sensors presented in Fig. 7. Historically,

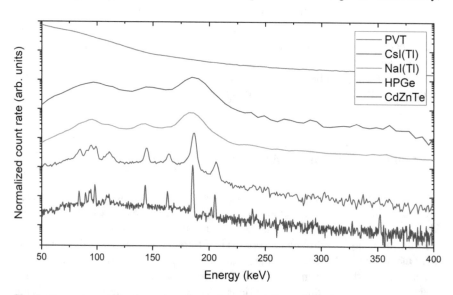

Fig. 7 Comparison of the spectral response of commercially available radiation sensors in measurements of a 93% ^{235}U source. Detectors comprised of NaI(Tl) (5.1 cm Ø × 5.1 cm), CsI (Tl) (5.1 cm × 2.5 cm × 1.3 cm), HPGe (8.5 cm Ø × 3.0 cm), CdZnTe (19 cm^3), and PVT (15,000 cm^3) were sourced from various COTS systems. Counts are vertically offset to arbitrary values

identification algorithms have required well-defined spectral features to provide high-confidence identification results to an operator. At NaI(Tl) resolutions of 7–10%, it can be exceedingly difficult to resolve radiation signatures that are confounded by nearby gamma-ray emission energies, down-scatter and attenuation caused by shielding materials between the radiation source and the detector, and presence of naturally occurring radioactive materials (NORM) in excess of background radiation. Spectral stability also has a major impact on the ability of an algorithm to identify a radiation source. Shifts in stabilization energy or calibration of a few percent caused by temperature variations or gain drift are often enough to prevent the correct identification of isotopes by automated routines or algorithms. Because of these challenges, "reachback support" efforts, where spectroscopists confirm the identity of a radioactive material, are currently (and may always be) an integral part of adjudicating radiation alarms in operational environments.

Although the ways algorithms process spectra to detect, identify, quantify, or localize radioactive materials in commercial radiation detectors are business proprietary information and vary by manufacturer, a few relatively common techniques are used broadly in isotope identification routines. While many dozens of algorithm methods exist in literature, the conventional isotope identification routines employed on spectrometers are types of photopeak "searching," template matching, region-of-interest (ROI) windows, and spectral deconvolution methods [185]. Photopeak "searching" is among the most rudimentary routines and is highly prone to error. The energy positions of local maxima in a spectrum are compared to the photopeak energies produced by various isotopes in a reference library. Peak searching algorithms are complicated when isotopes do not produce well-resolved photopeaks per the resolution of the sensor and when signals are close to background. In template matching, a measured spectrum is compared against a library of template spectra produced by individual or a combination of isotopes. A statistical test on the comparison of the measured spectra and the template provides a confidence value used to determine whether isotopes are present, and if so, what their identity is. ROI windowing is common on sensors with low spectroscopic quality and is performed by comparing ratios of counts within broad windows in energy regions across a spectrum. The ratios of counts within windows across different energy regions are compared to background ratios to determine if a source with emission characteristics apart from background radiation is present. Spectral deconvolution techniques are among the most powerful conventional algorithms and rely on unfolding the incident gamma-ray spectrum from the observed measurement through the detector response function. Deconvolution routines consider spectral contributions from scatter, shielding, background interference, and other factors to build a response matrix that is used to generate an estimation of the incident gamma-radiation spectrum. The incident spectrum is then compared to a library to identify isotopes present.

More recently, advancements in sensor materials, detector designs, and algorithms for detection and identification are enabling radiation sensor and algorithm combinations that are more accurate and reliable than ever before. Novel sensor materials are being identified that can be synthesized through easier growth routes such as solution processing and 3D printing [18, 186]. Perhaps the most interesting materials emerging in the field of radiation sensors are the inorganic and hybrid organic-inorganic perovskite compounds [18, 187–189]. Whereas most scintillator and RTSD compounds are grown from melt-based methods in high-temperature furnaces over periods up to the scale of weeks, the organic-inorganic perovskites are capable of being grown from solution in glassware at near-ambient temperatures. Yield, output rate, and cost per volume of these compounds make them compelling to further investigate for applications as large-volume detectors. If purity and optoelectronic stability can be improved in these compounds, hybrid organic perov-skites could possibly lead the field of radiation sensor research toward more effective radiation detectors in the upcoming years.

Low-resolution sensors equipped with advanced spectral identification algo-rithms are enabling "spectroscopy on a smear" where no prominent photopeaks are present to identify a source. Despite the signal quality in materials like PVT, algorithms now exist that allow isotopes to be identified even when no identifying characteristics are present to the human eye. A large portion of recent identification algorithm research has focused on the application of artificial neutral networks and deep learning to train more accurate isotope detection and identification capabilities [190–193]. Increasingly clever design factors are being employed such as embedded radioactive stabilization sources, LED pulsers, and stabilization routines that use background radiation to keep spectra calibrated and capable of identifying radioac-tive materials. Even in cases where energy calibration of a spectra is off, machine learning algorithms trained to identify and correct data quality are emerging that can detect and identify sources even in the most complex environmental conditions.

6 Conclusion

In summary, new hardware and software are constantly developed for more reliable and higher performing radiation detection systems. Generally, research and devel-opment efforts on radiation sensors drive to enable detection with higher energy resolution and high detection efficiency per unit volume. This typically manifests itself in detection technology in one of the two ways [97]:

1. Providing equal identification capability at smaller volumes as lower resolution spectroscopic detectors
2. Enabling higher identification capability at equivalent volumes as lower resolu-tion spectroscopic detectors

Sensor materials that utilize spectroscopy can, and should, be utilized in detection missions to reject nuisance alarms from benign radiation sources. One of the key challenges in radiation detection is the practical trade-off between sensitivity and selectivity. Compounds that can be produced and utilized in the largest, most sensitive radiation detectors are often the materials with the most limited spectro-scopic capability. An example is the application of large plastic scintillators such as PVT in detection systems used to monitor vehicles and cargo. These sensor materials are often able to detect radiation at interaction count rates much lower than background; however, because Compton scatters are the predominant interaction mechanism, the lack of spectroscopic signatures in energy spectra has historically rendered the largest detectors from easily identifying or categorizing sources. Despite these limitations, modern algorithms continue to emerge that enable selectivity in sensors with poor resolution.

As a final note, while the majority of research and development efforts in radiation detection science focus on improving how systems perform in the detection and identification of radiation, it is equally important to consider that radiation detection systems must be maintainable, sustainable, and easily operated by end users who are not experts in detection science. Amazing capabilities from a sensor material cannot be realized in a system that is not robust or reliable in field operations. A high-resolution spectrum that would excite a spectroscopist most often has no meaning to detection system operators, who are usually first responders or law enforcement personnel that need a simple "yes" or "no" response to the presence of radiation. Exceptionally performing sensors incorporated with burdensome calibration requirements, unoptimized receiver-operator characteristics, or low mean time between failure will not deliver success in operational environments. Many such practical requirements are equal to, or more important than, the performance characteristics of a sensor. Only with these considerations in mind can there be a future where radiation detection science truly becomes a field with a perfect sensor.

References

1. Dujardin, C., et al.: Needs, Trends, and Advances in Inorganic Scintillators. IEEE Trans. Nucl. Sci. **65**(8), 1977–1997 (2018)
2. Dhanaraj, G., et al.: Springer handbook of crystal growth. Springer Science & Business Media, New York (2010)
3. Pla-Dalmau, A., Bross, A.D., Mellott, K.L.: Low-cost extruded plastic scintillator. Nucl. Instrum. Methods Phys. Res., Sect. A. **466**(3), 482–491 (2001)
4. Meng, L.J., Ramsden, D.: An inter-comparison of three spectral-deconvolution algorithms for gamma-ray spectroscopy. IEEE Trans. Nucl. Sci. **47**(4), 1329–1336 (2000)
5. Knoll, G.F.: Radiation Detection and Measurement. John Wiley & Sons, Hoboken (2010)
6. Milbrath, B.D., et al.: Radiation detector materials: An overview. J. Mater. Res. **23**(10), 2561–2581 (2008)
7. Lempicki, A., Wojtowicz, A.J., Berman, E.: Fundamental limits of scintillator performance. Nucl. Instrum. Methods Phys. Res., Sect. A. **333**(2), 304–311 (1993)
8. Moszynski, M., et al.: Absolute light output of scintillators. IEEE Trans. Nucl. Sci. **44**(3), 1052–1061 (1997)
9. Bliss, M., Stave, J.A.: Final Report on Actinide Glass Scintillators for Fast Neutron Detection. Pacific Northwest National Lab.(PNNL), Richland, WA (2012)
10. Cherepy, N.J., et al.: Scintillators with Potential to Supersede Lanthanum Bromide. IEEE Trans. Nucl. Sci. **56**(3), 873–880 (2009)
11. Derenzo, S.E., et al.: Design and Implementation of a Facility for Discovering New Scintillator Materials. IEEE Trans. Nucl. Sci. **55**(3), 1458–1463 (2008)

12. Hofstadter, R.J.P.R.: The detection of gamma-rays with thallium-activated sodium iodide crystals. **75**(5), 796 1949
13. Hofstadter, R.: Alkali halide scintillation counters. Phys. Rev. **74**(1), 100–101 (1948)
14. Siciliano, E.R., et al.: Comparison of PVT and NaI(Tl) scintillators for vehicle portal monitor applications. Nucl. Instrum. Methods Phys. Res., Sect. A. **550**(3), 647–674 (2005)
15. Dorenbos, P.: (INVITED) The quest for high resolution γ-ray scintillators. Opt. Mater. X, **1**, 100021 (2019)
16. Nikl, M., et al.: Development of LuAG-based scintillator crystals – A review. Prog. Cryst. Growth Charact. Mater. **59**(2), 47–72 (2013)
17. Martin, T., Koch, A., Nikl, M.: Scintillator materials for X-ray detectors and beam monitors. MRS Bull. **42**(6), 451–457 (2017)
18. Zhou, Y., et al.: Metal halide perovskites for X-ray imaging scintillators and detectors. ACS Energy Lett. **6**(2), 739–768 (2021)
19. Hajagos, T.J., et al.: High-Z sensitized plastic scintillators: a review. Adv. Mater. **30**(27), 1706956 (2018)
20. Korzhik, M., Gektin, A.: Engineering of scintillation materials and radiation technologies. , Springer, New York (2017)
21. Stephen Derenzo M.B., Weber M., Brennan K.: Scintillation Properties. 2021 November 22, 2019 [cited 2021 March 1]; Available from: http://scintillator.lbl.gov/
22. Galunov, N.Z., Tarasenko, O.A., Tarasov, V.A.: Determination of the light yield of organic scintillators. Funct. Mater. **3**, 305 (2013)
23. Gundiah, G., et al.: Structure and scintillation of Eu2+–activated solid solutions in the BaBr2–BaI2 system. Nucl. Instrum. Methods Phys. Res., Sect. A. **652**(1), 234–237 (2011)
24. Yan, Z., et al.: Eu2+–activated BaCl2, BaBr2 and BaI2 scintillators revisited. Nucl. Instrum. Methods Phys. Res., Sect. A. **735**, 83–87 (2014)
25. Yan, Z., Shalapska, T., Bourret, E.D.: Czochralski growth of the mixed halides BaBrCl and BaBrCl:Eu. J. Cryst. Growth. **435**, 42–45 (2016)
26. Bizarri, G., et al.: Scintillation and Optical Properties of ${\rm BaBrI}\!:\!{\rm Eu}^{2+}$ and ${\rm CsBa}_{2}{\rm I}_{5}\!:\!{\rm Eu}^{2+}$. IEEE Trans. Nucl. Sci. **58**(6), 3403–3410 (2011)
27. Bourret-Courchesne, E.D., et al.: BaBrI:Eu2+, a new bright scintillator. Nucl. Instrum. Methods Phys. Res., Sect. A. **613**(1), 95–97 (2010)
28. Shirwadkar, U., et al.: Promising Alkaline Earth Halide Scintillators for Gamma-Ray Spectroscopy. IEEE Trans. Nucl. Sci. **60**(2), 1011–1015 (2013)
29. Yan, Z., Bizarri, G., Bourret-Courchesne, E.: Scintillation properties of improved 5% Eu2+-doped BaCl2 single crystal for X-ray and γ-ray detection. Nucl. Instrum. Methods Phys. Res., Sect. A. **698**, 7–10 (2013)
30. Sakai, E.: Recent Measurements on Scintillator-Photodetector Systems. IEEE Trans. Nucl. Sci. **34**(1), 418–422 (1987)
31. Haas, J.T.M.D., Dorenbos, P.: Advances in yield calibration of scintillators. IEEE Trans. Nucl. Sci. **55**(3), 1086–1092 (2008)
32. Michail, C., et al.: Absolute luminescence efficiency of europium-doped calcium fluoride (CaF2:Eu) single crystals under X-ray excitation. Crystals. **9**(5), 234 (2019)
33. Holl, I., Lorenz, E., Mageras, G.: A measurement of the light yield of common inorganic scintillators. IEEE Trans. Nucl. Sci. **35**(1), 105–109 (1988)
34. Plettner, C., et al.: CaF2(Eu): an "old" scintillator revisited. J. Instrum. **8**(06), P06010–P06010 (2013)
35. Hofstadter, R., O'Dell, E., Schmidt, C.: CaI2 and CaI2 (Eu) scintillation crystals. Rev. Sci. Instrum. **35**(2), 246–247 (1964)
36. Shah, K.S., et al.: CeBr/sub 3/ scintillators for gamma-ray spectroscopy. IEEE Trans. Nucl. Sci. **52**(6), 3157–3159 (2005)
37. Quarati, F., et al.: Scintillation and detection characteristics of high-sensitivity CeBr3 gamma-ray spectrometers. Nucl. Instrum. Methods Phys. Res., Sect. A. **729**, 596–604 (2013)

38. Alekhin, M.S., et al.: Optical and scintillation properties of CsBa2I5:Eu2+. J. Lumin. **145**, 723–728 (2014)
39. Stand, L., et al.: Crystal growth and scintillation properties of Eu2+ doped Cs4CaI6 and Cs4SrI6. J. Cryst. Growth. **486**, 162–168 (2018)
40. Yuan, D.: Air-Stable Bulk Halide Single-Crystal Scintillator Cs3Cu2I5 by Melt Growth: Intrinsic and Tl Doped with High Light Yield. ACS Appl. Mater. Interfaces. **12**(34), 38333–38340 (2020)
41. Wu, Y., et al.: Zero-dimensional Cs 4 EuX 6 (X= Br, I) all-inorganic perovskite single crystals for gamma-ray spectroscopy. J. Mater. Chem. C. **6**(25), 6647–6655 (2018)
42. Burger, A., et al.: Cesium hafnium chloride: A high light yield, non-hygroscopic cubic crystal scintillator for gamma spectroscopy. Appl. Phys. Lett. **107**(14), 143505 (2015)
43. Ariesanti, E., et al.: Improved growth and scintillation properties of intrinsic, non-hygroscopic scintillator Cs2HfCl6. J. Lumin. **217**, 116784 (2020)
44. Glodo, J., et al.: Selected properties of Cs $_{2}$LiYCl $_{6}$, Cs $_{2}$LiLaCl $_{6}$, and Cs $_{2}$LiLaBr $_{6}$ scintillators. IEEE Trans. Nucl. Sci. **58**(1), 333–338 (2011)
45. Samulon, E.C., et al.: Luminescence and scintillation properties of Ce3+-activated Cs2NaGdCl6, Cs3GdCl6, Cs2NaGdBr6 and Cs3GdBr6. J. Lumin. **153**, 64–72 (2014)
46. Gundiah, G., et al.: Structure and scintillation properties of Ce3+-activated Cs2NaLaCl6, Cs3LaCl6, Cs2NaLaBr6, Cs3LaBr6, Cs2NaLaI6 and Cs3LaI6. J. Lumin. **149**, 374–384 (2014)
47. Wei, H., et al.: Two new cerium-doped mixed-anion elpasolite scintillators: Cs2NaYBr3I3 and Cs2NaLaBr3I3. Opt.Mater. **38**, 154–160 (2014)
48. Yang, K., Zhuravleva, M., Melcher, C.L.: Crystal growth and characterization of CsSr1−xEux I3 high light yield scintillators. Physica Status Solidi (RRL)–Rapid Res. Lett. **5**(1), 43–45 (2011)
49. van Loef, E.V., et al.: Crystal growth and characterization of rare earth iodides for scintillation detection. J. Cryst. Growth. **310**(7), 2090–2093 (2008)
50. Stand, L., et al.: Scintillation properties of Eu2+-doped KBa2I5 and K2BaI4. J. Lumin. **169**, 301–307 (2016)
51. van Loef, E.V.D., et al.: Scintillation properties of K2LaX5:Ce3+ (X=Cl, Br, I). Nucl. Instrum. Methods Phys. Res., Sect. A. **537**(1), 232–236 (2005)
52. van't Spijker, J.C., et al.: Scintillation and luminescence properties of Ce3+ doped K2LaCl5. J. Lumin. **85**(1), 1–10 (1999)
53. Lindsey, A.C., et al.: Crystal growth and characterization of europium doped KCaI3, a high light yield scintillator. Opt. Mater. **48**, 1–6 (2015)
54. Stand, L., et al.: Crystal growth and scintillation properties of potassium strontium bromide. Opt. Mater. **46**, 59–63 (2015)
55. Stand, L., et al.: Potassium strontium iodide: A new high light yield scintillator with 2.4% energy resolution. in *2013 IEEE Nuclear Science Symposium and Medical Imaging Conference (2013 NSS/MIC)* (2013)
56. Glodo, J., et al.: LaBr/sub 3/:Pr/sup 3+/ - a new red-emitting scintillator. in IEEE Nuclear Science Symposium Conference Record, 2005 (2005)
57. Menge, P.R., et al.: Performance of large lanthanum bromide scintillators. Nucl. Instrum. Methods Phys. Res., Sect. A. **579**(1), 6–10 (2007)
58. Rozsa, C.M., P.R. M., Mayhugh M. R.: Scintillation Products technical note lanthanum bromide scintillators performance summary. pp. 0419 (2019). https://www.crystals.saint-gobain.com/sites/imdf.crystals.com/files/documents/labr-performance-summary-2019.pdf
59. Alekhin, M.S., et al.: Improvement of γ-ray energy resolution of LaBr3: Ce3+ scintillation detectors by Sr2+ and Ca2+ co-doping. Appl. Phys. Lett. **102**(16), 161915 (2013)
60. Soundara-Pandian, L., et al.: Lithium Alkaline Halides—Next Generation of Dual Mode Scintillators. IEEE Trans. Nucl. Sci. **63**(2), 490–496 (2016)

61. Porter-Chapman, Y.D., et al.: Scintillation and luminescence properties of undoped and cerium-doped ${\rm LiGdCl}_{4}$ and ${\rm NaGdCl}_{4}$. IEEE Trans. Nucl. Sci. **56**(3), 881–886 (2009)
62. Glodo, J., et al.: Mixed Lutetium Iodide Compounds. IEEE Trans. Nucl. Sci. **55**(3), 1496–1500 (2008)
63. Birowosuto, M.D., et al.: High-light-output scintillator for photodiode readout: Lu I 3: Ce 3+. J. Appl. Phys. **99**(12), 123520 (2006)
64. Dorenbos, P., et al.: Scintillation properties of RbGd2Br7 : Ce3+ crystals; fast, efficient, and high density scintillators. Nucl. Instrum. Methods Phys. Res., Sect. B. **132**(4), 728–731 (1997)
65. Rowe, E., et al.: A New Lanthanide Activator for Iodide Based Scintillators: ${\hbox {Yb}}^{2+}$. IEEE Trans. Nucl. Sci. **60**(2), 1057–1060 (2013)
66. Loef, E.V.V., et al.: Crystal growth and scintillation properties of strontium iodide scintillators. IEEE Trans. Nucl. Sci. **56**(3), 869–872 (2009)
67. Khan, A., et al.: Ce3+-activated Tl2GdCl5: Novel halide scintillator for X-ray and γ-ray detection. J. Alloys Compd. **741**, 878–882 (2018)
68. Khan, A., et al.: Scintillation properties of TlGd2Cl7 (Ce3+) single crystal. IEEE Trans. Nucl. Sci. **65**(8), 2152–2156 (2018)
69. Hawrami, R., et al.: Tl2LaCl5:Ce, high performance scintillator for gamma-ray detectors. Nucl. Instrum. Methods Phys. Res., Sect. A. **869**, 107–109 (2017)
70. Khan, A., et al.: Crystal growth and Ce3+ concentration optimization in Tl2LaCl5: An excellent scintillator for the radiation detection. J. Alloys Compd. **827**, 154366 (2020)
71. Kim, H.J., et al.: Scintillation performance of the TlSr2I5 (Eu2+) single crystal. Opt. Mater. **82**, 7–10 (2018)
72. Fujimoto, Y., et al.: New intrinsic scintillator with large effective atomic number: Tl2HfCl6 and Tl2ZrCl6 crystals for X-ray and gamma-ray detections. Sens. Mater. **30**(7), 1577–1583 (2018)
73. Phan, Q.V., et al.: Tl2ZrCl6 crystal: Efficient scintillator for X- and γ-ray spectroscopies. J. Alloys Compd. **766**, 326–330 (2018)
74. Yehuda-Zada, Y., et al.: Optimization of 6LiF:ZnS(Ag) scintillator light yield using GEANT4. Nucl. Instrum. Methods Phys. Res., Sect. A. **892**, 59–69 (2018)
75. Kim, Y.K., et al.: Properties of semiconductor scintillator ZnSe:O. Nucl. Instrum. Methods Phys. Res., Sect. A. **580**(1), 258–261 (2007)
76. Alekhin, M.S., et al.: Improvement of LaBr3: 5% Ce scintillation properties by Li+, Na+, Mg2+, Ca2+, Sr2+, and Ba2+ co-doping. J. Appl. Phys. **113**(22), 224904 (2013)
77. Glodo, J., et al.: Pulse shape discrimination with selected elpasolite crystals. IEEE Trans. Nucl. Sci. **59**(5), 2328–2333 (2012)
78. Roush, M.L., Wilson, M.A., Hornyak, W.F.: Pulse shape discrimination. Nucl. Inst. Methods. **31**(1), 112–124 (1964)
79. Thermo scientific RadEye SPRD-GN personal radiation detector. Thermo Fisher Scientific (2017). https://www.thermofisher.com/order/catalog/product/4250812#/4250812
80. Brubaker, E., Dibble D., Yang P.: Thermal neutron detection using alkali halide scintillators with 6Li and pulse shape discrimination. in 2011 IEEE Nuclear Science Symposium Conference Record (2011)
81. Yang, K., Menge, P.R., Ouspenski, V.: Li Co-Doped NaI:Tl (NaIL)—A large volume neutron-gamma scintillator with exceptional pulse shape discrimination. IEEE Trans. Nucl. Sci. **64**(8), 2406–2413 (2017)
82. Caifeng, L., et al.: Particle discrimination and fast neutron response for a NaI:Tl and a NaI:Tl scintillator detector. Nucl. Instrum. Methods Phys. Res., Sect. A. **978**, 164372 (2020)
83. Mayer, M., Bliss, M.: Optimization of lithium-glass fibers with lithium depleted coating for neutron detection. Nucl. Instrum. Methods Phys. Res., Sect. A. **930**, 37–41 (2019)
84. Bliss, M., et al.: Glass-fiber-based neutron detectors for high- and low-flux environments. SPIE's 1995 International Symposium on Optical Science, Engineering, and Instrumentation, vol. 2551 (1995). SPIE

85. Moore, M.E., et al.: Neutron imaging with Li-glass based multicore SCIntillating FIber (SCIFI). J. Lightwave Technol. **37**(22), 5699–5706 (2019)
86. Wieczorek, A.: Development of novel plastic scintillators based on polyvinyltoluene for the hybrid J-PET/MR tomograph. arXiv preprint arXiv:1710.08136 (2017)
87. BC-400,BC-404,BC-408,BC-412,BC-416 premium plastic scintillators, S.-G. Crystals, Editor (2018). https://www.crystals.saint-gobain.com/products/bc-408-bc-412-bc-416
88. Rose, P.B., et al.: Onset of fogging and degradation in polyvinyl toluene-based scintillators. IEEE Trans. Nucl. Sci. **67**(7), 1765–1771 (2020)
89. Lance, M.J., et al.: Nature of Moisture-Induced fogging defects in scintillator plastic. Nucl. Instrum. Methods Phys. Res., Sect. A. **954**, 161806 (2020)
90. Schlesinger, T.E., et al.: Cadmium zinc telluride and its use as a nuclear radiation detector material. Mater. Sci. Eng. R Rep. **32**(4), 103–189 (2001)
91. ORTEC, G., Series Coaxial HPGe Detector Product Configuration Guide. AM Technology (Ed.) (2009)
92. McGregor, D.S., Hermon, H.: Nucl. Instrum. Methods Phys. Res., Sect. A. **395**, 101 (1997)
93. Owens, A., Peacock, A.: Compound semiconductor radiation detectors. Nucl. Instrum. Methods Phys. Res., Sect. A. **531**(1), 18–37 (2004)
94. Luke, P.N., Amman, M.: Room-temperature replacement for Ge detectors - Are we there yet? IEEE Trans. Nucl. Sci. **54**(4), 834–842 (2007)
95. Sellin, P.J., Vaitkus, J.: New materials for radiation hard semiconductor detectors. Nucl. Instrum. Methods Phys. Res., Sect. A. **557**(2), 479–489 (2006)
96. Zaletin, V.M., Varvaritsa, V.P.: Wide-bandgap compound semiconductors for X- or gamma-ray detectors. Russ. Microelectron. **40**(8), 543–552 (2011)
97. Johns, P.M., Nino, J.C.: Room temperature semiconductor detectors for nuclear security. J. Appl. Phys. **126**(4), 040902 (2019)
98. Owens, A.: Compound semiconductor radiation detectors. Taylor & Francis, Milton Park (2012)
99. Awadalla, S.: Solid-state radiation detectors: technology and applications. CRC Press, Boca Raton, FL (2015)
100. Iniewski, K.: Semiconductor Radiation Detection Systems. CRC Press, Boca Raton, FL (2010)
101. Takahashi, T., Watanabe, S.: Recent progress in CdTe and CdZnTe detectors. IEEE Trans. Nucl. Sci. **48**(4), 950–959 (2001)
102. Amman, M., et al.: Evaluation of THM-Grown CdZnTe Material for Large-Volume Gamma-Ray Detector Applications. IEEE Trans. Nucl. Sci. **56**(3), 795–799 (2009)
103. Szeles, C., et al.: Fabrication of high-performance CdZnTe quasi-hemispherical gamma-ray CAPture plus detectors. In Hard X-Ray and Gamma-Ray Detector Physics and Penetrating Radiation Systems VIII (Vol. 6319, p. 631909). International Society for Optics and Photonics (2006)
104. Szeles, C., et al.: Development of the high-pressure electro-dynamic gradient crystal-growth technology for semi-insulating CdZnTe growth for radiation detector applications. J. Electr. Mater. **33**(6), 742–751 (2004)
105. Hitomi, K., et al.: Recent development of TlBr gamma-ray detectors. IEEE Trans. Nucl. Sci. **58**(4), 1987–1991 (2011)
106. Kim, H., et al.: Developing larger TlBr detectors detector performance. IEEE Trans. Nucl. Sci. **56**(3), 819–823 (2009)
107. Wu, W., et al.: Traveling heater method growth and characterization of CdMnTe crystals for radiation detectors. Physica Status Solidi (c) **13**(7–9), 408–412 (2016)
108. Cui, Y., et al.: CdMnTe in X-ray and gamma-ray detection: potential applications. No. BNL-81493-2008-CP. Brookhaven National Lab.(BNL), Upton, NY (United States) (2008)
109. Rafiei, R., et al.: High-Purity CdMnTe radiation detectors: a high-resolution spectroscopic evaluation. IEEE Trans. Nucl. Sci. **60**(2), 1450–1456 (2013)

110. Kim, K.H., et al.: Spectroscopic properties of large-volume virtual Frisch-grid CdMnTe detectors. J. Korean Phys. Soc. **66**(11), 1761–1765 (2015)
111. Burger, A., Shilo, I., Schieber, M.: Cadmium selenide: a promising novel room temperature radiation detector. IEEE Trans. Nucl. Sci. **30**(1), 368–370 (1983)
112. Roth, M.: Advantages and limitations of cadmium selenide room temperature gamma ray detectors. Nucl. Instrum. Methods Phys. Res., Sect. A. **283**(2), 291–298 (1989)
113. Kishore, V., et al.: Structural and electrical measurements of CdZnSe composite. Bull. Mater. Sci. **28**(5), 431–436 (2005)
114. Roy, U.N., et al.: Growth of CdTexSe1−x from a Te-rich solution for applications in radiation detection. J. Cryst. Growth. **386**, 43–46 (2014)
115. Roy, U., et al.: Growth and characterization of CdTeSe for room-temperature radiation detector applications. in SPIE Optical Engineering+ Applications (2013). International Society for Optics and Photonics
116. Sze, S.M., Ng, K.K.: Physics of semiconductor devices. John Wiley & Sons, Hoboken (2006)
117. Bor-Chau, J., et al.: Characterization of GaSb photodiode for gamma-ray detection. Appl. Phys. Express. **9**(8), 086401 (2016)
118. Manfredotti, C., Murri, R., Vasanelli, L.: GaSe as nuclear particle detector. Nucl. Inst. Methods. **115**(2), 349–353 (1974)
119. Nakatani, H., et al.: GaSe nuclear particle detectors. Nucl. Instrum. Methods Phys. Res., Sect. A. **283**(2), 303–309 (1989)
120. Mandal, K.C.: et al. GaSe and GaTe anisotropic layered semiconductors for radiation detectors (2007)
121. Zdansky, K., et al.: Evaluation of semi-insulating annealed InP:Ta for radiation detectors. IEEE Trans. Nucl. Sci. **53**(6), 3956–3961 (2006)
122. Yee, J.H., Swierkowski, S.P., Sherohman, J.W.: ALSB as a high-energy photon detector. IEEE Trans. Nucl. Sci. **24**(4), 1962–1967 (1977)
123. Li, H., et al.: TlHgInS3: an indirect-band-gap semiconductor with X-ray photoconductivity response. Chem. Mater. **27**(15), 5417–5424 (2015)
124. Li, H., et al.: Crystal Growth and Characterization of the X-ray and γ-ray Detector Material Cs2Hg6S7. Cryst. Growth Des. **12**(6), 3250–3256 (2012)
125. Li, H., et al.: Investigation of semi-Insulating Cs2Hg6S7 and Cs2Hg6-xCdxS7 alloy for hard radiation detection. Cryst. Growth Des. **14**(11), 5949–5956 (2014)
126. Li, H., et al.: CsHgInS3: a new quaternary semiconductor for γ-ray detection. Chem. Mater. **24**(22), 4434–4441 (2012)
127. Johnsen, S., et al.: Thallium chalcogenide-based wide-band-gap semiconductors: TlGaSe2 for radiation detectors. Chem. Mater. **23**(12), 3120–3128 (2011)
128. Roy, U.N.: et al. Crystal growth, characterization, and fabrication of AgGaSe2 crystals as novel material for room-temperature radiation detectors (2004)
129. Alekseev, I.V.: A neutron semiconductor detector based on TlInSe2. Instrum. Exp. Tech. **51**(3), 331–335 (2008)
130. Kilday, D.G., et al.: Electronic structure of the "chain" chalcogenide ${\mathrm{TlInSe}}_{2}$. Phys. Rev. B. **35**(2), 660–663 (1987)
131. Alekseev, I.: Application of TlInSe 2 crystals for the detection of hard radiation (1993)
132. Tupitsyn, E., et al.: Single crystal of LiInSe2 semiconductor for neutron detector. Appl. Phys. Lett. **101**(20), 202101 (2012)
133. Bell, Z.W., et al.: Neutron detection with LiInSe2. in SPIE Optical Engineering+ Applications (2015). International Society for Optics and Photonics
134. Stowe, A.C., et al.: Crystal growth in LiGaSe2 for semiconductor radiation detection applications. J. Cryst. Growth. **379**, 111–114 (2013)
135. Li, H., et al.: CsCdInQ3 (Q = Se, Te): new photoconductive compounds as potential materials for hard radiation detection. Chem. Mater. **25**(10), 2089–2099 (2013)
136. Kahler, D., et al.: Performance of novel materials for radiation detection: Tl3AsSe3, TlGaSe2, and Tl4HgI6. Nucl. Instrum. Methods Phys. Res., Sect. A. **652**(1), 183–185 (2011)

137. Androulakis, J., et al.: Dimensional reduction: a design tool for new radiation detection materials. Adv. Mater. **23**(36), 4163–4167 (2011)
138. Wang, P.L., et al.: Hard Radiation Detection from the Selenophosphate Pb2P2Se6. Adv. Funct. Mater. **25**(30), 4874–4881 (2015)
139. Wang, P.L., et al.: Refined synthesis and crystal growth of pb2p2se6 for hard radiation detectors. Cryst. Growth Des. **16**(9), 5100–5109 (2016)
140. Lin, W., et al.: Cu2I2Se6: a metal–inorganic framework wide-bandgap semiconductor for photon detection at room temperature. J. Am. Chem. Soc. **140**(5), 1894–1899 (2018)
141. Squillante, M.R., et al.: InI nuclear radiation detectors. in Nuclear Science Symposium and Medical Imaging Conference, 1992., Conference Record of the 1992 IEEE (1992)
142. Onodera, T., Hitomi, K., Shoji, T.: Fabrication of indium iodide X- and gamma-ray detectors. IEEE Trans. Nucl. Sci. **53**(5), 3055–3059 (2006)
143. Onodera, T., Baba, K., Hitomi, K.: Evaluation of antimony Tri-iodide crystals for radiation detectors. Sci. Technol. Nucl. Install. **2018** (2018)
144. Podraza, N.J., et al.: Band gap and structure of single crystal BiI3: Resolving discrepancies in literature. J. Appl. Phys. **114**(3) (2013)
145. Saito, T., et al.: BiI3 single crystal for room-temperature gamma ray detectors. Nucl. Instrum. Methods Phys. Res., Sect. A. **806**, 395–400 (2016)
146. Wibowo, A.C., et al.: An unusual crystal growth method of the chalcohalide semiconductor, β-Hg3S2Cl2: a new candidate for hard radiation detection. Cryst. Growth Des. (2016)
147. Stoumpos, C.C., et al.: Crystal growth of the perovskite semiconductor CsPbBr3: a new material for high-energy radiation detection. Cryst. Growth Des. **13**(7), 2722–2727 (2013)
148. Li, H., et al.: Mercury chalcohalide semiconductor Hg3Se2Br2 for hard radiation detection. Cryst. Growth Des. (2016)
149. Wibowo, A.C., et al.: Photoconductivity in the chalcohalide semiconductor, SbSeI: a new candidate for hard radiation detection. Inorg. Chem. **52**(12), 7045–7050 (2013)
150. Wang, S., et al.: Crystal growth of Tl4CdI6: a wide band gap semiconductor for hard radiation detection. Cryst. Growth Des. **14**(5), 2401–2410 (2014)
151. Nguyen, S.L., et al.: Photoconductivity in Tl6SI4: a novel semiconductor for hard radiation detection. Chem. Mater. **25**(14), 2868–2877 (2013)
152. Johnsen, S., et al.: Thallium chalcohalides for X-ray and γ-ray detection. J. Am. Chem. Soc. **133**(26), 10030–10033 (2011)
153. Lin, W., et al.: TlSn2I5, a robust halide antiperovskite semiconductor for γ-ray detection at room temperature. ACS Photonics. **4**(7), 1805–1813 (2017)
154. He, Y., et al.: Defect antiperovskite compounds Hg3Q2I2 (Q = S, Se, and Te) for room-temperature hard radiation detection. J. Am. Chem. Soc. **139**(23), 7939–7951 (2017)
155. McCall, K.M., et al.: α-Particle Detection and Charge Transport Characteristics in the A3M2I9 Defect Perovskites (A= Cs, Rb; M= Bi, Sb). ACS Photonics. **5**(9), 3748–3762 (2018)
156. Johnson, B.R., et al.: FY07 Annual Report: Amorphous Semiconductors for Gamma Radiation Detection (ASGRAD) (2008) Pacific Northwest National Laboratory
157. Montag, B.W., et al.: Device fabrication, characterization, and thermal neutron detection response of LiZnP and LiZnAs semiconductor devices. Nucl. Instrum. Methods Phys. Res., Sect. A. **836**, 30–36 (2016)
158. Kuriyama, K., Kato, T., Kawada, K.: Optical band gap of the filled tetrahedral semiconductor LiZnAs. Phys. Rev. B. **49**(16), 11452–11455 (1994)
159. Kuriyama, K., Katoh, T.: Optical band gap of the filled tetrahedral semiconductor LiZnP. Phys. Rev. B. **37**(12), 7140 (1988)
160. Nava, F., et al.: Silicon carbide and its use as a radiation detector material. Meas. Sci. Technol. **19**(10), 102001 (2008)
161. Friedl, M.: Diamond detectors for ionizing radiation. Austrian Academy of Sciences, Vienna (1999)
162. Conte, G., et al.: Temporal response of CVD diamond detectors to modulated low energy X-ray beams. Physica Status Solidi (a). **201**(2), 249–252 (2004)

163. Willig, W.R.: Large bandgap mercury and lead compounds for nuclear particle detection. Nucl. Inst. Methods. **101**(1), 23–24 (1972)
164. Li, H., et al.: Chem. Mater. **24**, 4434 (2012)
165. Yakunin, S., et al.: Detection of gamma photons using solution-grown single crystals of hybrid lead halide perovskites. Nat Photon, (2016). advance online publication
166. Mettan, X., et al.: Tuning of the thermoelectric figure of merit of CH3NH3MI3 (M, Pb, Sn) photovoltaic perovskites. J. Phys. Chem. C. **119**(21), 11506–11510 (2015)
167. He, Y., et al.: Resolving the energy of γ-ray photons with MAPbI3 single crystals. ACS Photonics. **5**(10), 4132–4138 (2018)
168. Han, Q., et al.: Single crystal formamidinium lead iodide (FAPbI3): Insight into the structural, optical, and electrical properties. Adv. Mater. **28**(11), 2253–2258 (2016)
169. Nazarenko, O., et al.: Single crystals of caesium formamidinium lead halide perovskites: solution growth and gamma dosimetry. Npg Asia Mater. **9**, e373 (2017)
170. Wei, H., et al.: Sensitive X-ray detectors made of methylammonium lead tribromide perovskite single crystals. Nat. Photonics. **10**(5), 333–339 (2016)
171. Ryu, S., et al.: Voltage output of efficient perovskite solar cells with high open-circuit voltage and fill factor. Energy Environ. Sci. **7**(8), 2614–2618 (2014)
172. Wei, H., et al.: Dopant compensation in alloyed CH 3 NH 3 PbBr 3− x Cl x perovskite single crystals for gamma-ray spectroscopy. Nat. Mater. **16**(8), 826 (2017)
173. Butler, J., Lingren, C., Doty, F.: Cd 1-x Zn x Te gamma ray detectors. IEEE Trans. Nucl. Sci. **39**(4), 605–609 (1992)
174. Del Sordo, S., et al.: Progress in the development of CdTe and CdZnTe semiconductor radiation detectors for astrophysical and medical applications. Sensors. **9**(5), 3491–3526 (2009)
175. Zhang, Q., et al.: Progress in the development of CdZnTe unipolar detectors for different anode geometries and data corrections. Sensors. **13**(2), 2447–2474 (2013)
176. Bolotnikov, A.E., et al.: Use of high-granularity CdZnTe pixelated detectors to correct response non-uniformities caused by defects in crystals. Nucl. Instrum. Methods Phys. Res., Sect. A. **805**, 41–54 (2016)
177. Streicher, M., et al.: A portable 2 x 2 digital 3D CZT imaging spectrometer system. in 2014 IEEE Nuclear Science Symposium and Medical Imaging Conference (NSS/MIC) (2014)
178. Hitomi, K., et al.: Polarization phenomena in TlBr detectors. IEEE Trans. Nucl. Sci. **56**(4), 1859–1862 (2009)
179. Hitomi, K., Shoji, T., Niizeki, Y.: A method for suppressing polarization phenomena in TlBr detectors. Nucl. Instrum. Methods Phys. Res., Sect. A. **585**(1–2, 102), –104 (2008)
180. Kim, H., et al.: Continued development of thallium bromide and related compounds for gamma-ray spectrometers. Nucl. Instrum. Methods Phys. Res., Sect. A. **629**(1), 192–196 (2011)
181. Hitomi, K., et al.: TlBr Capacitive Frisch Grid Detectors. IEEE Trans. Nucl. Sci. **60**(2), 1156–1161 (2013)
182. Baciak, J.E., He, Z.: Comparison of 5 and 10 mm thick HgI2 pixelated γ-ray spectrometers. Nucl. Instrum. Methods Phys. Res., Sect. A. **505**(1–2), 191–194 (2003)
183. Barrett, H.H., Eskin, J.D., Barber, H.B.: Charge transport in arrays of semiconductor gamma-ray detectors. Phys. Rev. Lett. **75**(1), 156–159 (1995)
184. McNeil, W.J., et al.: Single-charge-carrier-type sensing with an insulated Frisch ring CdZnTe semiconductor radiation detector. Appl. Phys. Lett. **84**(11), 1988–1990 (2004)
185. Stinnett, J.B.: Automated isotope identification algorithms for low-resolution gamma spectrometers. University of Illinois at Urbana-Champaign, Champaign, IL (2016)
186. Son, J., et al.: Improved 3D printing plastic scintillator fabrication. J. Korean Phys. Soc. **73**(7), 887–892 (2018)
187. Wei, H., Huang, J.: Halide lead perovskites for ionizing radiation detection. Nat. Commun. **10**(1), 1–12 (2019)

188. Pan, W., Wei, H., Yang, B.: Development of halide perovskite single crystal for radiation detection applications. Front. Chem. **8** (2020)
189. Gao, L., Yan, Q.: Recent advances in lead halide perovskites for radiation detectors. Solar RRL. **4**(2), 1900210 (2020)
190. Koo, B.T., et al.: Development of a radionuclide identification algorithm based on a convolutional neural network for radiation portal monitoring system. Radiat. Phys. Chem. **180**, 109300 (2021)
191. Gomez-Fernandez, M., et al.: Isotope identification using deep learning: An explanation. Nucl. Instrum. Methods Phys. Res., Sect. A. **988**, 164925 (2021)
192. Kamuda, M., Sullivan, C.J.: An automated isotope identification and quantification algorithm for isotope mixtures in low-resolution gamma-ray spectra. Radiat. Phys. Chem. **155**, 281–286 (2019)
193. Daniel, G., et al.: Automatic and real-time identification of radionuclides in gamma-ray spectra: A new method based on convolutional neural network trained with synthetic data set. IEEE Trans. Nucl. Sci. **67**(4), 644–653 (2020)

Inorganic Perovskite CsPbBr$_3$ Gamma-Ray Detector

Lei Pan, Praneeth Kandlakunta, and Lei R. Cao

Abstract Wide-bandgap (WBG) semiconductors offer stability, endurability, and room-temperature operation than narrow-bandgap semiconductors such as Si and Ge detectors. One example of the most developed WBG is the well-known CZT detector. One emerging WBG is lead halide perovskites that show a great potential in X- and gamma-ray detection with a desirable high attenuation coefficient, low leakage current, and a large mobility-lifetime product. The all-inorganic CsPbBr$_3$ perovskite has its own advantages due to the structural stability over the organic-inorganic perovskites. This chapter starts with a brief introduction of WBG for radiation detection, with a focus on gamma rays. We then describe the performance of gamma-ray detectors made of solution-grown CsPbBr$_3$ single crystals. The device structure, fabrications, surface processing, nuclear instrumentation, and eventually performance evaluation are discussed. The reported Cs-137 energy spectrum with FWHM resolution at 5.5–11% @ 662 keV and peak-to-Compton valley ratio at 2 are presented. Low leakage current has been consistently achieved (~2–5nA@-200V), comparable to a commercial CZT detector. Excellent reproducibility and stability are also demonstrated.

1 Basics of a Semiconductor Radiation Detector

1.1 Introduction

The all-inorganic perovskite (cesium lead tribromide, CsPbBr$_3$) has recently emerged as a promising candidate for gamma-ray detector due to its low-cost growth method, lead (Pb) bearing, and premium charge transport properties. This chapter focuses on the effects of material preparation, metal deposition, and device architecture on the energy resolution (ER) of CsPbBr$_3$.

L. Pan · P. Kandlakunta · L. R. Cao (✉)
The Ohio State University, Columbus, OH, USA
e-mail: Pan.707@osu.edu; Kandlakunta.1@osu.edu; cao.152@osu.edu

© The Author(s), under exclusive license to Springer Nature Switzerland AG 2022
K. Iniewski (ed.), *Advanced Materials for Radiation Detection*,
https://doi.org/10.1007/978-3-030-76461-6_2

Although showing the properties of a good solar cell material, a major distinction of gamma-ray detection medium from solar cell material is that the former requires a large detection volume to intercept gamma-ray photons, which further necessitates a much higher crystal quality for the liberated free carriers to travel a long distance without getting trapped. Applying high voltage (up to several thousand volts) is a necessity, however, at the expense of an increased dark current. Materials with high electron density ($N*Z$), which is the product of the atomic density N and the Z number (effective atomic number, Z_{eff}) of the atom, are preferred for a gamma-ray detector, since the detection of gamma rays is essentially the conversion of gamma-ray photons to fast electrons, through three major interaction mechanisms (photoelectric effect, Compton scattering, and pair production). A high-Z property relaxes the requirement on thickness or volume of a gamma-ray detector, although only to a certain extent. $CsPbBr_3$ is a wide-bandgap semiconductor with 2.3 eV bandgap, which makes it less susceptible to the interference from thermal noise, i.e., thermally excited free e-h pairs, and enabling its ability for room-temperature operation. The reduced thermal noise, though shown as a low-leakage current, is at the expense of an increased Fano noise. It simply takes more energy to liberate one electron-hole (e-h) pair in a wide-bandgap semiconductor sensor; thus the total number of free e-h pairs available for charge collection is reduced, when compared to other narrow-bandgap semiconductors such as silicon (Si) or germanium (Ge).

While the statistical nature of the gamma-ray photons and electronic noise of the spectroscopy system extrinsically affect the ER, it is primarily affected by factors intrinsic to the material itself. These include material defects such as those formed by vacancies and interstitials acting as traps, and leakage current, all of which impact the charge collection efficiency, thereby the ER. The most recent work by He et al. has demonstrated an impressive ER of 1.4% with $CsPbBr_3$ grown by high-temperature melt method [1], while the solution growth $CsPbBr_3$ indicated the best achievable resolution of 5.5% [2].

When a charged particle enters the active region of a detector, for example, with parallel planar structure as shown in Fig. 1, it creates a cloud of e-h pairs equivalent to a total charge Q based on the amount of energy deposited and the average e-h pair creation energy of the detector material. The e-h pair creation energy, also known as W-value, may be estimated by the empirical formula, $W = 1.43 + 2 E_g$ (where E_g is the bandgap energy and W and E_g are both in units of eV). For example, a ^{137}Cs gamma photon of 662 keV when fully absorbed by a $CsPbBr_3$ detector ($E_g = 2.3$ eV

Fig. 1 The illustration of a planar detector with charge produced by interaction of radiation with the detection media

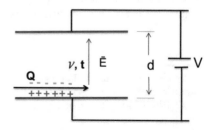

and $W = 6.03$ eV) would produce only about 110,000 e-h pairs, equivalent to merely 0.0176 pC of charge. In all measures, radiation detection can be understood as the acquisition of a small random signal from a detector using a simple model where the detector has a thickness d, and an applied bias V that creates an electric field E, as shown in Fig. 1.

The drifting of these electrons to the positive electrode and holes towards the negative electrode induces the charge movement on the electrode. It is the movement of this charge that sets off the onset of the current flow in the external circuit. By the time the carriers reach the electrode or get lost in the trapping sites, the current flow will be stopped. If there is no blocking effect from metal contacts, carriers will also be injected into the detector to eventually neutralize the free carriers produced by the radiation:

$$\nu = \mu \cdot \bar{E} \tag{1}$$

where ν is the carrier drift velocity in units of cm/s, $\bar{E}$ is the electric field in units of V/cm, and μ is the mobility of hole or electron in units of cm^2/V/s. From this definition in Eq. (1), the mobility is nothing more than a proportionality constant. We could also rearrange to have

$$\tau/t = \nu \cdot \tau/\nu \cdot t = \mu \cdot \tau \cdot \bar{E} \, (\mathbf{cm})/d \, (\mathbf{cm}) \tag{2}$$

where τ is the carrier lifetime in unit of s, $\mu\tau$ has the unit of cm^2/V, $d = \nu\,t$ is the detector thickness, and t is the time for carrier to drift to electrode. From Eq. (2), it is seen that the charge collection is the competition of $\mu\cdot\tau\cdot\bar{E}$, the mean free path of electrons or holes, with the detector thickness. While the E could be controlled manually by adjusting the applied bias voltage and d remaining a constant, the ER performance eventually is dictated by the material property, i.e., the $\mu\tau$ product.

1.2 Inorganic Perovskite CsPbBr3 for Gamma-Ray Detection

1.2.1 Review of Material Properties

Lead halide perovskites are a family of semiconductor materials with molecular formula APbX$_3$ [where A$^+$ = cesium (Cs), methylammonium (MA or CH$_3$NH$_3$), or formamidinium (FA or CH(NH$_2$)$_2$) and X$^-$ = chlorine (Cl), bromine (Br), or iodine (I)] showing attractive properties for gamma-ray detection. The perovskites can be further classified into an organic-inorganic group with organic component (i.e., MA + or FA+ or a mixture of them) as the cations and an all-inorganic group with Cs+ as the cations. Table 1 lists some typical properties of interest for a few room-temperature gamma-ray detectors.

Table 1 Typical properties of interest for some room-temperature semiconductor gamma-ray detectors [1–3, 6, 13–16]

	Constituent elements and atomic numbers	Bandgap (eV)	$\mu\tau$ product (cm^{2}V^{-1})	ER @662 keV
Cd$_{1-x}$Zn$_x$Te	Cd: 48, Zn: 30, Te, 52; $Z_{eff} = 49.9$ when $x = 0.1$	1.5–1.6	0.004–0.01	0.5%
TlBr	Tl:81, Br: 35, $Z_{eff} = 72.6$	2.68	0.0005	1.0%
Organic-inorganic perovskites	H:1, C:6, N:7, Pb:82, Bi:83, Cl:17, Br:35, I: 53, e.g., MAPbBr$_3$: $Z_{eff} = 62$	1.5–3.1	0.001–0.018	6.5% for MAPbBr$_{2.94}$Cl$_{0.06}$
All-inorganic perovskites	Cs:55, Pb:82, Bi:83, Cl:17, Br:35, I: 53, e.g., CsPbBr$_3$: $Z_{eff} = 62$	1.7–2.8	0.0001–0.001	1.4% for melt-growth CsPbBr$_3$ 5.5% for solution-growth CsPbBr$_3$

As the most technologically mature room-temperature gamma-ray detector, cadmium zinc telluride (CZT or Cd$_{1-x}$Zn$_x$Te with x ≈ 0.1) has achieved an ER of less than 1.0% at 662 keV with a signal readout correction algorithm applied [3]. However, the raw material growth issues and the associated high costs with CZT sustain the demand for an alternative detector technology [4]. The thallium bromide (TlBr) detectors have been reported to suffer from ionic conduction-induced polarization phenomena, which leads to the performance deterioration over time under an applied bias [5].

Perovskites have a large effective atomic number (Z_{eff}), a large and tunable bandgap, and high $\mu\tau$ product, which makes them promising for high-ER gamma-ray spectroscopy [6]. Additionally, the perovskites can be grown into a large size from low-cost solution-grown method compared to the high-temperature melting-grown method with a relatively higher initial cost. Among the members of the perovskite family, the organic-inorganic perovskites typically have larger $\mu\tau$ product than the all-inorganic perovskites. However, this advantage is counteracted by their sensitivity to moisture and hysteresis due to the presence of organic cations [7, 8]. Their thermal and thermodynamic instability towards decomposition is problematic too, given the presence of the toxic Pb [9]. On the contrary, the all-inorganic perovskites show a superior long-term stability both chemically and mechanically, while maintaining the desirable electric and photovoltaic properties, which is realized by the replacement of organic cation by its inorganic counterpart [10, 11]. The lower Z_{eff} of CsPbCl$_3$ and the relatively higher instability of CsPbI$_3$ in ambient air make them both inferior detector candidates compared to CsPbBr$_3$ [12].

1.2.2 X-Ray Detection Performance of a Perovskite Detector

Because of the much higher photon flux (>10^5 cm^{-2} s^{-1} with a small X-ray tube) than the sparse gamma-ray photons from a gamma-ray source, the evaluation of

X-ray detection performance can usually serve as a good starting point to a gamma-ray detector's evaluation. Wei et al. reported X-ray detectors made of MAPbBr$_3$ single crystals with a lowest detectable X-ray dose rate of 0.5 μGy$_{air}$s^{-1}, and a sensitivity of 80 μC Gy$_{air}^{-1}$ cm^{-2} that is four times higher than the sensitivity of the α-Se X-ray detectors [17]. MAPbBr$_3$ single crystals integrated onto Si substrates have been reported to produce a sensitivity of 2.1 $\times$ 10^4μC Gy$_{air}^{-1}$ cm^{-2} and a lowest detectable dose rate of 247 nGy$_{air}$ s^{-1} [18]. With dopant compensation, a MAPbBr$_{2.94}$Cl$_{0.06}$ crystal achieved a sensitivity of 8.4 $\times$ 10^4μC Gy$_{air}^{-1}$ cm^{-2} and a lowest detectable dose rate of 7.6 nGy$_{air}$ s^{-1} [15]. An alloyed MAPbI$_3$ X-ray detector, such as GAMAPbI3 (GA=guanidinium) single crystal, has achieved a sensitivity of 2.3 $\times$ 10^4μC Gy$_{air}^{-1}$ cm^{-2} with a lowest detectable dose rate of 16.9 nGy$_{air}$ s^{-1} [19]. A hot-pressed CsPbBr$_3$ quasi-monocrystalline film was reported to have a sensitivity of 5.6 $\times$ 10^4μCGy$_{air}^{-1}$ cm^{-2} and a lowest detectable dose rate of 215 nGy$_{air}$ s^{-1} [20].

1.2.3 Perovskite Gamma-Ray Detector Performance Review

X-ray detectors typically operate in current mode whereas a gamma-ray spectroscopy detector is an event-by-event counting instrument operated in pulse mode with histogram binning of pulse heights. This difference really stems from their respective application thrust. While X-ray imaging demands a high fluence of X-ray photons, the gamma-ray spectrum acquisition is mostly for quantitative analysis of isotopes or elements of interest, often in trace level, or astrophysics, with sparse photons to spare. Resolving the gamma-ray energy spectrum requires the charge generated by a single photon to be collected with a high collection efficiency that poses great challenges right from the synthesis of detector materials up to the device fabrication. Consequently, although various perovskite X-ray detectors have been reported with a wide range of good X-ray sensitivity, only a few have demonstrated discernable peaks in their gamma-ray energy spectra. Wei et al. demonstrated a dopant-compensated CH$_3$NH$_3$PbBr$_{3-x}$Cl$_x$ gamma-ray detector with an ER of 6.5% at 662 keV [15]. A Schottky-type CH$_3$NH$_3$PbI$_3$ gamma-ray detector was reported by He et al. to achieve an ER of 6.8% at 122 keV and 12% at 59.5 keV [14]. He et al. also reported detectors made of high-temperature melt-grown CsPbBr$_3$ single crystal with an ER of 3.9% and 3.8% at 122 keV and 662 keV, respectively [13]. Pan et al. demonstrated solution-grown CsPbBr$_3$ single-crystal perovskite detectors that produced ^{137}Cs energy spectrum with an ER of 5.5% at 662 keV [2]. A recent breakthrough was achieved by He et al. where a melting-growth CsPbBr$_3$ single crystal demonstrated 1.4% ER at 662 keV applying a resolution improvement technique.

2 Signal-to-Noise Ratio: Key to the Performance of a CsPbBr3 Gamma-Ray Detector

A high signal-to-noise ratio (SNR) of the detector signal is indispensable to achieve a good performance from a gamma-ray detector. High SNR implies a high charge collection efficiency for the charge generated by an incident gamma-ray photon, which requires a high schubweg distance $\mu\tau E$ (the mean free path of electron or hole). A CsPbBr$_3$ single crystal with high bulk quality, e.g., low trap/defect densities, is necessary to achieve a high $\mu\tau$ value. A large electric field E further increases the $\mu\tau E$ value. However, as the applied electric field E increases, the leakage current also increases and the ion migration might be an issue too.

2.1 Metal/Perovskite CsPbBr3/Metal Gamma-Ray Detector

The Schottky junction structures have been applied to perovskite radiation detector architecture due to their ease of fabrication that involves a straightforward metal deposition process. The perovskite radiation detectors with Schottky junction typically have a metal/perovskite/metal structure. Depending on the majority charge carrier type of the perovskite single crystal, i.e., n- or p-type, and the metal work function, the metal/perovskite/metal structure can be classified into three categories in terms of their device current-rectifying behavior or three current-voltage (I-V) characteristics: ohmic-ohmic, Schottky-ohmic, and Schottky-Schottky.

2.1.1 The Ohmic-Ohmic Structure

A perovskite radiation detector showing an ohmic-ohmic current-voltage behavior has a metal (low work function)/n-type perovskite/metal (low work function) structure or metal (high work function)/p-type perovskite/metal (high work function) structure. In this case, the potential barrier at the metal/perovskite interface is negligible resulting in a large leakage current through charge injection. No current-rectifying behavior exists in either direction of applied electric field, which is manifested in a linear symmetric I-V curve (Fig. 2). A detector with ohmic-ohmic structure works in resistive mode with its entire volume acting as the active region for radiation detection. A gamma-ray detector working in resistive mode will have a relatively high leakage current at an increased bias voltage due to the lack of current-suppressing behavior, unless the material is ultrapure.

Fig. 2 Qualitative
illustration of I-V curves for
ohmic-ohmic, Schottky-
ohmic, and Schottky-
Schottky structures

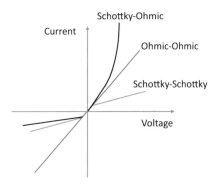

2.1.2   The Schottky-Ohmic Structure

Contrary to the resistive mode, a perovskite gamma-ray detector with Schottky-
ohmic or Schottky-Schottky structure is preferred due to their current-suppressing
behavior. The Schottky junction forms at the interface of metal (low work function)/
p-type perovskite or at the interface of metal (high work function)/n-type perovskite,
while the other side of the detector is made into ohmic. The Schottky-ohmic
structure exhibits an asymmetric I-V curve, indicating its current-rectifying behavior
(Fig. 2). Reverse bias voltage needs to be applied with electric field direction from
n-type perovskite to the high-work-function metal (or from low-work-function metal
to p-type perovskite). A depletion region is formed at the Schottky junction side,
which serves as the active volume for gamma-ray detection. The charge carriers (i.e.,
e-h pairs) generated inside the depletion region by the incident gamma-ray photons
would drift towards the electrodes, which induces charge at the electrode that can be
sensed by the external circuit. Outside the depletion region is the charge-neutral
region that is essentially inactive for gamma-ray detection. This is because the excess
e-h pairs created in this region by incident gamma-ray photons recombine or get
trapped due to the absence of an electric field, and hence do not drift in the electric
field and are not collected by the electrode, unless the charge carrier lifetimes are
sufficiently long to allow carriers at a diffusion length's distance from the depletion
region to diffuse towards it.

The energy band diagrams of a Schottky-ohmic structure at zero reverse bias,
moderate reverse bias, and full depletion reverse bias are shown in Fig. 3. At zero
bias, there is a built-in electric field at Schottky junction side, corresponding to the
band-bending region. In contrast, there is essentially no or negligible built-in electric
field at ohmic junction, corresponding to no band bending. Outside the depletion
region is the charge-neutral region depicted by a flat energy band. As the applied
reverse bias increases, the built-in electric field in the depletion region is reinforced
by the applied voltage, which is indicated by the stronger band bending. When the
reverse bias reaches the full depletion voltage, the entire detector volume is depleted.
The hole injection from anode that contributes to the bulk leakage current is
suppressed by the energy barrier created by the low-work-function metal. Similarly,

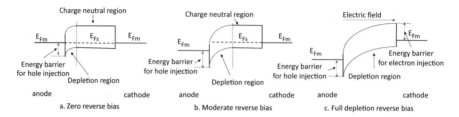

Fig. 3 Energy band diagram of a Schottky-ohmic structure at zero reverse bias (left), moderate reverse bias (middle), and full-depletion reverse bias (right). A p-type semiconductor is used for illustration. E_{Fs} and E_{Fm} stand for the Fermi level of semiconductor and metal, respectively

the electron injection from cathode, another contribution to the bulk leakage current, is prohibited by the energy barrier created by the high-work-function metal.

Unlike the ohmic-ohmic structure where charge carriers can recirculate inside the detector volume, no charge carrier recirculation exists for the Schottky-ohmic structure under reverse bias due to the energy barriers at either boundary of the detector volume. As a consequence, the maximum charge that can be collected is equal to the charge generated by the incident gamma-ray photons, which yields a maximum photoconductive gain of 1. The hole-induced photocurrent due to the drifting of the holes under surface illumination condition, e.g., laser excitation, can be obtained by Many Eq. (3) [21], where e is the elementary charge of electron; G_s is the surface charge carrier generation rate; A and L are the semiconductor cross-section area and length, respectively; s is the charge carrier surface recombination rate; τ_p and μ_p are the hole lifetime and mobility, respectively; and E is the applied electric field. The electron-induced photocurrent under cathode surface illumination can be obtained similarly by replacing τ_p and μ_p with τ_n and μ_n, respectively, in Eq. (3). Multiplying both sides of Eq. (3) by a time interval Δt gives Eq. (4), where Q_L is the charge collected in Δt and $eG_sA\Delta t$ is the charge generated by photons in Δt. As the applied bias voltage increases, the collected charge approaches the generated charge and finally saturates:

$$I_L = \frac{eG_sA}{1 + \frac{s}{\mu_p E}} \left(\tau_p \mu_p E/L \right) \left(1 - \exp\left(-\frac{L}{\tau_p \mu_p E} \right) \right) \tag{3}$$

$$Q_L = I_L \Delta t = \frac{eG_sA\Delta t}{1 + \frac{s}{\mu_p E}} \left(\tau_p \mu_p E/L \right) \left(1 - \exp\left(-\frac{L}{\tau_p \mu_p E} \right) \right) \tag{4}$$

2.1.3 The Schottky-Schottky Structure

A perovskite gamma-ray detector with Schottky-Schottky structure has the metal (high work function)/n-type perovskite/metal (high work function) or metal (low work function)/p-type perovskite/metal (low work function) structure. Schottky

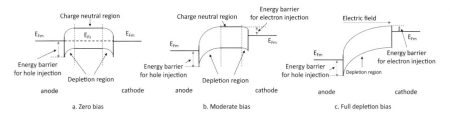

Fig. 4 Energy band diagrams of a Schottky-Schottky structure at zero bias (left), moderate bias (middle), and full-depletion bias (right). A p-type semiconductor is used for illustration. E_{Fs} and E_{Fm} represent the Fermi levels of semiconductor and metal, respectively

junction exists at both sides of the detector. The Schottky-Schottky structure exhibits symmetric I-V curve (Fig. 2). However, it is worth noting that there is a large difference between the symmetric I-V curves shown by an ohmic-ohmic structure and a Schottky-Schottky structure. The I-V curve of an ohmic-ohmic structure is linear over the entire range, whereas the I-V curve of a Schottky-Schottky structure shows current-suppressing effect, indicated by the I-V curve slope reduction at low voltage range. For the same detector material, the magnitude of leakage current with an ohmic-ohmic structure is typically much larger than that with a Schottky-Schottky structure. A perovskite gamma-ray detector with Schottky-Schottky structure can work in either bias voltage polarity.

The energy band diagrams of a Schottky-Schottky structure at zero bias, moderate bias, and full-depletion bias are shown in Fig. 4. The hole injection from the anode is suppressed by the energy barrier provided by the low-work-function metal. On the cathode side, the Schottky-ohmic structure displays an advantage over the Schottky-Schottky structure regarding bulk current suppression as a high-work-function metal provides larger barrier height for electron injection from cathode against a low-work-function metal. However, this advantage may be practically insignificant, since at a high reverse-bias voltage the surface leakage current may dominate the total leakage current.

2.2 Electron Transport Layer and Hole Transport Layer

In addition to the metal/perovskite/metal structure with a proper selection of metals, an electron transport layer (ETL) and/or a hole transport layer (HTL) may also be used to form a junction with perovskite, which could show current-rectifying behavior and suppress bulk leakage current. The ETL/HTL are semiconductor materials, either inorganic or organic, typically used to ensure proper operation of a perovskite solar cell, by enabling for example good charge carrier mobility and compatible energy levels with respect to perovskite absorber [22–24]. The junction between ETL/HTL and perovskite essentially forms a p-n heterojunction structure.

The ETL/HTL generates selectivity for electrons or holes, necessary to drive the charge collection in a perovskite solar cell in the absence of an external bias [25]. In a radiation detector where an external bias voltage is applied for charge extraction, the ETL/HTL is in principle inessential. A depletion layer may be formed in ETL and perovskite when a junction between ETL (e.g., n-type) and perovskite (e.g., p-type) is reverse biased, which may introduce nonlinearity between energy deposition and charge generation. The same applies to HTL and a n-type perovskite. The different bandgaps of ETL/HTL and perovskite, and thus different e-h pair creation energies, would lead to different amounts of charge generated from the same amount of energy deposited in the respective depletion regions, which may deteriorate the detector energy spectrum resolution. However, such degradation of energy resolution may be minimal in certain practical scenarios such as (1) the ETL/HTL is much thinner compared to the perovskite crystal, (2) the ETL/HTL doping concentration is sufficiently high to minimize the depletion region spread in ETL/HTL, and (3) the Z_{eff} of ETL/HTL is relatively low such that its interaction with gamma rays is negligible. Thus, the effect of Z_{eff}, doping, and thickness of the ETL/HTL on the linearity between energy deposition and charge production must be understood to make a suitable choice of ETL/HTL for use in a perovskite gamma-ray detector.

2.3   Perovskite CsPbBr3 Gamma Detector Surface Processing and Fabrication

While a good design architecture of perovskite gamma-ray detector is critical to reducing the bulk leakage current, surface processing of raw single crystals with good polishing and cleaning is key to reducing the surface leakage current. The organic cations used in perovskite are hygroscopic so that any medium used for surface processing must not contain water to avoid damage to the perovskite crystals. The perovskite crystals are also heat sensitive as it has been reported that organic perovskites ($MAPbX_3$, X = Cl, Br, I) decompose at moderate temperatures (~60 °C), which may lead to their long-term instability issues and health concerns due to the release of toxic decomposition product [9]. Although the inorganic perovskite $CsPbBr_3$ has a melting temperature of ~560 °C, it has two phase transitions that occur at ~80 °C and ~130 °C [10]. Any phase transition should be carefully avoided during the crystal surface processing as that may induce stress or even deformation of the $CsPbBr_3$ crystal leading to degradation of detector performance.

2.3.1   Surface Polishing, Cleaning, Passivation, and Metal Deposition

Polishing paper or polishing paste may be used to reduce surface roughness, while any polishing medium containing water must be avoided. In addition, chemical-mechanical polishing may be used with suitable chemicals, which may potentially reduce surface defects. Nonpolar organic solvent, such as toluene, is recommended for surface cleaning. Surface passivation/oxidation may be used to further reduce surface leakage current. However, the passivation layer could potentially bring the risk of charge accumulation at surface, which may degrade gamma-ray detector performance. The passivation layer is essentially a resistive layer between perovskite and metal contact. The exact resistance of such a layer is material specific and depends on several factors such as its chemical composition, thickness, and surface morphology. Although a resistive layer creates a physical isolation between metal and perovskite CsPbBr$_3$, it is not detrimental to charge induction process. The metal electrode does not have to be in an immediate physical contact with the semiconductor surface to allow charge induction and the linearity between generated charge and induced signal amplitude still holds according to the Shockley-Ramo theorem [26], which is also known as proximity charge sensing and experimentally demonstrated by P. Luke [27]. If the resistive layer has a resistance small enough that the charge carriers from metal and CsPbBr$_3$ can easily tunnel through its energy barrier, no significant charge accumulation occurs at the interface between resistive layer and CsPbBr$_3$ and its negative effects on ER of gamma spectrum are negligible. When the resistive layer has a resistance large enough, a considerable charge accumulation occurs at the resistive layer-CsPbBr$_3$ interface, which could cause polarization and distortion of electric field inside the detector leading to a severe ER degradation. The ultimate effect of oxide/passivation layer on gamma-spectrum ER is a balancing act between surface defect passivation and charge accumulation, which depends quantitatively on the exact resistance of the oxide/passivation layer and its defect passivation effects. In practice, a passivation/oxidation layer less than tens of nanometers in thickness is recommended during the detector fabrication to achieve a well-controlled detector performance.

After polishing, cleaning, and any surface passivation, metal electrodes may be deposited onto the crystal surface. Physical vapor deposition methods such as thermal evaporation and E-beam evaporation could be used for the metal contact deposition. The samples in the E-beam evaporators can reach up to 150 °C. While this is a much lower temperature to decompose CsPbBr$_3$, it is more of a concern with organic perovskites, the decomposition of which produces toxic lead and carcinogen. Furthermore, the samples are under high-vacuum environment in the metal evaporator, which could be below the vapor pressure of the material. Thus, although no material may be lost at atmospheric pressure, a significant amount of the material may be lost at vacuum and the material decomposition under such conditions may contaminate the evaporator chamber. Thus, the metal deposition process for perovskite must be carefully monitored and controlled so that the temperature of the deposition environment would not lead to decomposition or phase transition of the perovskite crystals.

Any other metal deposition methods that involve water or high temperatures should also be avoided. After metal deposition, the metal electrodes may be connected to the external circuit. Since the perovskite crystals are fragile and pressure sensitive, proper encapsulation, for example using epoxy or wax, and packaging should be employed to avoid any crystal and electrode damage.

2.3.2   Guard-Ring Electrode for Surface Leakage Current Reduction

Guard-ring electrode structure is another effective technique to reduce the surface leakage current in addition to the aforementioned techniques. A guard-ring-based detector structure has a planar electrode on one side, and on the other side has a central and a guard electrode. The central electrode is usually made in a circular shape and surrounded by the guard electrode in an annulus shape, which are both biased at essentially the same voltage. Consequently, the surface leakage current that normally flows from the surface edge to the central electrode in a direction parallel to detector signal current is bypassed by the guard electrode. In other words, the surface leakage current would flow through the guard electrode, and not the central electrode. Since only the central electrode is connected to the detector readout circuitry, the leakage current signal from guard electrode is not collected and is successfully eliminated from the detector current to provide a much cleaner signal with higher SNR for energy spectrum acquisition.

3   Perovskite CsPbBr3 Gamma-Ray Detector Performance Characterization

With a perovskite $CsPbBr_3$ device successfully fabricated and electrically characterized using, for example, I-V and C-V (or $1/C^2$-V) measurements, the next step would be the evaluation of its radiation detection capability. A typical sequence of tests could involve using multiple different radiation types and sources such as starting with a UV laser source, an X-ray source, and a weak alpha button-sized source, followed by the laboratory-scale standard gamma-ray sources. The laser and X-ray measurements, due to the stable and high intensity of the radiation involved, are relatively easy in terms of generating a current response in the detector. The alpha- and gamma-ray tests are typically encountered by a more challenging event-by-event-based particle counting with or without an energy measurement. An alpha source such as [241]Am is typically employed for detector evaluation due to its highly energetic and strongest alpha particle emission at 5.486 MeV and the limited penetration depth of such heavily charged particles in a perovskite crystal. In addition to alpha particles, [241]Am also emits low-energy gamma rays with 59.5 keV energy at an emission intensity of ~36%.

3.1 Perovskite CsPbBr₃ Gamma-Ray Detector Response to Alpha Particles

3.1.1 Electron/Hole-Induced Alpha Energy Spectra

Alpha particle energy spectra acquired by a CsPbBr$_3$ detector have been demonstrated by Pan et al. [28]. The energy spectra of a 1.0 μCi ^{241}Am alpha particle source were acquired by a perovskite detector with a structure of Ti(titanium)/CsPbBr$_3$/Au (gold) (Fig. 5). Essentially all charge carriers are generated near the surface of the crystal due to the limited penetration depth of the incident alpha particles (range at 28.8 μm for 5.486 MeV alpha particles simulated by SRIM). Reverse-bias voltage is applied with electric field direction from Ti electrode to CsPbBr$_3$ crystal. If the charge carriers are generated near the Ti/CsPbBr$_3$ interface, the collected signals are mainly hole-induced signals and thus the acquired energy spectra are regarded as hole-induced spectra. Similarly, if the charge carriers are generated near the Au/CsPbBr$_3$ interface, the acquired energy spectra are regarded as electron-induced spectra.

As the applied reverse-bias voltage increases, the energy spectrum shifts to the right, indicating a higher charge collection efficiency at a higher bias voltage. As the alpha particle measurement was conducted not in the vacuum, a 2 cm air gap between the alpha source and the electrode causes energy loss of alpha particles, which is manifested in the broadened peak. A comparison of the hole- and electron-induced energy spectra reveals a more refined spectrum shape and higher detector pulse amplitude (represented in # ADC channels) of the hole-induced spectra, which indicate superior hole transport characteristics compared to that of electrons in this CsPbBr$_3$ detector.

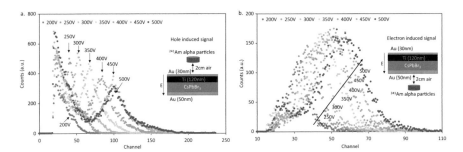

Fig. 5 ^{241}Am alpha energy spectra acquired in air by a perovskite CsPbBr$_3$ detector with a structure of Ti/CsPbBr$_3$/Au. On the left are shown the hole-induced energy spectra and the right figure is the electron-induced energy spectra. Same gamma-ray spectroscopy system settings were used in the acquisition of all energy spectra

3.1.2 Charge Carrier Mobility Estimation by Time-of-Flight Method

Charge carrier mobility can also be estimated through alpha particle response evaluation, which is known as time-of-flight method. The charge carriers generated near the surface will drift towards the electrode of opposite polarity along the trajectory of the electric field inside a detector. The time it takes for carriers to drift from one side to the other side of the detector is defined as the charge-drifting time, which is assumed to be the same as the preamplifier output pulse-rising time (time taken for the pulse to rise from 10% to 90% of its amplitude). Therefore, when the applied bias voltage and crystal thickness are known, the charge carrier mobility can be estimated through the equation $\mu = d^2/(t_r \times V)$, where μ is the mobility, d is the crystal thickness, t_r is the pulse-rising time, and V is the applied bias voltage. Hole or electron mobility can be estimated from signals that are primarily hole induced or electron induced, respectively. The hole mobility estimation of a $CsPbBr_3$ detector is demonstrated in Fig. 6. The histogram of pulse-rising time confirms that, as the applied bias voltage increases, the pulse-rising time as well as their deviation decreases. The hole mobility value is obtained as 1.78 cm^2/V-s for this particular device from the linear fit of drift velocity $v = d/t_r$ against the electric field $E = V/d$. The estimated mobility values may be different for different $CsPbBr_3$ detectors, owing to a crystal-to-crystal quality variation.

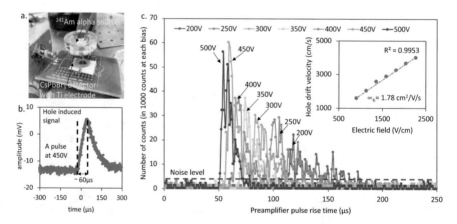

Fig. 6 Hole mobility estimation of a perovskite $CsPbBr_3$ detector with a structure of Ti/CsPbBr$_3$/Au using [241]Am alpha particle measurements. (**a**) A picture of the detector-source geometry setup. (**b**) A single hole-induced pulse of alpha particle. (**c**) Histogram of preamplifier pulse-rising time at different applied reverse-bias voltages

3.1.3   μτ Product and Surface Recombination Velocity Evaluation by Hecht Equation and Many Equation

The μτ product is normally overestimated when a laser or a high-intensity X-ray beam is used for its evaluation. This is because the overpopulated charge carriers tend to fill the trapping centers, resulting in a detector response that is better than when collecting the minute charge release by gamma rays. The alpha particles are used to estimate single-polarity μτ product and its energy spectra are used for μτ product evaluation through a fit to the single-charge-carrier Hecht Eq. (5):

$$Q = Q_0 \frac{V \mu\tau}{d^2} \left(1 - \exp\left(-\frac{d^2}{V \mu\tau} \right) \right) \tag{5}$$

where Q/Q_0 = CCE is the charge collection efficiency that can be obtained with the peak centroid channel at each applied bias voltage divided by the centroid channel of the saturation peak (i.e., the channel corresponding to maximum charge collection), again V is the applied bias voltage, and d is the detector thickness. He et al. demonstrated μτ product evaluation of CsPbBr$_3$ detectors through single-charge-carrier Hecht equation fitting [29].

While it is straightforward to apply Hecht equation for μτ product evaluation, there are some limitations that restrict its use. The derivation of the Hecht equation assumes a uniform electric field inside the detector, no charge de-trapping, and negligible surface recombination effect. A modification to Hecht equation considering the effect of surface recombination is the Many Eq. (6):

$$CCE = \frac{V \mu\tau}{d^2} \left(1 - \exp\left(-\frac{d^2}{V \mu\tau} \right) \right) \frac{1}{1 + \frac{d}{V\mu}S} \tag{6}$$

where S is the charge carrier surface recombination velocity in the units of cm/s. The surface recombination velocity indicates the effect of surface treatment on charge collection efficiency. From Eq. (6), it can be seen that a smaller surface recombination velocity is desirable for higher CCE, which may be achieved by proper surface treatment.

3.2   *Perovskite CsPbBr3 Detector Gamma Spectrum Performance*

Resolving the photopeak produced by a weak gamma-ray source (a few μCi or less) requires the detector working in the single-photon counting mode. To maintain the proportionality between the energy deposited by an incident gamma-ray photon and the charge induced, the charge collection efficiency of the CsPbBr$_3$ gamma-ray detector has to be as high as possible. The doping concentration or defect density

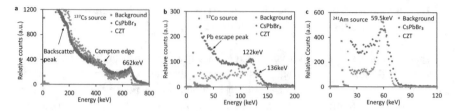

Fig. 7 Gamma-ray energy spectra acquired by solution-grown CsPbBr$_3$ single-crystal gamma-ray detector (2.53 × 7.08 × 8.29 mm^3) in comparison with that of a commercial CZT detector (5 × 5 × 4 mm^3) under same experimental conditions (bias 500 V, shaping time 10μs). a. ^{137}Cs. b. ^{57}Co. c. ^{241}Am gamma-ray spectra

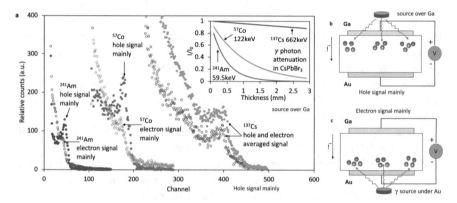

Fig. 8 (a) ^{137}Cs, ^{57}Co, ^{241}Am gamma-ray energy spectra acquired under different source-to-detector configurations to demonstrate and compare the effects of electron and hole transport properties on spectral performance. Source-detector geometry setup for **(b)** hole-induced and **(c)** electron-induced signal collection

of the CsPbBr$_3$ single crystal has to be as low as possible to reduce charge carrier trapping. A high resistivity, typically >10^9 Ω * cm, of the perovskite single crystal is required. Gamma-ray energy spectra acquired by CsPbBr$_3$ gamma-ray detector have been demonstrated by He et al. [13], using melt-grown CsPbBr$_3$ single crystal and by Pan et al. using solution-grown CsPbBr$_3$ single crystal, as shown in Fig. 7 [2].

3.2.1 Electron/Hole-Induced Gamma Energy Spectra

The attenuation of gamma-ray photons of different energies in CsPbBr$_3$ is thickness dependent, as shown in Fig. 8a, inset. It is thus possible, using a proper choice of gamma-ray energies, to demonstrate the difference in contribution of hole and electron transport to the gamma-ray energy spectrum. By switching the relative position of the gamma-ray source with respect to the Ga (gallium)/CsPbBr$_3$/Au detector (Fig. 8b, c), the electron-induced and hole-induced gamma spectra can be acquired. The spectrum resulted primarily from hole transport clearly presents a stronger and more symmetric 59.5 keV full energy peak compared to that resulted primarily from electron transport. This phenomenon is also observable with the 122 keV full energy peak from ^{57}Co source due to relatively low attenuation thickness for 122 keV gamma rays. In case of the 662 keV gamma-ray photons from ^{137}Cs, the longer attenuation thickness indicates a uniform generation of charge carriers in the detector volume, which results in a negligible difference between hole- and electron-contributed energy spectra, regardless of the source position relative to the detector.

3.2.2 CsPbBr$_3$ Gamma-Ray Detector Energy Spectrum Linearity and Stability

A linear energy response is also an important characteristic for a gamma-ray detector. CsPbBr$_3$ gamma-ray detector made of solution-grown CsPbBr$_3$ single crystal is shown to have excellent energy linearity (Fig. 9).

The long-term stability of the solution-grown CsPbBr$_3$ single-crystal gamma-ray detector was evaluated by Pan et al. [28]. The dark currents measured for both a Bi (bismuth)/CsPbBr$_3$/Au detector and a Ga/CsPbBr$_3$/Au detector remained stable for several hours under a constant bias voltage of -200 V (Fig. 10a), which shows little or no ion migration issues. In addition, the ^{137}Cs gamma-ray spectra of Ga/CsPbBr$_3$/Au detector acquired at the beginning and end of a 24-h period during which a bias voltage of -200 V was continuously applied showed no difference (Fig. 10b). This result further indicates the good stability of Ga/CsPbBr$_3$/Au detector under a long-period bias condition.

The solution-grown CsPbBr$_3$ single crystals are also shown to be robust and the same crystal can be reutilized multiple times for device fabrication. A CsPbBr$_3$

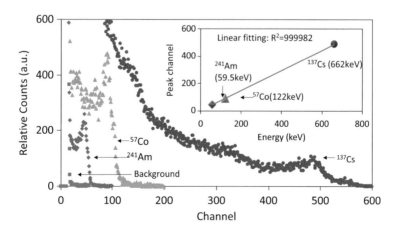

Fig. 9 ^{137}Cs, ^{57}Co, and ^{241}Am gamma-ray energy spectra acquired separately with solution-grown CsPbBr$_3$ detector under same experimental conditions (bias voltage 500 V, shaping time 10µs). Shown in the inset is the linear fit of photopeak channel to photopeak energy

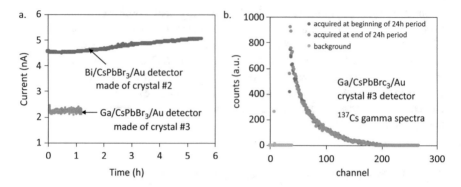

Fig. 10 Solution-grown CsPbBr₃ gamma-ray detector stability test. (**a**) Dark current as a function of time measured for a Bi/CsPbBr₃/Au detector (thickness 2.58 mm, planar electrode area: 3.5×3.5 mm²) and a Ga/CsPbBr₃/Au detector (thickness 2.7 mm, planar electrode area: $\pi \times 2.1 \times 2.1$ mm²) under -200 V bias. (**b**) ¹³⁷Cs gamma-ray energy spectra acquired with a Ga/CsPbBr₃/Au detector at the beginning and end of a 24-h period under continuously applied -200 V bias (shaping 10μs, acquisition time 7000 s for each spectrum)

single crystal was repeatedly used with deposition of different metal contacts to evaluate its gamma spectrum performance for each metal/perovskite/metal configuration while eliminating crystal-to-crystal variation [28]. Leakage current as low as 2 nA to 3 nA was consistently achieved at -200 V bias with detector thickness of ~2–3 mm and Zr(zirconium), Bi, Ti, and Ga as the Schottky contact metals, which is comparable to the leakage current of a commercial CZT detector (Fig. 11). Gamma-ray energy spectra with a prominent photopeak at 662 keV were consistently acquired with all the studied metal contacts and device structures (Fig. 12). The best ER at 662 keV was 11% with a peak-to-Compton valley ratio of 2 (Fig. 12d). The Bi/CsPbBr₃/Bi detector was stored in ambient air and the energy spectrum acquired after 1 month showed no significant difference from that acquired using the as-fabricated device (Fig. 12b).

3.2.3 Long Pulse-Rising Time of CsPbBr3 Gamma-Ray Detector and Digital Pulse Processing

Conventional instrumentation for semiconductor-based detectors is designed to process charge-drifting times shorter than 10 μs, for example, 10 s of ns for Si detectors and 100 s of ns for CZT detectors. The preamplifier, often charge-sensitive preamplifier, has an output pulse-rising time assumed to be equal to the charge-drifting time in the detector. Perovskite CsPbBr₃ detector poses a new challenge to the traditionally applied gamma-ray spectroscopy in terms of its long charge-drifting times. For example, a perovskite CsPbBr₃ gamma-ray detector can have a charge-drifting time longer than 10 μs [2, 30], which would lead to ballistic deficit issues during charge processing, subsequently resulting in the distortion of gamma-ray energy spectrum.

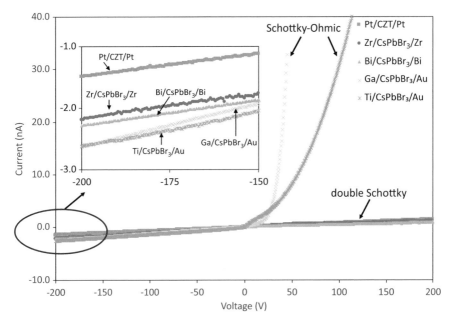

Fig. 11 I-V curves acquired with CsPbBr₃ detector made of a repeatedly used solution-grown single crystal with different metal electrodes

Digital pulse processing (DPP) may be applied to address the potential ballistic deficit issue resulting from the long charge-drifting times of perovskite gamma-ray detectors. DPP may be used to reconstruct energy spectra free from ballistic deficit. Pan et al. developed a custom DPP algorithm for perovskite CsPbBr₃ gamma-ray detector energy spectrum acquisition [2]. The DPP algorithm applies a digital deconvolution to the preamplifier output signal, which eliminates the ballistic deficit at preamplifier charge collection stage. Subsequently, the energy spectrum is reconstructed by calculating the deconvolved signal amplitudes directly from the difference between the signal peak and its baseline.

3.2.4 Gamma Energy Spectrum Resolution Improvement

Energy spectrum resolution degradation due to poor electron transport in the CsPbBr₃ single crystal could be reduced using single charge carrier collection technique based on the Shockley-Ramo theorem [26]. The electric field distribution inside the detector volume can be modified by modifying the electrode geometry. Hence, the contribution of one type of charge carrier to the total induced charge is elevated, while the contribution of the charge carrier with the opposite polarity is

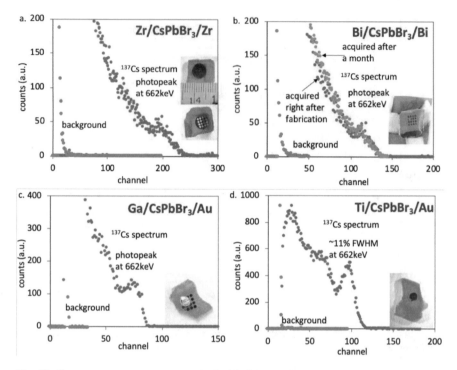

Fig. 12 Gamma-ray energy spectra acquired with CsPbBr$_3$ detector made of a repeatedly used solution-grown single crystal with different metal electrodes

reduced. Electrode geometries other than planar electrodes can be used to implement single-charge-carrier sensing, e.g., pixelated electrode and coplanar grid electrode.

A successful demonstration of using such techniques to improve the ER of the CsPbBr$_3$ gamma detectors made of melt-grown CsPbBr$_3$ single crystals is reported by He et al. [1], where 1.4% ER at 662 keV has been achieved with planar electrodes using a melt-grown CsPbBr$_3$ single crystal with small thickness and ER at ~4.5–7% at 662 keV with a larger crystal volume. To achieve a high ER while maintaining the high peak-to-Compton ratio enabled by a CsPbBr$_3$ crystal with relatively large volume, He et al. used quasi-hemispherical electrode and pixelated electrode, which yields 1.8% ER at 662 keV for quasi-hemispherical electrode and 1.6% ER at 662 keV for pixelated electrode, respectively [1].

Acknowledgments We thank the financial support of the U.S. Defense Threat Reduction Agency and the collaboration with Dr. Jinsong Huang, with whom our cited works are published together. We also thank Dr. John Carlin and Paul Steffen at the Nanotech West Lab, The Ohio State University, for the helpful discussions on metal deposition process on perovskite single crystals.

References

1. He, Y., Petryk, M., Liu, Z. et al. CsPbBr3 perovskite detectors with 1.4% energy resolution for high-energy γ-rays. Nat. Photonics. **15**, 36–42 (2021). https://doi.org/10.1038/s41566-020-00727-1
2. Pan, L., Feng, Y., Kandlakunta, P., Huang, J., Cao, L.R.: Performance of perovskite CsPbBr3 single crystal detector for gamma-ray detection. IEEE Trans. Nucl. Sci. (2020). https://doi.org/10.1109/TNS.2020.2964306
3. Zhang, F., Herman, C., He, Z., De Geronimo, G., Vernon, E., Fried, J.: Characterization of the H3D ASIC readout system and 6.0 cm ^{3} 3-D position sensitive CdZnTe detectors. IEEE Trans. Nucl. Sci. **59**(1), 236–242 (2012)
4. Roy, U.N., Weiler, S., Stein, J.: Growth and interface study of 2 in diameter CdZnTe by THM technique. J. Cryst. Growth. (2010). https://doi.org/10.1016/j.jcrysgro.2010.05.046
5. Hitomi, K., Kikuchi, Y., Shoji, T., Ishii, K.: Polarization phenomena in TlBr detectors. IEEE Trans. Nucl. Sci. **56**(4), 1859–1862 (2009)
6. Wei, H., Huang, J.: Halide lead perovskites for ionizing radiation detection. Nat. Commun. **10**(1), 1–12 (2019). https://doi.org/10.1038/s41467-019-08981-w
7. Conings, B., et al.: Intrinsic thermal instability of methylammonium lead trihalide perovskite. Adv. Energy Mater. (2015). https://doi.org/10.1002/aenm.201500477
8. wook Park, B., Il Seok, S.: Intrinsic instability of inorganic–organic hybrid halide perovskite materials. Adv. Mater. (2019). https://doi.org/10.1002/adma.201805337
9. Brunetti, B., Cavallo, C., Ciccioli, A., Gigli, G., Latini, A.: On the thermal and thermodynamic (in)stability of methylammonium lead halide perovskites. Sci. Rep. (2016). https://doi.org/10.1038/srep31896
10. Stoumpos, C.C., et al.: Crystal growth of the perovskite semiconductor CsPbBr3: A new material for high-energy radiation detection. Cryst. Growth Des. (2013). https://doi.org/10.1021/cg400645t
11. Kang, J., Wang, L.W.: High defect tolerance in lead halide perovskite CsPbBr3. J. Phys. Chem. Lett. (2017). https://doi.org/10.1021/acs.jpclett.6b02800
12. Deng, J., Li, J., Yang, Z., Wang, M.: All-inorganic lead halide perovskites: A promising choice for photovoltaics and detectors. J. Mater. Chem. C. (2019). https://doi.org/10.1039/c9tc04164h
13. He, Y., et al.: High spectral resolution of gamma-rays at room temperature by perovskite CsPbBr3 single crystals. Nat. Commun. **9**(1), 1–8 (2018). https://doi.org/10.1038/s41467-018-04073-3
14. He, Y., et al.: Resolving the energy of γ-ray photons with MAPbI3 single crystals. ACS Photonics. (2018). https://doi.org/10.1021/acsphotonics.8b00873
15. Wei, H., et al.: Dopant compensation in alloyed CH3NH3PbBr3-x Clx perovskite single crystals for gamma-ray spectroscopy. Nat. Mater. (2017). https://doi.org/10.1038/nmat4927
16. Hitomi, K., Kikuchi, Y., Shoji, T., Ishii, K.: Improvement of energy resolutions in TlBr detectors. Nucl. Instrum. Methods Phys. Res. Sect. A. **607**(1), 112–115 (2009)
17. Wei, H., et al.: Sensitive X-ray detectors made of methylammonium lead tribromide perovskite single crystals. Nat. Photonics. **10**(5), 333–339 (2016). https://doi.org/10.1038/nphoton.2016.41
18. Wei, W., et al.: Monolithic integration of hybrid perovskite single crystals with heterogenous substrate for highly sensitive X-ray imaging. Nat. Photonics. (2017). https://doi.org/10.1038/nphoton.2017.43
19. Huang, Y., et al.: A-site cation engineering for highly efficient MAPbI3 single-crystal X-ray detector. Angew. Chem. Int. Ed. (2019). https://doi.org/10.1002/anie.201911281
20. Pan, W., et al.: Hot-pressed CsPbBr3 quasi-monocrystalline film for sensitive direct X-ray detection. Adv. Mater. (2019). https://doi.org/10.1002/adma.201904405
21. Many, A.: High-field effects in photoconducting cadmium sulphide. J. Phys. Chem. Solids. (1965). https://doi.org/10.1016/0022-3697(65)90133-2

22. Mohamad Noh, M.F., et al.: The architecture of the electron transport layer for a perovskite solar cell. J. Mater. Chem. C. **6**(4), 682–712 (2018). https://doi.org/10.1039/c7tc04649a
23. Ansari, M.I.H., Qurashi, A., Nazeeruddin, M.K.: Frontiers, opportunities, and challenges in perovskite solar cells: A critical review. J. Photochem. Photobiol. C Photochem. Rev. **35**, 1–24 (2018). https://doi.org/10.1016/j.jphotochemrev.2017.11.002
24. Mahmood, K., Sarwar, S., Mehran, M.T.: Current status of electron transport layers in perovskite solar cells: materials and properties. RSC Adv. **7**(28), 17044–17062 (2017). https://doi.org/10.1039/c7ra00002b
25. Zhou, Y., Gray-Weale, A.: A numerical model for charge transport and energy conversion of perovskite solar cells. Phys. Chem. Chem. Phys. **18**(6), 4476–4486 (2016). https://doi.org/10.1039/c5cp05371d
26. He, Z.: Review of the Shockley-Ramo theorem and its application in semiconductor gamma-ray detectors. Nucl. Instruments Methods Phys. Res. Sect. A. (2001). https://doi.org/10.1016/S0168-9002(01)00223-6
27. Luke, P.N., Tindall, C.S., Amman, M.: Proximity charge sensing with semiconductor detectors. IEEE Trans. Nucl. Sci. **56**(3), 808–812 (2009). https://doi.org/10.1109/TNS.2008.2011483
28. Pan, L., Feng, Y., Huang, J., Cao, L.R.: Comparison of Zr, Bi, Ti, and Ga as metal contacts in inorganic perovskite CsPbBr gamma-ray detector. IEEE Trans. Nucl. Sci. (2020). https://doi.org/10.1109/TNS.2020.3018101
29. He, Y., et al.: Perovskite CsPbBr 3 single crystal detector for alpha-particle spectroscopy. Nucl. Instruments Methods Phys. Res. Sect. A. **922**(January), 217–221 (2019). https://doi.org/10.1016/j.nima.2019.01.008
30. Lukosi, E., et al.: Methylammonium lead tribromide semiconductors: Ionizing radiation detection and electronic properties. Nucl. Instruments Methods Phys. Res. Sect. A. (2019). https://doi.org/10.1016/j.nima.2019.02.059

The Impact of Detection Volume on Hybrid Halide Perovskite-Based Radiation Detectors

Pavao Andričević

Abstract Ionizing radiation, wherever present, e.g., in medicine, nuclear environment, or homeland security, due to its strong impact on biological matter, should be closely monitored. Availability of semiconductor materials with distinctive characteristics required for an efficient high-energy photon detection, especially with high atomic numbers (high Z), in sufficiently large, single-crystalline forms, which would also be both chemically and mechanically robust, is still very limited.

In this chapter, we introduce the new metal halide perovskite material, which meets all aforementioned key requirements, at an extremely low cost. In particular, γ-ray detectors based on crystals of methylammonium lead tribromide (MAPbBr3) equipped with carbon electrodes were fabricated, allowing radiation detection by photocurrent measurements at room temperatures with record sensitivities (333.8 $\mu C\ Gy^{-1}\ cm^{-2}$). Importantly, the devices operated at low bias voltages (<1.0 V), which may enable future low-power operation in energy-sparse environments, including space. The detector prototypes were exposed to radiation from a ^{60}Co source at dose rates up to 2.3 Gy h^{-1} under ambient and operational conditions for over 100 h, without any sign of degradation. We postulate that the excellent radiation tolerance stems from the intrinsic structural plasticity of the organic-inorganic halide perovskites, which can be attributed to a defect-healing process by fast ion migration at the nanoscale level.

Furthermore, since the sensitivity of the γ-ray detectors is proportional to the volume of the employed MAPbBr3 crystals, a unique crystal growth technique is introduced, baptized as the "oriented crystal-crystal intergrowth" or OC2G method, yielding crystal specimens with volume and mass of over 1000 cm^3 and 3.8 kg, respectively. Large-volume specimens have a clear advantage for radiation detection; however, the demonstrated kilogram-scale crystallogenesis coupled with future cutting and slicing technologies may have additional benefits, for instance, enable the development for the first time of crystalline perovskite wafers, which may challenge the status quo of present and future performance limitations in all optoelectronic applications.

P. Andričević (✉)
École Polytechnique Fédérale de Lausanne, Lausanne, Switzerland

© The Author(s), under exclusive license to Springer Nature Switzerland AG 2022
K. Iniewski (ed.), *Advanced Materials for Radiation Detection*,
https://doi.org/10.1007/978-3-030-76461-6_3

1 Importance of New Materials for Radiation Detection

Ionizing radiation, besides all its negative effects, has found to benefit greatly the humankind in many ways. It is used nowadays in a wide range of domains, from medicine, academic research, and industry to energy generation, specifically in various applications in agriculture, archeology, space exploration, law enforcement, geology, and many others [1, 2]. Accompanying all these applications is always radiation detection, precisely because of its damaging effects on different materials, especially the human body. Therefore, there is a need to control and monitor radiation constantly, particularly as humans do not have a direct way of recognizing it.

The key process by which radiation is detected is ionization, where the incident ray gives part of its energy to an electron, which then collides with other atoms liberating many more electrons. These electrons are then collected either directly using proportional counters or solid-state semiconductor detectors, or indirectly via scintillation detectors, in order to register the presence of radiation and measure its energy [3]. The final result is an electrical pulse whose voltage is proportional to the deposited energy.

In this work, we will concentrate on solid-state detectors based on semiconductor materials, which first started to appear in the 1970s. Their main advantage towards commonly used scintillator detectors is that the charge produced by the photon interaction is collected directly. Therefore, the energy resolution of these detectors is dramatically better, enabling great spectral detail to be measured. Moreover, the need of photomultipliers and other accompanying electronics is not needed which drastically lowers their fabrication costs.

Multiple distinct characteristics of a semiconductor material are required for an efficient high-energy photon detection. For example, a large detecting volume to intercept radiation, a large linear attenuation coefficient, large and balanced carrier mobilities (μ), and concomitant long charge carrier lifetimes (τ), thus resulting in high values of their product ($\mu\tau$), are all important for efficient charge collection. And finally, a high resistivity accompanied by a low charge trap density is needed to avoid charge trapping under a single-event analysis [4]. The product $\mu\tau$ attains the highest values in crystalline semiconductor materials, in which carrier transport is not limited by scattering and trapping at grain boundaries. However, the availability of such semiconductors, especially with high atomic numbers (high Z), in sufficiently large, single-crystalline forms which would also be both chemically and mechanically robust, is still limited [5].

High-purity germanium crystals possess the ideal electronic characteristics in this regard. They are the most widely used semiconductor materials in solid-state-based detectors. However, due to a small bandgap of germanium, to perform a high-resolution spectroscopy a liquid nitrogen cooling system is necessary for germanium-based solid-state detectors. Therefore, to achieve a high-resolution detection at room temperature, there have been applications of semiconducting materials other than germanium and silicon, such as CdTe, HgI_2, and GaAs [3].

Zinc-alloyed CdTe ($Cd_{1-x}Zn_xTe$, denoted CZT for $0 < x < 0.2$) single crystals produce the best resolution of γ-ray spectra among the non-cooled semiconductor radiation detectors, due to their large bandgap of above 1.6 eV, a high resistivity up to $(10^8–10^9)$ Ω cm at room temperature, and a large $\mu\tau$ product [6]. However, important limitations can be identified for the application of CZT detectors, such as the cost-restricted crystal manufacturing at a scaled-up level, the incompatibility of high-temperature crystal growth with readout circuits, and a low hole mobility.

The recently rediscovered hybrid halide perovskites (ABX_3) have been found to meet all aforementioned key requirements for high-energy radiation detection. The high-Z chemical elements, such as lead (Pb), iodine (I), and bromine (Br), concomitant with relatively large densities of perovskite (4.0 g cm^{-3}), allow for a substantial attenuation of high-energy photons. They possess large proper bandgaps to reduce thermal noise, and high resistivity to suppress dark current and device noise. Moreover, a great mobility-lifetime product for efficient charge collection has been calculated (e.g., for $Cs_{0.1}FA_{0.9}PbI_{2.8}Br_{0.2}$ SCs the $\mu\tau$ product is of about 1.2×10^{-1} cm^2/V) [7]. The linear attenuation coefficient and its corresponding penetration depth of all aforementioned radiation detector materials as well as that of the methylammonium lead tribromide single crystals ($MAPbBr_3$ SCs), which will be discussed more in detail in this work, are shown in Fig. 1. Furthermore, a comparison of all the fundamental figures of merit for radiation detectors of these materials is given in Table 1.

However, unlike the market leading crystals, these perovskite SCs may grow from abundant and low-cost raw materials in solutions at near room temperature

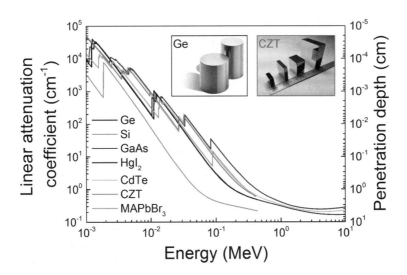

Fig. 1 Linear attenuation coefficient and corresponding penetration depth as a function of photon energy for different radiation detection materials from X- to γ-ray. Inset: Optical images of high-purity Ge [8] and CZT [9] single crystals. Values taken from [10]

Table 1 Comparative table of the fundamental figures of merit of radiation detection for different materials

Material	Bandgap (ev)	Bulk resistivity (Ω cm)	$\mu\tau$ product (cm^2 V^{-1})	Energy resolution (%)	Temperature (K)	Reference
Ge	0.67	10^2–10^3	>1	0.2	77	[11]
GaAs	1.42	10^6–10^7	0.1	5.3–12	255–300	[12]
HgI$_2$	2.13	10^{10}	0.01	0.85–1.3	300	[13]
CZT	1.5–1.6	10^{10}	0.004–0.01	0.5	300	[14]
Perovskite	1.5–3.1	10^7–10^{10}	0.001–0.01	3.9	300	[6]

without using high-capital-demanding infrastructure, therefore making them a good candidate for fabrication of the next generation of high-energy radiation detectors.

2 X-Ray Detection by Perovskite

The first perovskite-based radiation detectors were polycrystalline thin films of methylammonium lead triiodide (MAPbI$_3$). These very common devices in the photovoltaic field were repurposed for X-ray detection [15]. Yakunin et al. successfully demonstrated an X-ray image of a leaf using this device. Despite the still imperfect image it showed great potential for application in medical diagnostics, especially if better quality thin films or single crystals will be used [16].

Solution-growth perovskite SCs offered even more advantages when compared to their polycrystalline counterparts. Náfrádi et al. [17] first demonstrated the high mass attenuation coefficient of 14 ± 1.2 cm^2 g^{-1}, and large charge collection efficiency of $75 \pm 6\%$ for a MAPbI$_3$ bulk single crystal under unfiltered X-ray radiation in the 20–35 keV range. This work was closely followed by X-ray detectors fabricated from 2 to 3 mm thick MAPbBr$_3$ SCs with record high $\mu\tau$ products (1.2×10^{-1} cm^2 V^{-1}) able to detect dose rates as low as 0.5 μGy$_{air}$ s^{-1} with a sensitivity of 80 μC mGy$_{air}^{-1}$ cm^{-2} [18]. With the required dose rate for medical diagnostics being 5.5 μGy$_{air}$ s^{-1} [16], this would allow to reduce the radiation dose applied to the human body for many medical and security operations. Later, the same group of Huang and coworkers [19] monolithically integrated these perovskite SCs onto different substrates, from silicon wafers to glass, through facile, low-temperature, solution-processed molecular bonding. The dipole of this bonding molecule at the interface of perovskite and Si significantly reduces the dark current at higher optimum bias, enabling the Si-integrated MAPbBr$_3$ single-crystal detector to reach a sensitivity of $2.1 \times 10^4 \mu$C Gy$_{air}^{-1}$ cm^{-2} under 8 keV X-ray radiation. In addition, the low-noise current led to a decrease in the lowest detectable X-ray dose rate (36 nGy$_{air}$ s^{-1}) [19].

Most recently, we demonstrated a promising method for building X-ray detector units by 3D aerosol jet printing perovskite into electric circuits [20]. With aerosol jet printing, the material can be selectively deposited at any location of interest with μm precision. The success of this technique in writing well-defined 3D structures is linked to the existence of intermediate phases of $MAPbX_3$ formed with polar aprotic solvents in the form of elongated crystallites. These solvatomorph phases are preformed in the nozzle, grow during the time of flight, and land on the substrate as already-formed crystalline nanowires containing little solvent. This prevents the spreading of the solution on the surface and/or the dissolving of the underlying layers in the case of defining high-aspect-ratio 3D structures. Once at the surface, with further evaporation of the solvent, the intermediate phases transform into $MAPbX_3$. This deposition feature is very specific to $MAPbX_3$ chemistry with polar aprotic solvents and allows the designing of a well-defined network, necessary for an X-ray photodetector. In this work, a device was fabricated depositing $MAPbI_3$ on graphene, as seen in Fig. 2. The detector exhibited ultrahigh sensitivity of $2.2 \times 10^8 \mu C \, Gy_{air}^{-1} \, cm^{-2}$, for detecting 8 keV X-ray photons at dose rates below 1 μGy/s (detection limit 0.12 μGy/s), which was a fourfold improvement of the best-in-class devices [21].

Therefore, perovskites have already been studied substantially as X-ray photo-detectors, even reaching industrial players. Samsung Electronics developed a detector that can decrease radiation exposure to less than 1/10th of the normal amount typical for medical imaging such as fluoroscopy, digital radiography, CT, and other radiology equipment [22].

3   Gamma Detection by Perovskite

Compared to X-rays, gamma photons have a much higher energy, meaning stronger penetration capabilities. Although perovskite materials have high linear attenuation coefficients, thicker devices will be needed for sufficient stopping of γ-rays. Stoumpos et al. [5] initially proposed centimeter-size SCs of all inorganic $CsPbBr_3$ for X- and γ-ray detection due to their high attenuation, high resistivity, and significant photoconductivity response. Dong et al. [23] then introduced, for the first time, a photodetector based on a $MAPbI_3$ SC to monitor an intense γ-ray radiation from a ^{137}Cs source, with *photon-to-electron* conversion efficiency of 3.9%.

Kovalenko and coworkers [4] followed that by demonstrating the γ-energy resolution capabilities with various hybrid lead halide perovskite SCs, mainly with $FAPbI_3$ SCs exhibiting high $\mu\tau$ products in the range of $10^{-2} \, cm^2 \, V^{-1}$. However, despite the record $\mu\tau$ product of their $Cs_xFA_{1-x}PbI_{3-y}Br_y$ ($x = 0$–0.1, $y = 0$–0.6) SC gamma dosimeter ($0.12 \, cm^2 \, V^{-1}$), a poor signal-to-noise ratio was visible, probably due to the low bulk resistivity and large dark current. Wei et al. [6] attempted to solve this issue by using a p-type $MAPbBr_3$ SC via dopant compensation of Cl^-. By further suppressing the crystal surface/edge leakage current with a guard-ring electrode, a tenfold improved bulk resistivity to $3.6 \times 10^9 \, \Omega \, cm$ was documented for the $MAPbBr_{2.94}Cl_{0.06}$ sample. The best photopeak energy resolution reached 6.5% as

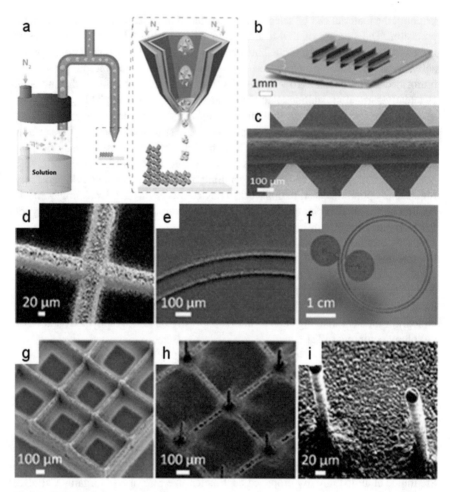

Fig. 2 Three-dimensional MAPbX$_3$ aerosol jet printing. (**a**) Schematic of the AJP system. The jet-focusing nitrogen flow helps the fast evaporation of the solvent and the growth of the intermediate phases, which is important for the creation of well-defined 3D structures. (**b**) 1 cm^2 sensing chip with 3D printed MAPbI$_3$ walls about 600μm in height. (**c**) False-colored SEM image of the 3D printed MAPbI$_3$ wall on the Ti/Au electrodes (graphene in blue, MAPbI$_3$ in purple, and metal electrodes in yellow). (**d–i**) Various printed patterns of MAPbI$_3$ lines, spirals, grids, and pillars written on the glass substrate [20]

the first demonstration of γ-ray energy spectrum collection was demonstrated by halide perovskite. He et al. [24] improved that by growing centimeter-sized single crystals of CsPbBr$_3$ with extremely low impurity levels (below 10 ppm for total 69 elements) that can resolve the 59.5 keV (^{241}Am γ-ray), 122 and 136 keV (^{57}Co γ-ray), 511 keV (^{22}Na γ-ray), and 662 keV (^{137}Cs γ-ray) lines with best spectral resolution of 3.8%.

Nevertheless, to date the volume of laboratory-grown metal halide perovskite dedicated for gamma-detecting purposes did not exceed 1.2 cm^3. In this chapter [25], the detection of γ-rays with MAPbBr$_3$ SCs with different volumes ranging from 0.1 to 1000 cm^3 is presented, as well as the influence of different electrode designs from silver epoxy to various carbon allotropes. In addition, the capability of dose rate measurements with great long-term stability is discussed.

3.1   Device Architecture

MAPbBr$_3$ single crystals, used in this work, were synthesized using the inverse temperature crystallization method. In 2015 Kadro et al. [26] and Saidaminov et al. [27] showed that MAPbX$_3$ (X = I, Br, Cl) perovskites exhibit inverse temperature solubility behavior in certain solvents. This important observation led to an innovative crystallization method, baptized as the inverse temperature crystallization route. The method allowed the growth of high-quality size- and shape-controlled single crystals rapidly, at a rate that is an order of magnitude faster than previously reported methods.

0.8 g MAPbBr$_3$ was dissolved per cm^3 of DMF at room temperature. Crystal growth was initialized by increasing the temperature of the solution from room temperature to 40 °C with a heating rate of 5 °C/h. Nice cubic like, centimeter-size single crystals can be harvested after only a couple of hours of crystallogenesis.

For resistively detected γ-radiation, the elaboration of stable and low-resistance contacts is very important. Our detector devices were initially assembled using copper wires as electrical contacts glued to the surface of MAPbBr$_3$ SCs with conductive adhesives, like Dupont 4929 silver epoxy (Fig. 3a). However, it has been recently recognized that perovskite-based optoelectronic devices implementing silver electrodes exhibited stability issues over long-term operation. Notably, silver and other noble metal-based electrodes undergo electro-corrosion, thus degrading and weakening the electronic properties. Therefore, lately, carbon-derived components such as carbon nanotubes, graphene, reduced graphene oxide, fullerenes, and graphite have been proposed.

Here, three types of carbon electrodes were investigated: vertically aligned carbon nanotube (VACNT) forests, where the perovskite single crystal engulfed the individual nanotubes as protogenetic inclusions (Fig. 3b), thus leading to the formation of a three-dimensionally enlarged interface [28]; graphite-spray electrodes covering the whole surface of the MAPbBr$_3$ single crystals, allowing large electrode surfaces (Fig. 3c); and lastly, graphite paper in pressed MAPbBr$_3$ polycrystalline sample in the form of pallets (Fig. 3d).

Gamma irradiation measurements were done inside an irradiation cavity "LOTUS" in the Laboratory for Reactor Physics and System Behavior at EPFL [29]. The experimental cavity is 3.6 m long, 2.4 m wide, and 3 m high, with a 2.2 m concrete shielding, which makes it adapted to irradiation with strong sources. It is equipped with a ^{60}Co radiative source and a shielded irradiator that allows safe usage

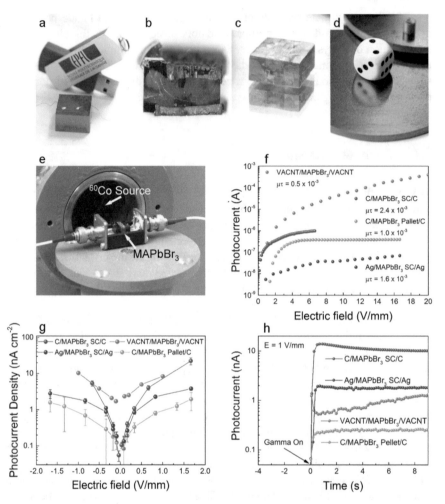

Fig. 3 Images of the samples different electrode strategies: (**a**) silver paste contacts, (**b**) aligned carbon nanotube contacts engulfed by the crystal, (**c**) graphite-spray contacts, and (**d**) pressed carbon paper on the surface of compressed polycrystalline sample. (**e**) Image of the detector device with graphite contacts inserted in the cavity and exposed to a ^{60}Co γ-source. (**f**) Photocurrent versus voltage measured at room temperature for the extraction of mobility-lifetime product for holes according to the Hecht equation. (**g**) The electric field dependence of photocurrent under a 2.3 Gy h^{-1} dose rate for the four device architectures, showing an excellent resolution even at low electric fields. (**h**) Time dependence of the photocurrent responses at 1 V mm^{-1} electric field. For simplicity we will use gamma on-source out, direct exposure

of the source and irradiation with a horizontal collimated beam. The ^{60}Co source has an activity of 269 GBq and produces γ-rays with energies of 1.173 and 1.332 MeV. It is doubly encapsulated in stainless steel and hermetically sealed. It is held by a steel and tungsten source rod, and incorporated in the irradiator. In the storage position, the lead and tungsten shielding is made so that the dose rate at 30 cm of the irradiator stays under 50 μSv h^{-1}. Therefore, users can safely enter the cavity and approach the irradiator. The tungsten rod, driven by a pneumatic system, travels inside the irradiator, allowing it to lift the source into the top exposed position in less than a second. In this position, the source is held in front of the beam port allowing a flow of γ-rays to directly enter the cavity. A removable collimator, made of several tapped tungsten and lead disks, is inserted in the beam port in order to limit scattering and provide a horizontal radiation beam with a 15-degree opening angle.

The perovskite γ-ray detector devices were positioned in the beam line (Fig. 3e) at various distances from 25 cm to 125 cm from the source. By varying the distance, we were able to change the exposed dose rate of the samples, due to the inverse-square law, from 2.3 to 0.07 Gy h^{-1}. Long coaxial cables were used to connect the detectors through tunnels in the concrete shielding with the measuring setup outside the cavity. The cavity is kept in ambient conditions; thus the samples were encapsulated in polydimethylsiloxane (PDMS), an optically clear and inert silicon-based organic polymer, to preserve them from humidity. Additionally, this polymer coating layer allows easier handling of the detector device, lowering the risk of lead toxicity. Lastly, all light sources were removed from the inner part of the cavity so as not to have any interference to the signal while under gamma exposure.

To compare the different devices, the fundamental figure of merit of γ-radiation detectors, i.e., the $\mu\tau$ product, was calculated from the γ-photocurrent-voltage curves showed in Fig. 3f. The obtained $\mu\tau$ products range from 0.5×10^{-3} to 0.3×10^{-2} cm^2 V^{-1}, which is comparable to the highest quality CZT SCs of 0.91×10^{-2} cm^2 V^{-1} [30]. However, the actual advantage of perovskite over CZT is that it possesses comparable $\mu\tau$ products for both holes and electrons [31] even though MAPbBr$_3$ has shown to be a hole transport semiconductor with electrons mostly affected by traps [32]. CZT, on the other hand, has one order of magnitude lower hole mobility [33]. Furthermore, these values are in range with $\mu\tau$ products of other hybrid halide perovskite detectors, keeping in mind that they were extracted from photocurrent generated from high-energy photon irradiation.

To further compare the different device types, J-E (current density vs. electric field) curves as well as photocurrent responses were plotted as seen in Fig. 3g, h. Typical current-voltage curves were not used as samples are very different in volume and electrode distance. The two best-performing devices are fabricated with the VACNT and graphite-spray electrodes. The VACNT/MAPbBr$_3$/VACNT device configuration offers the highest photocurrent density at low electric fields, presumably due to the large contact area between the CNTs and the perovskite, as well as the field enhancement factor at the nanotube tips [34]. However, due to the complexity of the device fabrication larger devices with VACNT electrodes are hard to obtain. Therefore, the absolute value of photocurrent response is much higher in the device fabricated with graphite-spray electrodes, because of larger volume allowing more interactions of high-energy photons with the device active area. In addition, the VACNT electrode device has unstable currents in time as seen in Fig. 3h. These VACNTs have shown to enhance ion migration in perovskite SCs [34]. This property utilized to an advantage for light emission is here, unfortunately, causing an unwanted current drift, which should be reduced in detectors as much as possible.

One can notice that even the "low-budget" device consisting of the pressed and sintered polycrystalline perovskite with graphite paper can perform properly. During all measurements the devices were positioned precisely 25 cm from the source, exposing them to a 2.3 Gy h^{-1} dose rate.

3.2 Dose Rate Measurements

Dose rate-dependent measurements, usually neglected in most of the reports, are important for the characterization of the device performance. In the case of a fixed-activity γ-source, the dose could be changed by the distance between the device and the irradiation source, this way varying the number of γ-photons reaching the detector in a unit time. Hence, our detector was positioned at different distances from the ^{60}Co source, ranging from 25 to 125 cm. Corresponding values of dose rates ranged from 2.30 to 0.07 Gy h^{-1}, which were measured by a calibrated γ-sonde.

On-off characteristics were measured for all aforementioned device architectures, for multiple distances from the source. The bias voltage for each device was chosen to give a predominantly stable and constant baseline current. This is crucial for calibrating the device properly to perform dosimetry measurements, and will be one of the main challenges to overcome. Furthermore, the value of the photocurrent at the leading edge of the γ-ray on was plotted as a function of distance from the source. These two types of curves are shown in Fig. 4a and b for the MAPbBr$_3$ SCs with graphite-spray electrodes.

The photocurrent measurements plotted as a function of distance (Fig. 4b) are in good agreement with the theoretical inverse-square law behavior. All curves are fitted to the $\dot{D}_0/r^x$ function, where $\dot{D}_0$ is the dose rate at the distance r and x is the power constant. The extracted value of the exponent ranged from (1.0 ± 0.3) to (2.45 ± 0.06) for the pressed pallet to the vertically aligned CNT electrode config-uration, respectively.

However, it is important to mention that because the irradiator with the ^{60}Co radioactive source is placed in a concrete irradiation cavity of reduced dimensions (3.6 × 2.4 × 3 m^3), the shielding geometry has to be taken into account. An intense scattering of gamma particles from the concrete walls affects the theoretically estimated inverse-square law behavior of dose rates for a point-like source in open space.

Therefore, to further prove that MAPbBr$_3$ γ-ray detectors can perform dose rate measurements, the devices were compared with two types of commercially available detectors for gamma dose rate measurements: a thermoluminescent dosimeter (Harshaw TLDs-700, 4.38% ^{7}Li-doped), and a calibrated γ-sonde (Berthold LB 6414). The commercial devices were positioned at the same distances from the source as our detectors, and measured in the same conditions. The distance-dependent curves are shown in comparison to our graphite spray and VACNT electrode samples in Fig. 4c. All measured values show a strong correspondence.

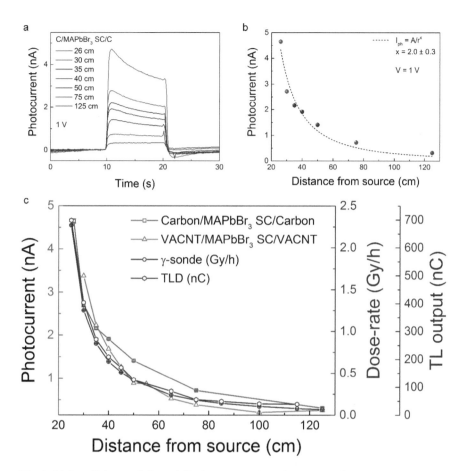

Fig. 4 (**a**) On-off characteristics and (**b**) photocurrent dependence on the distances from the source fitted to the $\dot{D}_0/r^x$ function for the sample with graphite electrodes. (**c**) Comparison of the detector photocurrents at different distances from the source (graphite and VACNT electrodes) with commercially available and calibrated gamma detectors

This is once more a confirmation of the advantages of using MAPbBr$_3$ SC devices with carbon electrodes, proving that they can be applied in dosimetry applications.

From the dose rate measurements, one can calculate the sensitivity of detecting gamma irradiation for the devices, a figure of merit more commonly used for X-ray detection. As mentioned previously, the dose rates corresponding to the distances of the detectors from the ^{60}Co source were read by the γ-sonde. The sensitivity was estimated from the slope of the liner fit of the photocurrent dose rate dependence. Values of 112 ± 3 μC Gy^{-1} cm^{-2} and 49 ± 2 μC Gy^{-1} cm^{-2} were determined, for our two best samples, MAPbBr$_3$ SC with VACNTs and graphite-spray contacts (Fig. 5a and c), respectively. It is important to point out how these sensitivities are achieved for low bias voltages (<2 V) and are enhanced by increasing the voltage. In Fig. 5b and d the dependence of the photocurrent density and sensitivity of the two

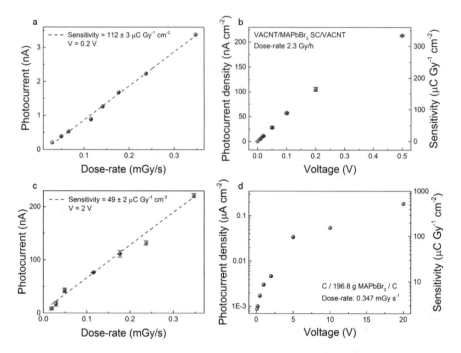

Fig. 5 Calculated sensitivity from the photocurrent dose rate dependence and its change with voltage for the γ-ray photodetector devices with (**a, b**) VACNT and (**c, d**) graphite-spray electrodes

devices with different bias voltages can be seen, reaching sensitivity values as high as 531 μC Gy^{-1} cm^{-2} for the MAPbBr$_3$ SC gamma detector device with graphite-spray electrodes (334 μC Gy^{-1} cm^{-2} for VACNTs at 0.5 V) when exposed to our strongest dose rate of 2.3 Gy/h.

These values are one order of magnitude higher than the best reported so far sensitivity of perovskite detectors when exposed to gamma rays, of 41 μC Gy^{-1} cm^{-2} [35], as well as in the same order as that of CZTs used as X-ray detectors for medical imaging [36].

3.3 Long-Term Operational Stability

Radiation damage is one of the major concerns of commercially available state-of-the-art detectors. It is well known that neutrons, protons, electrons, and even γ-radiation can lead to displacement damage in solid-state detectors (Si, CZT, etc.) [37]. In the case of high-energy particles, the displacement damage is a cascade-type process involving multiple interactions, resulting in an extended damage region or defect clusters. In contrast, in the case of γ-radiation, the displacement damage is caused by high-energy Compton electrons (around 1 MeV) that only produce point defects. These displacement damages lead to an increase of the leakage current and damage the sensor. Conventional materials for γ-ray detection can be made more radiation resistant by a purposeful introduction of known impurities. Conceivably, certain impurities, by their interaction with the primary defects caused by radiation, vacancies, and interstitials, can neutralize them without significantly altering the material properties. For example, silicon detectors with enhanced oxygen content exhibited better radiation hardness, being almost insensitive to γ-ray radiation up to 6 MGy [37].

As compared to the robust Si and CZT, hybrid halide perovskites are considered even more prone to degradation when exposed to various environmental factors. Consequently, performance deterioration of the perovskite-based γ-ray detector can clearly be expected in long-term operation. Therefore, since the detector stability is in general an important factor for perovskite optoelectronic devices, we performed test measurements on different timescales ranging from 1 h up to 100 h of operation.

The first γ-ray detector that was investigated was the simplest configuration, a MAPbBr$_3$ single crystal with silver epoxy electrodes. The device was positioned inside the irradiation cavity in front of the γ-beam exposing it to a constant dose rate of 2.3 Gy h^{-1} for 24 h, resulting in a total dose on the SC device of around 55 Gy. Moreover, a constant bias voltage of 5 V was applied constantly during the 24 h, simulating continuous operational conditions.

On-off measurements done in the same way as previously to test the γ-detection performance of the device were performed before and after the 24-h irradiation. Immediately after the long-term exposure the gamma response degrades to 10% of its initial value. These results did not come as a surprise, because as we mentioned previously, silver and other noble metal electrodes have shown many stability issues in all perovskite-based optoelectronic devices. This is quite logical as halides are widely used to etch metal layers as copper, silver, aluminum, and gold [38].

However, recently it has been documented that degradation in the performance of perovskite solar cells, due to illumination and load, can be recovered when leaving the device in the dark for a comparable amount of time. We attempted the same for gamma detection, measuring the on-off curves at different time intervals after the 24-h irradiation. During the first hour, no regeneration was visible. Yet, when the sample was left for 60 h in the dark under no external bias voltage, an almost complete recovery was made. The baseline current did not return to its initial value, but the photocurrent response fully restored.

Nevertheless, due to the initial degradation, our interest moved the samples fabricated with carbon electrodes. Carbon compared to other metal electrodes lacks electro- and photocorrosion, is cheap and abundant, and provides excellent chemical stability and tunability offering the possibility of selective electrodes. We concentrated mainly on the MAPbBr$_3$ SCs with graphite-spray contacts. Long-term operation measurements were performed in the same way as for the previous silver electrode device, exposing it to a constant 2.3 Gy h^{-1} dose rate and under a bias voltage of 1 V. The measurements were done for different irradiation intervals from 10 min to 100 h. One distinctive characteristic, which can be visible with all-time intervals, is that immediately after exposing the detector to the source we see an

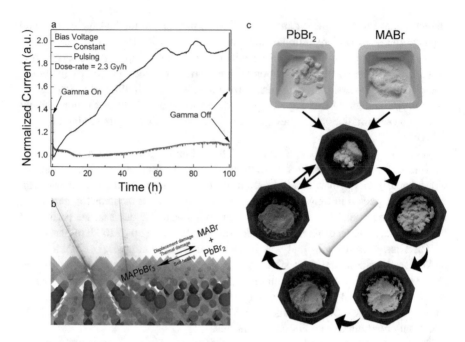

Fig. 6 Long-term stability improvement of the C/MAPbBr₃/C γ-ray photodetector by pulsing the bias voltage. (**a**) Comparison of photocurrent stability in time for different bias voltage schemes under a 2.3 Gy h⁻¹ dose rate exposure for 100 h. (**b**) Schematic representation of the self-repairing behavior of the perovskite under gamma irradiation. (**c**) Demonstration of a solvent-free, solid-state PbBr₂ and MABr reaction and a formation of MAPbBr₃ obtained by mechanical grinding, as a possible reason for a fast self-healing of the material after radiation damage

almost instantaneous increase in current (<1 s), followed by a decrease in current: fast in the first couple minutes, later saturating to a more or less constant value. This effect appears as well with visible light, and can be associated with the capacitive effect due to the pre-applied voltage before exposure. However, in the case of gamma irradiation, this drop in current can last up to 1 h and more. After this drop, the current recovers to its initial baseline, but also continues to increase further in time. This is probably due to the ion migration that starts to govern the change in baseline current for longer time intervals. The 24-h and 100-h exposures confirm this, as the current in their case continues to increase constantly, doubling in value for the longest duration (Fig. 6a, black curve).

Although the baseline current changes during these long-term exposures under operational conditions, no visible degradation of the sample or its detection properties is visible. Photocurrent response measurements were done before and after each of the irradiation intervals. When normalized to the baseline current, because it changes during operation, the detectors exhibit a very similar response before and after the exposure (<5% difference).

The constant bias voltage applied to the detector is the main reason for the substantial current drift in long-term operation. Ions inside the single crystal migrate due to the external electric field governing the current characteristics. This effect is even more pronounced for carbon electrodes, especially CNTs, which is the reason why we opted for graphite-spray electrodes. However, if one could apply a pulsing voltage with time intervals in between at open-circuit conditions, this poling effect could be suppressed. A pulsing voltage scheme, applying 1 V every 60 s, was tested for the C/MAPbBr$_3$/C device. For all cases (1, 12, and 100 h) the dark current variation was reduced substantially allowing changes of 10% or even less for the 100-h exposure as seen in Fig. 6a (red curve). After the initial decrease in photocurrent, commonly seen in perovskite devices, the value saturates and shows remarkable stability, most likely even beyond 100 h, which was just the chosen timescale of our experiment. Very small peaks of decrease in current visible throughout the curves could appear from different electronic effects during the pulsing of the voltage. These could later be removed with even more improved electronics. Therefore, the weak variation of the photocurrent in time, which is caused by ion migration inside the perovskite single crystal, can in principle be suppressed by using a low-frequency pulsed voltage source to operate the detector.

It is important to remember that all devices are encapsulated into PDMS, which protects the perovskite single crystal from ambient conditions, most notably humidity and oxygen. Therefore, all changes in current are in direct correlation with either the bias voltage or the gamma radiation. Photodetector devices were furthermore tested 1 and 2 years after their fabrication and initial measurements. They continue to perform in the same manner, confirming an excellent shelf life stability of 2 years and more. During all measurements the same γ-detector device was exposed to more than 300 h accumulating an absorbed dose of more than 600 Gy, and it still remains enacted with no degradation either visible or in device performance. This demonstrates that hybrid halide perovskites could be a viable radiation detection material with great radiation hardness.

However, the mechanism behind this phenomenon is currently unknown. It has been reported that perovskites possess natural defects, which are known not to interfere with detection sensitivity. These defects may act in the same way as the aforementioned purposely introduced impurities in Si, likewise improving the radiation hardness of perovskite single crystals. Another more important feature is that, unlike the robust semiconductors Si and CZT, perovskites are "soft mixed ionic and electronic semiconductors." When atoms are knocked out from their equilibrium positions, they form point defect like vacancies. This can, for example, create bromine-rich and bromine-deficient parts, which can further lead to a complete separation, resulting in the production of the MAPbBr$_3$ degradation compounds, MABr and PbBr$_2$ [39], accompanied by a phase transition and decreased absorption, which have already been observed in perovskites after long-term irradiation [40]. This would result in deterioration of the device performance. However, perovskites have shown the possibility to revert to their initial conditions after or even during these long-term operations, known as the self-healing mechanism [41–43]. Migration of the charged defects [44] or their (trans)formation by chemical reactions [41] is postulated to be the cause of this behavior (see the sketch of Fig. 6b).

As a proving example, experiments showed that perovskite solar cells survived accumulated dose levels up to 10^{16} and 10^{15} particles cm^{-2} of electrons (1 MeV) and protons (50 keV), respectively. These are known to completely destroy crystalline Si-, GaAS-, and InGaP/GaAs-based solar cells [45]. Besides, perovskite solar cells have shown to retain 96.8% of their initial power conversion after more than 1500 h of continuous gamma irradiation, with an accumulated dose of 23 kGy [42]. Yet, the mechanism behind this long-term γ-radiation stability under operational conditions and very high irradiation dose rates has not been addressed. The structural plasticity of organic-inorganic halide perovskites, fast ion migration, and thermodynamically favorable chemical reaction of "radiation-damage-phase-separated" components (MABr and PbBr$_2$) to perovskite product stimulate the reversible self-healing process on the nanoscale. The affinity of these components to enter into a reaction in solid state, without a solvent, just by steric proximity promoted by mechanical mixing is shown in Fig. 6c. One can see that after several minutes of grinding macroscopic quantities, the orange-yellow color of MAPbBr$_3$ is obtained. Therefore, even if a complete degradation of the material would occur, in a confined space of the crystal (at a microscopic scale), this mechanism might induce the self-healing of the radiation damage.

4 Effect of Size on Detector Capabilities

One of the main challenges of the perovskite photovoltaic field is upscaling, managing to retain the same cell efficiency while increasing the active area. However, for ionizing radiation in general, but especially γ-rays, it is not the surface of the crystal that is important but actually the thickness or volume. This can be seen nicely from Fig. 1 as the penetration depths for photons with energies above 100 keV become larger than 1 cm.

To test the real effect of crystal size and orientation and detection capabilities, perovskite devices of the same architecture but different sizes of MAPbBr$_3$ SCs were placed in the irradiation cavity. Initially, a detector device with silver epoxy electrodes was exposed to both γ-rays and visible light illumination at different crystal orientations, thus allowing for exposure of different crystal surfaces. As expected, under visible light illumination, the photocurrent notably increased when a larger active surface of the MAPbBr$_3$ SCs was exposed. In contrast, the photocurrent response to γ-irradiation exhibited a negligible difference between the two orientations.

However, when detectors of different volumes were exposed to the gamma radiation a vastly different outcome was measured. The devices were fabricated around larger MAPbBr$_3$ SCs with silver epoxy electrodes, keeping the same spacing between the electrodes, as schematically shown in Fig. 7a. The detectors were then positioned in the gamma beam keeping their first edge at a fixed distance from the source. This allowed us to measure the responsivity of γ-ray detectors as a function of volume of the MAPbBr$_3$ SCs in the range of 141–16,800 mm^3.

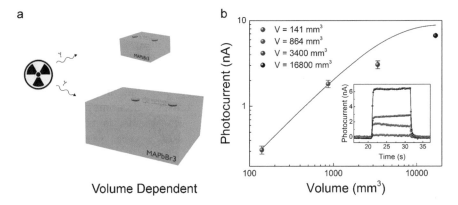

Fig. 7 (**a**) Schematic illustration of the configuration of volume dependence measurements of the photocurrent. (**b**) Photocurrent dependence as a function of the crystal volume. Blue line = guide for the eye. Inset: On-off characteristics at different crystal volumes

As shown in Fig. 7b, the photocurrent is markedly growing with the increasing crystal volumes of MAPbBr$_3$ SCs. Initially, for MAPbBr$_3$ SCs of smaller sizes up to about 2000 mm^3, the photocurrent increases linearly with growing crystal volumes, thus confirming both a high photocarrier extraction efficiency and the sensitivity dependence on the crystal volume. As might be expected, however, due to the finite value of the $\mu\tau$ product, as well as simplicity of our detector which employs just two electrodes placed on one crystal facet, this dependence becomes sublinear with further increasing of the crystal volumes. Clearly, the herein observed loss of photocarriers, which cannot be harvested by the electrodes due to recombination or trapping within the active crystal volume, can be eliminated by designing a more volumetric detection pattern in the future.

To estimate what thickness of SCs is needed for efficient interaction of high-energy photons with the active material the penetration depth was calculated. For the ^{60}Co gamma source, a linear attenuation coefficient of 0.1917 cm^{-1} was estimated which corresponds to a penetration depth of 5.22 cm. This proves that multiple-centimeter crystal sizes would be needed for efficient detection of photons with above 1 MeV energies.

4.1 Oriented Crystal-Crystal Intergrowth: OC2 Growth

Therefore, to estimate the limitations of the aforementioned volume dependence, a goal of growing a perovskite crystal with a thickness of over 10 cm was set. Upscaling of perovskite-based optoelectronic devices remains a challenge for the whole community. There is still no reliable technique to grow perovskite crystals in a controlled manner on a large scale. Nevertheless, laboratory prototypes of perovskite optoelectronic devices are posting record performance numbers, surpassing some well-established semiconductor materials such as silicon. Before 1915, monocrystalline silicon was at this stage of development. About 100 years ago, it had started to be implemented in various electrical applications with great success; however there were some limitations regarding the growth and purification methods and producible quantities. Nowadays, in the well-established silicon industry, up to 2 m large ingots of monocrystalline silicon are routinely grown and sliced into thin wafers. Today these are the basis of most of our modern solid-state electronic devices, from transistors to solar cells. The question then arises whether or not one day a Czochralski-type growth will be a routine method of large-scale crystallogenesis for perovskite single crystals as well.

In this chapter, the "oriented crystal-crystal intergrowth" method is introduced [25], hereinafter referred to as the OC2G technique, which can yield solution-grown MAPbBr$_3$ crystals with volumes and masses over 1000 cm^3 and close to 4 kg, respectively.

In the OC2G technique, nice cubic shape SCs of MAPbBr$_3$, that were previously solution grown by the inverse temperature crystallization method, serve as building blocks for the final large crystal. They are then precisely aligned side by side along their facets, like a mosaic or a 3D Rubik's cube pattern, as illustrated in Fig. 8a. It is not required to have SCs of the same size; however, attention should be on the perfect alignment of the edges to achieve a nice and fast intergrowth. After immersing them into a solution of MAPbBr$_3$ in DMF, they can be fused together by an inverse temperature crystallization. Under controlled and slowly increased temperature, the solution will gradually reach saturation, and the growing crystals fill up the space between the aligned facets (Fig. 8b), resulting in a very firm connection without inducing any unwanted polycrystallinity.

From two up to four crystals can be intergrown by OC2G at the same time, depending on the size and weight of the individual single crystals. Each additional added crystal increases the time of a single growth cycle, which is limited due to the concentration of MAPbBr$_3$ in the solution. Repeatable cycles, from room temperature to for example 80 °C, can be done to achieve a better junction. It is interesting to point out that under an applied force the final crystal will break at random places, and not along the merged facets. This breaking behavior of the OC2G crystal is an indirect proof of crystallinity and grain boundary quality, if it exists at all at the created junction.

After studying in detail the OC2G technique and learning its possibilities and limitations, a growth of a 10 cm thick crystal with a mass of over 1 kg was initiated. Over 35 small MAPbBr$_3$ seed single crystals, from a couple of millimeters to centimeter large, were first grown using the inverse temperature crystallization method. Together, their weight reached about 500 grams. The seed single crystals were then fused together in a multiple-step OC2G to form the large crystal. A series of time-lapse images depicting every cycle during the growth are showed in Fig. 8c.

Three-by-three seed single crystals were firstly grown together to make crystal "lines." Subsequently, three of these lines were grown together to obtain crystal plates weighting approximately 200 g and having a top surface area of about 8 × 8 cm^2. Four of these plates containing nine seed single crystals were merged

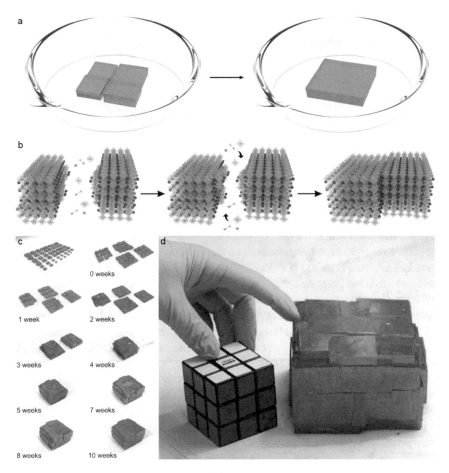

Fig. 8 (**a**) Schematic representation of the OC2G technique, fusing together individual single crystals into a larger crystal. (**b**) Illustration of the intergrowth at the molecular view. (**c**) Photographs of consecutive steps of the OC2G method took at different milestones. Full process starting from 36 single crystals of a total mass of 500 grams to a (**d**) 3.8 kg crystal

simultaneously over 2 weeks. Two-by-two crystal plates were then fused together by precisely aligning their flat surfaces. Due to the increased contact surface, a longer period was needed for this stage. Similarly, the same step was repeated to attain a full-connected MAPbBr$_3$ crystal with more than 5 cm long edges, weighting close to 2 kg (about 1 month of total growth). Repeating the growth cycles for additional weeks, while simultaneously introducing a fresh solution, the crystal continued to grow, increasing in size and weight. More than 300 g/week would be added to the mass of the crystal. Finally, the product of 10 weeks of total growth was a 1000 cm^3 and 3.82 kg crystal of MAPbBr$_3$, as shown in Figs. 8d.

To the best of our knowledge, this is by far the largest hybrid perovskite crystal produced, reaching the average baby birth weight of 3.8 kg. It is also worth mentioning that, in principle, there are no technological limitations to further exceed these impressive crystal sizes in the future. Likewise, with a complete understanding of the OC2G technique, the growth time could be cut at least in half, if not even more.

4.2   Increased Detection and Potential Shielding

A device was fabricated utilizing the large crystal to test if in fact the enlarged volume will benefit greatly the radiation detection signal. Firstly, silver epoxy contacts were deposited, as in our first size dependence measurements. However, it is obvious that a more volumetric electrode design is needed to be able to collect all generated charges. Therefore, the focus was moved to the graphite-spray electrode architecture (Fig. 9a). The 1 dm^3 crystal was positioned in the beam path as close as possible to the source, corresponding to a dose rate of 1.25 Gy h^{-1}. Photocurrents of over 100 nA were acquired at relatively low bias voltages, 2–5 V. These values were compared to other graphite-spray electrode samples based on $MAPbBr_3$ SCs, grown by inverse temperature crystallization. As seen in Fig. 9b, an increase in the photoresponse of over 100 times is present when utilizing crystals grown by the OC2G method, even despite a lower dose rate of irradiation due to setup limitations.

Therefore, once again the importance of an increased volume is confirmed, as well as the fact that the gain is not linear. Clearly, a loss of photocarriers, which cannot be harvested by the electrodes due to recombination or trapping within the active crystal volume, is present. This problem can be solved by designing a truly volumetric charge collection pattern. Nevertheless, the OC2G crystal, even with this simple electrode design, exhibited a significant signal increase and is proving to be a viable method for future large-volume perovskite radiation detectors.

Additionally, due to the high density of perovskite, large-volume $MAPbBr_3$ crystals could offer other applications, such as radiation shielding. For example, the 12 cm thick $MAPbBr_3$ crystal should attenuate 93.5% of the 269 GBq ^{60}Co (1.25 MeV) source. To prove this assumption, another detector was designed based on the large OC2G $MAPbBr_3$ crystal (volume $\sim$1000 cm^3). Two copper wires, embedded in silver epoxy contact pads on one facet of the crystal, served as biasing electrodes. The detector was then positioned in the beam of gamma radiation in three different orientations. Firstly, the crystal facet containing the biasing electrodes was exposed directly to the source from a distance of 35 cm (orientation "1," Fig. 9c). In this configuration, the "active volume," within which the charges are collected most likely, was directly exposed. Next, the detector was rotated by 90°, thus moving the contact area further from the radiation source (orientation "2"). Lastly, the detector was turned by an additional 90°, thus positioning the contact area entirely opposite to the source (orientation "3"). In this latter arrangement, the crystal facet containing the biasing electrodes was at the distance of $\sim$45 cm from the radiation source, thus also being "shielded" by the whole thickness of the $MAPbBr_3$ crystal (of $\sim$10 cm). The photocurrent across the detector drops markedly when the detector's spatial

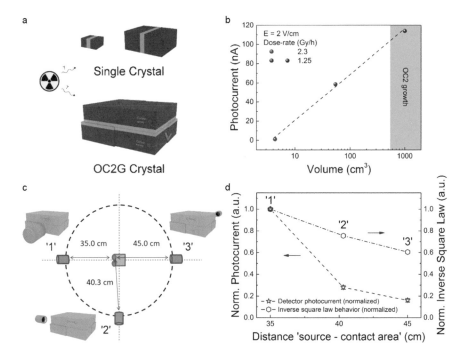

Fig. 9 The size effect of perovskite-based gamma detectors. (**a**) Schematic illustration of the configuration of volume dependence measurements of the photocurrent of MAPbBr$_3$ detectors based on single crystals attained by the inverse temperature crystallization method and crystals attained by the OC2G method. (**b**) Photocurrent dependence as a function of the crystal volume. (**c**) Sketchy representation of the different mutual orientations of the radiation source and the gamma-ray detector. (**d**) Comparison of the evolution of the photocurrent across the gamma-ray detector designed around a large MAPbBr$_3$ OC2G crystal (~1000 cm^3) and the "inverse square law" corresponding to three different mutual orientations of the crystal facet containing the biasing electrodes and the radiation source (Cobalt-60)

position is being switched from the orientation "1" to "2." This pronounced drop of the photocurrent (of ~73%, Fig. 9d) can be associated with strong "shielding" of the contact area provided by the outermost portion of the MAPbBr$_3$ crystal, which, for the orientation "2," is directly facing the radiation source. The photocurrent continues to decrease (by dropping up to ~82%) also for the orientation "3," in which, as mentioned above, the contact area of the detector is "shielded" by the whole thickness of the MAPbBr3 crystal (~10 cm). It is worth noting that the herein reported drop of the photocurrent is definitely more pronounced than the one, which could be expected for an "inverse distance squared" dependence (typically observed for exposure from a point radiation source, dark red circle in Fig. 9d). Altogether, these findings confirm strong attenuation of gamma rays by large-volume MAPbBr$_3$ crystals.

5   Conclusion

Detectors based on MAPbBr$_3$ SCs demonstrated detection of γ-rays from a ^{60}Co (1.173 and 1.332 MeV) source. Four different device configurations have been studied: simple copper wires as electrical leads glued to the SC by silver epoxy contacts; VACNTs engulfed inside the perovskite SC itself as a three-dimensionally enlarged electrode; graphite spray around the whole MAPbBr$_3$ SC as a large surface electrode; and "low-budget" device consisting of pressed and sintered polycrystal-line perovskite with graphite paper. All detector configurations were positioned inside an irradiation cavity directly in line with a gamma beam of the 269 GBq cobalt source, exposing them to a dose rate of 2.3 Gy h^{-1}. They managed to clearly detect a photocurrent response while moving the source from shielding to the exposed position and back. Directly from the photocurrent under γ-exposure the $\mu\tau$ products were obtained ranging from 0.5×10^{-3} to 0.3×10^{-2} cm^2 V^{-1}, which are comparable to those of the highest quality CZT SCs.

Furthermore, γ-radiation dose rate measurements were performed by varying the distance of the detectors to the gamma source and in that way changing the amount of γ-photons reaching the crystals in a unit of time. For distances from 25 to 125 cm, dose rates of 2.30 to 0.07 Gy h^{-1} were achieved on the sample. All devices exhibited an expected decrease of photocurrent response with the increase of distance from the source, while in the case of the SC device with carbon electrodes, the decrease of photocurrent of the leading edge of an on-off signal was in good agreement with the theoretical inverse-square law behavior, when fitted with a $\dot{D}_0/r^x$ function, where $\dot{D}_0$ is the dose rate at the distance r and x is the power constant. The extracted value of the exponent ranged from (1.95 ± 0.05) to (2.0 ± 0.3) for the VACNT and graphite-spray electrode configuration, respectively. Moreover, the measured dose rates showed a good correlation with commercially available thermoluminescent dosim-eters (Harshaw TLDs-700, 4.38% ^{7}Li-doped), and a calibrated γ-sonde (Berthold LB 6414), exhibiting the possibility of utilizing perovskites in dosimeter device.

Although the VACNT electrode device configuration showed the best responsivity in sample volume dependence, the design, which we believe could have the best commercial application potential, is the graphite-spray electrode configuration, as it requires less device fabrication complexity, and has a much more stable baseline current under long-term operation. Whereas the current of the VACNT detector increases constantly in time, due to the strong ion migration already present in all perovskite optoelectronic devices, it was further enhanced when using carbon nanotubes. The simplest architecture of just silver epoxy elec-trodes not only allows a big enough surface for increased charge collection, but more importantly has also proven to have very weak stability degrading the device γ-detection performance.

The MAPbBr$_3$ SC device with graphite-spray electrodes demonstrated good stability when irradiated from the ^{60}Co source under ambient and operational conditions for a testing period of 100 h. Utilizing a pulsing voltage scheme of applying 1 V every 60 s, the baseline current was stabilized, not changing more than 10% during this long-term operation. No degradation or deterioration of γ-radiation detection capabilities is observed during the testing period. We attribute the observed excellent radiation tolerance to the intrinsic structural plasticity of hybrid halide perovskites, as well as the fast ion migration accompanying a reversible component-phase separation-based defect-healing process, on the nanoscale.

Lastly, device photocurrent responses to γ-radiation are directly proportional to the active volume of $MAPbBr_3$ SCs, contrary to the active surface, as is the case for visible-light illumination. For this reason, a new growth technique was developed called oriented crystal-crystal intergrowth, or OC2G. This new technique successfully produced a very large $MAPbBr_3$ crystal of over 1 dm^3, and close to 4 kg. In the OC2G $MAPbBr_3$, single crystals are aligned side by side along their facets, similar to a Rubik's cube pattern, and fused together by the inverse temperature crystallization method. The formed crystals showed great solidity, unable to break at the junction.

An increase in the photoresponse of over 100 times is present when utilizing crystals grown by the OC2G method. However, for these higher crystal volumes the increase is sublinear, partially due to the finite value of the $\mu\tau$ product, but even more so because of the simplicity of the detector electrodes. A more volumetric detection pattern design would be needed in the future for better charge collection. It is important to point out that the large crystals obtained with the OC2G technique are, technologically, not limited in size. Furthermore, mostly due to their large thickness and attenuation, these crystals could be used as shielding, besides for detection, for highly radioactive sources.

In conclusion, the simplicity, low-cost solution-based fabrication process, and operational stability under γ-irradiation of the crystalline $MAPbBr_3$ make this material a good candidate for a new generation of high-energy radiation detectors.

Acknowledgement I would like to acknowledge László Forró and Endre Horváth for envisioning this project as well as the contributions from Márton Kollár, Anastasiia Glushkova, Pavel Frajtag, Vincent Pierre Lamirand, Andreas Pautz, Bálint Náfrádi, Andrzej Sienkiewicz, and Tonko Garma. The work was supported by the Swiss National Science Foundation (No. 513733) and the ERC advanced grant "PICOPROP" (Grant No. 670918).

References

1. Uses of Radiation. https://www.nrc.gov/about-nrc/radiation/around-us/uses-radiation.html. Accessed 01 July 2019
2. Radiation Answers. https://www.radiationanswers.org/radiation-sources-uses.html. Accessed 01 July 2019
3. Reilly, D., Ensslin, N., Smith Jr. H., Kreiner S.: Passive nondestructive assay of nuclear materials (No. NUREG/CR--5550). Nuclear Regulatory Commission, 1991
4. Yakunin, S., Dirin, D.N., Shynkarenko, Y., Morad, V., Cherniukh, I., Nazarenko, O., Kreil, D., Nauser, T., Kovalenko, M.V.: Detection of gamma photons using solution-grown single crystals of hybrid lead halide perovskites. Nat. Photonics. **10**(9), 585 (2016)
5. Stoumpos, C.C., Malliakas, C.D., Peters, J.A., Liu, Z., Sebastian, M., Im, J., Chasapis, T.C., Wibowo, A.C., Chung, D.Y., Freeman, A.J., Wessels, B.W.: Crystal growth of the perovskite

semiconductor CsPbBr3: A new material for high-energy radiation detection. Cryst. Growth Des. **13**(7), 2722–2727 (2013)

6. Wei, H., DeSantis, D., Wei, W., Deng, Y., Guo, D., Sevenije, T.J., Cao, L., Huang, J.: Dopant compensation in alloyed CH3NH3PbBr3−xClx perovskite single crystals for gamma-ray spectroscopy. Nat. Mater. **16**(8), 826 (2017)

7. Nazarenko, O., Yakunin, S., Morad, V., Cherniukh, I., Kovalenko, M.V.: Single crystals of caesium formamidinium lead halide perovskites: Solution growth and gamma dosimetry. NPG Asia Mater. **9**(4), e373 (2017)

8. High Purity Germanium Single Crystal. http://www.taikunchina.com/en/show-product-74. html. Accessed 05 July 2019

9. Manufacturing cadmium zinc telluride CZT and turning it into cutting-edge radiation detectors. https://www.kromek.com/cadmium-zinc-telluride-czt/. Accessed 05 July 2019

10. X-Ray Mass Attenuation Coefficients. https://physics.nist.gov/PhysRefData/XrayMassCoef/ tab4.html. Accessed 05 July 2019

11. Owens, A., Peacock, A.: Compound semiconductor radiation detectors. Nucl. Instrum. Methods Phys. Res., Sect. A. **531**(1–2), 18–37 (2004)

12. Andrea, Š., Zlatko, B., Nečas, V., Dubecký, F., Anh, T.L., Sedlačková, K., Boháček, P., Zápražný, Z.: From single GaAs detector to sensor for radiation imaging camera. Appl. Surf. Sci. **461**, 3–9 (2018)

13. Meng, L.J., He, Z., Alexander, B., Sandoval, J.: Spectroscopic performance of thich HgI2 detectors. IEEE Trans. Nucl. Sci. **53**(3), 1706–1712 (2006)

14. Limousin, O.: New trends in CdTe and CdZnTe detectors for X- and gamma-ray applications. Nucl. Instrum. Methods Phys. Res., Sect. A. **504**(1–3), 24–37 (2003)

15. Yakunin, S., Sytnyk, M., Kriegner, D., Shrestha, S., Richter, M., Matt, G.J., Azimi, H., Brabec, C.J., Stangl, J., Kovalenko, M.V., Heiss, W.: Detection of X-ray photons by solution-processed lead halide perovskites. Nat. Photonics. **9**(7), 444–449 (2015)

16. Wangyang, P., Gong, C., Rao, G., Hu, K., Wang, X., Yan, C., Dai, L., Wu, C., Xiong, J.: Recent advances in halide perovskite photodetectors based on different dimensional materials. Adv. Opt. Mater. **6**(11), 1701302 (2018)

17. Náfrádi, B., Náfrádi, G., Forró, L., Horváth, E.: Methylammonium lead iodide for efficient X-ray energy conversion. J. Phys. Chem. C. **119**(45), 25204–25208 (2015)

18. Wei, H., Fang, Y., Mulligan, P., Chuirazzi, W., Fang, H.H., Wang, C., Ecker, B.R., Gao, Y., Loi, M.A., Cao, L., Huang, J.: Sensitive X-ray detectors made of methylammonium lead tribromide perovskite single crystals. Nat. Photonics. **10**(5), 333 (2016)

19. Wei, W., Zhang, Y., Xu, Q., Wei, H., Fang, Y., Wang, Q., Deng, Y., Li, T., Guverman, A., Cao, L., Huang, J.: Monolithic integration of hybrid perovskite single crystals with heterogenous substrate for highly sensitive X-ray imaging. Nat. Photonics. **11**(5), 315 (2017)

20. Glushkova, A., Andričević, P., Smajda, R., Náfrádi, B., Kollár, M., Djokić, V., Arakcheeva, A., Forró, L., Pugin, R., Horváth, E.: Ultrasensitive 3D aerosol-jet-printed perovskite X-ray photodetector. ACS Nano. **15**(3), 4077–4084 (2021)

21. Sun, Q., Xu, Y., Zhang, H., Xiao, B., Liu, X., Dong, J., Cheng, Y., Zhang, B., Jie, W., Kanatzidis, M.G.: Optical and electronic anisotropies in perovskitoid crystals of Cs3Bi2I9 studies of nuclear radiation detection. J. Mater. Chem. A. **6**(46), 23388–23395 (2018)

22. Kim, Y.C., Kim, K.H., Son, D.Y., Jeong, D.N., Seo, J.Y., Choi, Y.S., Han, I.T., Lee, S.Y., Park, N.Y.: Printable organometallic perovskite enables large-area, low-dose X-ray imaging. Nature. **550**(7674), 87 (2017)

23. Dong, Q., Fang, Y., Shao, Y., Mulligan, P., Qiu, J., Cao, L., Huang, J.: Electron-hole diffusion lengths >175 um in solution-grown CH3NH3PbI 3 single crystals. Science. **347**(6225), 967–970 (2015)

24. He, Y., Matei, L., Jung, H.J., McCall, K.M., Chen, M., Stoumpos, C.C., Liu, Z., Peters, J.A., Chung, D.Y., Wessels, B.W., Wasielewski, M.R.: High spectral resolution of gamma-rays at room temperature by perovskite CsPbBr3 single crystals. Nat. Commun. **9**(1), 1609 (2018)

25. Andričević, P., Frajtag, P., Lamirand, V.P., Pautz, A., Kollár, M., Náfrádi, B., Sienkiewicz, A., Garma, T., Forró, L., Horváth, E.: Kilogram-scale crystallogenesis of halide perovskites for gamma-rays dose rate measurements. Adv. Sci. **8**(2), 2001882 (2021)

26. Noda, S., Hasegawa, K., Sugime, H., Kakehi, K., Zhang, Z., Maruyama, S., Yamaguchi, Y.: Millimeter-thick single-walled carbon nanotube forests: Hidden role of catalyst support. Jpn. J. Appl. Phys. **46**(5L), L399 (2007)
27. Saidaminov, M.I., Abdelhady, A.L., Murali, B., Alarousu, E., Burlakov, V.M., Peng, W., Dursun, I., Wang, L., He, Y., Muculan, G., Goriely, A.: High-quality bulk hybrid perovskite single crystals within minutes by inverse temperature crystallization. Nat. Commun. **6**(May), 7586 (2015)
28. Andričević, P., Kollár, M., Mettan, X., Náfrádi, B., Sienkiewicz, A., Fejes, D., Hernádi, K., Forró, L., Horváth, E.: Three-dimensionally enlarged photoelectrodes by a protogenetic inclusion of vertically aligned carbon nanotubes into $CH_3NH_3PbBr_3$ single crystals. J. Phys. Chem. C. **121**(25), 13549–13556 (2017)
29. Theler: Gamma irradiation in LOTUS with SILC. EPFL, student report, 2016
30. Kim, K., Kim, S., Hong, J., Lee, J., Hong, T., Bolotnikov, A.E., Camarda, G.S., James, R.B.: Purification of CdZnTe by electromigration. J. Appl. Phys. **117**(14), 145702 (2015)
31. He, Y., Ke, W., Alexander, G.C., McCall, K.M., Chica, D.G., Liu, Z., Hadar, I., Stoumpos, C. C., Wessels, B.W., Kanatzidis, M.G.: Resolving the energy of γ-ray photons with MAPbI3 single crystals. ACS Photonics. **5**(10), 4132–4138 (2018)
32. Pan, L., Feng, Y., Kandlakunta, P., Huang, J., Cao, L.R.: Performance of perovskite CsPbBr 3 single crystal detector for gamma-ray detection. IEEE Trans. Nucl. Sci. **67**(2), 443–449 (2020)
33. Musiienko, P.M., Grill, R., Praus, P., Vasylchenko, I., Pekarek, J., Tisdale, J., Ridzonova, K., Belas, E., Landová, L., Hu, B.: Deep levels, charge transport and mixed conductivity in organometallic halide perovskites. Energy Environ. Sci. **12**(4), 1413–1425 (2019)
34. Andričević, P., Mettan, X., Kollár, M., Náfrádi, B., Sienkiewicz, A., Garma, T., Rossi, L., Forró, L., Horváth, E.: Light-emitting electrochemical cells of single crystal hybrid halide perovskite with vertically aligned carbon nanotubes contacts. ACS Photonics. **6**(4), 967–975 (2019)
35. Wang, X., Wu, Y., Li, G., Wu, J., Zhang, X., Li, Q., Wang, B., Chen, J., Lei, W.: Ultrafast ionizing radiation detection by p-n junctions made with single crystals of solution-processed perovskite. Adv. Electr. Mater. **4**(11), 1800237 (2018)
36. Wei, H., Huang, J.: Halide lead perovskites for ionizing radiation detection. Nat. Commun. **10** (1), 1066 (2019)
37. Li, Z.: Radiation damage effects in Si materials and detectors and rad-hard Si detectors for SLHC. J. Instrum. **4**(03), P03011 (2009)
38. Choi, T.S., Hess, D.W.: Chemical etching and patterning of copper, silver, and gold films at low temperatures. ECS J. Solid State Sci. Technol. **4**(1), N3084–N3093 (2015)
39. Boyd, C.C., Cheacharoen, R., Leijtens, T., Mcgehee, M.D.: Chem. Rev. **119**, 3418 (2019)
40. Yang, K., Huang, K., Li, X., Zheng, S., Hou, P., Wang, J., Guo, H., Song, H., Li, B., Li, H., Liu, B., Zhong, X., Yang, J.: Org. Electron. **71**, 79 (2019)
41. Lang, F., Nickel, N.H., Bundesmann, J., Seidel, S., Denker, A., Albrecht, S., Brus, V.V., Rappich, J., Rech, B., Landi, G., Neitzert, H.C.: Adv. Mater. **28**, 8726 (2016)
42. Yang, S., Xu, Z., Xue, S., Kandlakunta, P., Cao, L.: Adv. Mater. **31**(1805547), 4 (2019)
43. Ceratti, D.R., Rakita, Y., Cremonesi, L., Tenne, R., Kalchenko, V., Elbaum, M., Oron, D., Alberto, M., Potenza, C., Hodes, G., Cahen, D.: Adv. Mater. **30**(1706273), 10 (2018)
44. Domanski, K., Roose, B., Matsui, T., Saliba, M., Turren-Cruz, S.H., Correa-Baena, J.P., Carmona, C.R., Richardson, G., Foster, J.M., De Angelis, F., Ball, J.M.: Migration of cations induces reversible performance losses over day/night cycling in perovskite solar cells. Energy Environ. Sci. **10**(2), 604–613 (2017)
45. Miyazawa, Y., Ikegami, M., Chen, H.W., Ohshima, T., Imaizumi, M., Hirose, K., Miyasaka, T.: Tolerance of perovskite solar cell to high-energy particle irradiations in space environment. iScience. **2**, 148–155 (2018)

Inorganic Halide Perovskite Thin Films for Neutron Detection

Leunam Fernandez-Izquierdo, Martin G. Reyes-Banda,
Jesus A. Caraveo-Frescas, and Manuel Quevedo-Lopez

Abstract The need for high-efficiency radiation detectors with wide-area coverage is essential in applications such as nuclear medicine, industrial imagining, environmental radioactivity monitoring, spacecraft applications, and homeland security, among others. For these applications, the detector material should interact strongly with high-energy particles or photons, must be able to operate at high electric fields with negligible leakage current, must possess high resistivity, and must be scalable. Cesium lead bromide ($CsPbBr_3$) possesses excellent electric, electronic, and spectroscopic properties while showing endurance to humidity and good stability under extreme operating conditions. These properties make it an ideal material for high-energy radiation detectors. The use of $CsPbBr_3$ for heavily charged particle sensing is normally limited to single crystals due to the lack of deposition techniques for thick $CsPbBr_3$ films, which is necessary for efficient radiation and neutron sensing. This chapter shows methods that allow the deposition of perovskite thin films with controlled thickness. The close-space sublimation (CSS) process allows for the deposition of stoichiometric and high-quality $CsPbBr_3$ films with reduced defects and large grains with high deposition rates. Alpha and neutron particle sensing using a p-n diode is discussed. This chapter demonstrates the potential of inorganic perovskite films for alpha and neutron detectors in planar and micro-structured perovskite thin films.

1 Introduction

Cesium lead bromide ($CsPbBr_3$) is a promising material for radiation sensing used as a scintillator or direct detector [1–3]. Besides its high stability, $CsPbBr_3$ possesses interesting electronic and optoelectronic properties such as high attenuation above the bandgap, good photoresponse, large electron and hole mobility, long lifetimes,

L. Fernandez-Izquierdo · M. G. Reyes-Banda · J. A. Caraveo-Frescas · M. Quevedo-Lopez (✉)
Department of Material Science & Engineering, University of Texas at Dallas, Richardson, TX, USA
e-mail: mquevedo@utdallas.edu

© The Author(s), under exclusive license to Springer Nature Switzerland AG 2022   81
K. Iniewski (ed.), *Advanced Materials for Radiation Detection*,
https://doi.org/10.1007/978-3-030-76461-6_4

low excitation binding energy, halogen self-passivation, defect tolerance, and luminosity [1, 4–12]. Device-quality single crystals have been prepared using many methods, including a high-temperature process [4], solution-based methods such as antisolvent vapor crystallization (AVC) [8, 9, 11], and inverse temperature crystallization (ITC) [13]. The carrier concentration (holes) of these crystals varies in the range of 5×10^7 to 1×10^8 cm^{-3} [10, 11] and about 1×10^9 cm^{-3} for electrons [10], resulting in nearly intrinsic crystals with resistivities in the range of 1–3 GΩ cm [11]. As a reference, Bridgman-grown crystals show resistivities as high as ~340 GΩ cm [4, 14] and mobility-lifetime (μτ) product for electrons and holes in the range of 1.7×10^{-3} to 4.5×10^{-4} cm^{2}V^{-1} and 1.3×10^{-3} to 9.5×10^{-4} cm^{2}V^{-1}, respectively [4, 15, 16]. These μτ values are better than those of CdZnTe (CZT) and CdTe. It is important to note that the electron μτ product of CZT and CdTe [12] is in the lower range of the corresponding values for $CsPbBr_3$, while the hole μτ product is only 0.1% that of $CsPbBr_3$.

The reported use of $CsPbBr_3$ for radiation sensing is limited to single crystals, except for its use as a thin film in X-ray scintillators [1, 3]. There are no reports for direct radiation detection using thin-film diodes based on $CsPbBr_3$. This is related to the lack of deposition techniques for thick $CsPbBr_3$ films. The close-space sublimation (CSS) process reported here addresses this problem. Although the intrinsic phase purity and crystal quality of single crystals offer improved optoelectronic properties, the high cost of the single-crystal approach renders this option nonviable for portable and large-area applications; hence thin films are a better option. Furthermore, the photon attenuation coefficient of $CsPbBr_3$ is linear and comparable to that of CZT for energies up to 1000 keV [4]. Still, no exhaustive studies exist for the interaction of $CsPbBr_3$ with charged particles (alphas, betas, etc.) or for higher photon energies such as gamma rays. The interaction of $CsPbBr_3$ 2D nanosheets with ionizing radiation has been reported recently, showing scintillation performance comparable to commercial crystals [1]. The observed luminosity of ~21,000 photons/MeV is comparable with commercial Cs_2LiYCl_6 (CLYC) crystals, but the luminescence decay time of <15 ns is much shorter than that of NaI:Tl (~200 ns), CLYC (>50 ns), and $LaBr_3$(Ce) (>16 ns) [17]. Recently gamma and alpha particle spectroscopy using $CsPbBr_3$ single crystals was reported [15].

There are a few reports for $CsPbBr_3$ thin-film deposition using solution processing [1, 18, 19], chemical vapor deposition (CVD) [20], vacuum evaporation [21], and even a hybrid vacuum solution process [22]. Although a solution process approach is "simple," economic, and more flexible, the stability and electronic properties of the resulting materials might be compromised by impurities and solvents incorporated from the precursor solution. On the other hand, physical vapor deposition methods, such as CVD and vacuum processes, eliminate solvents and yield higher quality materials, but precursor utilization is low and not practical for depositing thick films, necessary for efficient high-energy electromagnetic radiation and neutron sensing [23]. To overcome the aforementioned deficiencies, it has worked on developing a solution-free, simple, high-growth-rate, scalable, and inexpensive CSS process to deposit high-quality $CsPbBr_3$ films. Furthermore, this method can also be used for other material systems. $CsPbBr_3$ melts congruently at

~570 °C [4]; therefore, sublimation from $CsPbBr_3$ powder or crystals can produce stoichiometric films. The fact that CSS is a near-thermal-equilibrium deposition process results in films with reduced defects, large grain, high material utilization, and high growth rates. The films are used to demonstrate a thin-film solid-state neutron detector using an $ITO/Ga_2O_3/CsPbBr_3/Au$ diode.

Planar $CsPbBr_3/Ga_2O_3$ p-n junction diodes coupled with ^{10}B neutron conversion film have been demonstrated by showing a thermal neutron detection efficiency of 2.5% [24]. The theoretical neutron detection efficiency of planar detectors is limited to <4.5% due to the self-absorption of reaction products within the neutron conversion layer [25]. Higher efficiencies have been demonstrated when using micro-structured silicon diodes backfilled with a neutron conversion material [26–28]. Micro-structured detectors fabricated using thin-film semiconductors can be easily scaled for large-area applications than detectors using single-crystal semiconductors. Neutron detection efficiency using micro-structured detectors has also been demonstrated recently [28].

2 $CsPbBr_3$ Film Deposition Using Close-Space Sublimation

Figure 1a shows a diagram of the process to obtain perovskite thin films. $CsPbBr_3$ films with thicknesses between 8 and 20 μm can be deposited by the CSS process from $CsPbBr_3$ crystallites grown using the AVC method. Figure 1b shows the XRD patterns of the $CsPbBr_3$ precursor crystals and the resulting $CsPbBr_3$ film. The CSS-deposited $CsPbBr_3$ films have the same crystallographic nature and phase purity as the precursor crystals. High-incident-angle (2°) XRD analyses confirmed the phase purity in the bulk of the CSS-deposited $CsPbBr_3$ films with the ortho-rhombic perovskite structure [9, 10, 13, 29]. No diffraction peaks for polymorphs Cs_4PbBr_6 and $CsPb_2Br_5$ [30] were observed, further demonstrating the phase purity of the deposited $CsPbBr_3$ films. The crystallite size of the as-deposited $CsPbBr_3$ was in the range of ~245 nm with lattice constants $a = 8.205$, $b = 11.694$, and $c = 8.268$ Å, consistent with the orthorhombic $CsPbBr_3$ phase [10].

Surface morphology of as-deposited films (Fig. 1c), evaluated by scanning electron microscopy (SEM), shows grains with an average size of ~2.5 × ~6.5 μm with dense columnar growth (Fig. 1d). Recrystallization is evident after annealing at 450 °C for 30 min. This annealing was introduced further to increase the grain size and density of the films. The grain size and columnar growth of the CSS-deposited $CsPbBr_3$ are in sharp contrast with the smaller grains observed in films deposited by solution process or physical methods such as CVD and co-evaporation [19, 31]. EDXS analysis was performed across the film cross section at five points to examine any possible variation in stoichiometry. The composition is maintained constant throughout the film thickness. The surface roughness of the $CsPbBr_3$ film may result in discontinuities at the interface due to poor coverage of the contact films. To avoid this, a polishing process to reduce the surface roughness was introduced, reducing the roughness from ~270 nm to <50 nm.

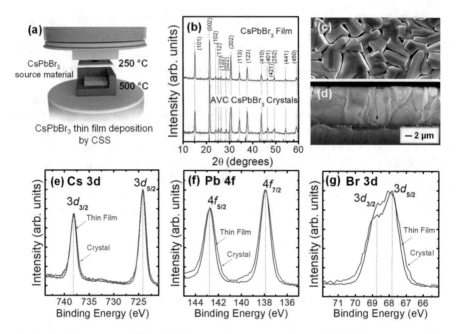

Fig. 1 (**a**) CSS process using the CsPbBr$_3$ crystals as the source material; (**b**) XRD patterns of the CsPbBr$_3$ film and crystals used as the source material, the reference XRD pattern corresponds to ICSD #97851; (**c**) top view and (**d**) cross-section SEM images of the as-deposited CsPbBr$_3$ thin film; (**e–g**) corresponding XPS spectra of Cs 3d, Pb 4f, and Br 3d showing identical chemical bonding for crystals and films

The X-ray photoelectron spectroscopy (XPS) spectra of the CsPbBr$_3$ film and crystals are shown in Fig. 1e–g. The binding energies for the Cs 3d, Pb 4f, and Br 3d regions are consistent with CsPbBr$_3$ [32]. Furthermore, the binding energies match for both the source crystals and the thin films, indicating that the CsPbBr$_3$ stoichiometry/composition and crystalline structure of the crystals are maintained during the CSS deposition process.

3   Ga$_2$O$_3$/CsPbBr$_3$ Diodes and Performance Characteristics

A schematic of the device configuration along with the band diagram is shown in Fig. 2a, b. Conduction (E_c) and valence band (E_v) edge energies of −3.0 and −5.3 eV, respectively, were estimated for the CsPbBr$_3$ film using data from photoelectron spectroscopy in air (PESA), Kelvin probe, and optical transmittance. From Hall measurements, the resistivity, mobility, and carrier concentration for the CsPbBr$_3$ film were determined as 1×10^{11} Ω cm, 0.013 cm^2 (V s)$^{-1}$, and 5×10^9 cm^{-3}, respectively. The corresponding values for the n-Ga$_2$O$_3$ are ~5×10^6 Ω cm, 1 cm^2 (V s)$^{-1}$, and ~10^{15} cm^{-3}, respectively.

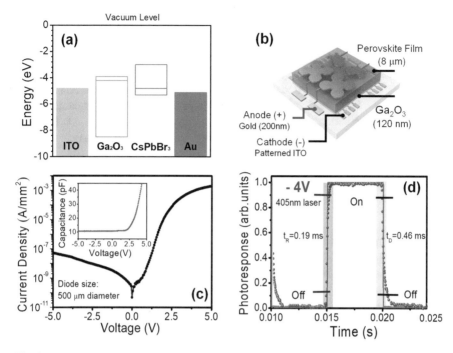

Fig. 2 (**a**) The band edge positions of different layers as determined experimentally and (**b**) schematic of the $ITO/Ga_2O_3/CsPbBr_3/Au$ device; (**c**) current-voltage characteristics (dark) for the $Ga_2O_3/CsPbBr_3$ diode; the inset shows the complete depletion of the device with an ~8 μm thick $CsPbBr_3$ film at low reverse bias; (**d**) rise/decay time estimation

To demonstrate charged particle sensing, a p-n junction diode with the configuration $ITO/Ga_2O_3/CsPbBr_3/Au$ was fabricated. The resulting diode showed leakage (reverse bias) current of 5×10^{-8} A/mm^2 and is fully depleted at -5 V (Fig. 2c and inset) with negligible I-V hysteresis and rectification of 10^4. The favorable band alignments and the high bandgap of Ga_2O_3 enabled low leakage current and diode rectification $>10^4$, which is significantly superior to that reported for $CsPbBr_3/ZnO$ [33] and $ZnO/CsPbBr_3/MoO_3$ diodes [34].

The rise and decay times for the $CsPbBr_3$ diode are shown in Fig. 2d. The device shows an on/off ratio of 10^2 at -4 V (E = 5 kV/cm) when exposed to a 405 nm laser (Fig. 2d). The estimated rise/decay time for the diode was 190/460 μs, which is much shorter than previously reported for $CsPbBr_3$-based diodes [33, 34] and photoresistors [8, 10, 35]. The superior performance of the fabricated devices, even without a hole transport layer, is attributed to the phase purity, large columnar grains of the CSS-deposited $CsPbBr_3$ films, as well as large bandgap of the Ga_2O_3 n-layer.

4   Alpha and Neutron Sensing with CsPbBr₃ Diodes

The diodes were evaluated as charged particle detectors, specifically for thermal neutron sensing using ^{10}B layer as conversion film. Neutron detectors normally include a "conversion" material such as ^{10}B in which a transmutation reaction produces an ionizing particle which in turn creates e-h pairs in the detector material [36]. The transmutation reaction includes a neutron captured by the ^{10}B that then becomes ^{11}B in an excited state that splits into ^{7}Li and an alpha particle (^{4}He) as shown in the reaction from Eq. (1) [36]:

$$^{10}B + n \rightarrow \left\{ \begin{array}{c} 94\%\,^{7}Li\,(0.840\text{ MeV}) + \alpha(1.470\text{ MeV}) + \gamma\,(0.48\text{ MeV}) \\ 6\%\,^{7}Li\,(1.02\text{ MeV}) + \alpha(1.77\text{ MeV}) \end{array} \right\} \quad (1)$$

The results for alpha and neutron response are shown in Fig. 3. The alpha peak is clearly observed after exposing the diodes to ^{210}Po (5.3 MeV) (Fig. 3a), demonstrating the Ga₂O₃/CsPbBr₃ diode sensitivity to alpha particles. The measured alpha response corresponds to 16% of that of a commercial OPF480 Si diode when exposed to the same ^{210}Po source. The peak broadening (Fig. 3a) can be due to various factors, such as the attenuation of alpha particles by the air and the top electrode, charge trapping in semiconductors, and electronic noise [37]. Next, the diode was exposed to thermal neutrons using a ^{252}Cf source. Results show neutron counts one order of magnitude higher than the background noise. Background noise measurements were performed before the exposure to the ^{252}Cf source. No increase in the background counts was observed for alpha particle or neutron exposure, indicating that the increase in the counts with time is a result of energetic particles absorbed in the Ga₂O₃/CsPbBr₃ diode, as seen in Fig. 3a (inset). No counts were detected above LLD. This confirms that the counts are originating from the alphas generated in the ^{10}B layer after neutron interaction and absorbed in the diode, and not due to gammas from the ^{252}Cf source. Figure 3b shows neutron response of the CsPbBr₃ diodes when exposed to a ^{252}Cf source. Calculated efficiency shows approx. 1% for CsPbBr₃ detectors representing 38.3% of the obtained silicon

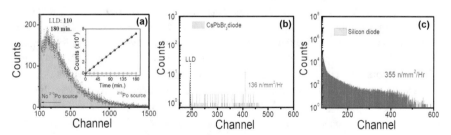

Fig. 3 (**a**) Alpha particle response when exposed to the ^{210}Po source, data collected for 180 min with a shaping time constant of 3 microseconds; inset: counts above LLD for alpha particles (black) and noise (green) recorded every 15 min; (**b**) normalized neutron response of CsPbBr₃-fabricated devices; (**c**) silicon-based neutron detector

efficiency (Fig. 3c). To the best of our knowledge, the demonstrated results are the first report of neutron detection using a thin-film CsPbBr$_3$-based device.

5 Composition Tuning in CsPbBr$_3$ Films

The diffractograms for films before and after solid-state ion exchange with chlorine (Cl) are shown in Fig. 4a. For films treated in PbCl$_2$, the characteristic XRD patterns of CsPbBr$_3$ shift depending on the annealing time. When annealed in PbCl$_2$ vapor, the characteristic peaks (101) and (002) shift towards higher 2θ values (Figs. 4a), indicating a smaller lattice constant resulting from the smaller Cl atoms substituting Br in the CsPbBr$_3$ films. This further indicates that Cl effectively incorporates in the crystalline structure of CsPbBr$_3$ [38, 39]. For PbCl$_2$-treated films, the lattice parameter is inversely proportional to the Cl content and shows a linear relationship.

The XPS spectra of the CsPbBr$_3$ films before and after annealing with PbCl$_2$ are shown in Fig. 4b, c. The binding energies of the Cs 3d region do not change for as-deposited or ion-exchanged films. This is demonstrated in the 723.66 and 737.66 eV peaks, corresponding to Cs 3d$_{5/2}$ and 3d$_{3/2}$ core levels [32, 40]. The shifting of Pb 4 f core levels to higher binding energies for PbCl$_2$-treated samples indicates stronger binding of the halogen with Pb, which is attributed to the higher electronegativity of Cl [41].

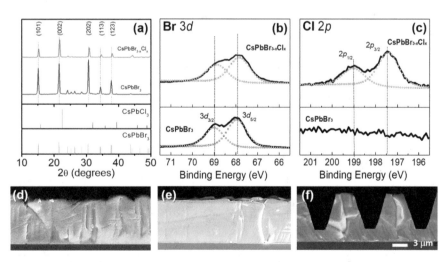

Fig. 4 (**a**) CsPbBr$_3$ films after thermal processing under the vapors of PbCl$_2$ for 120 min. The two bottom panes in each figure correspond to the ICSD patterns used as reference; (**b**) XPS spectra revealing Br 3d at 67.96 and 68.96 eV; (**c**) XPS spectra of Cl 2p at 197.3 and 198.9 eV for films treated in PbCl$_2$ vapors; (**d**) cross section of as-deposited CsPbBr$_3$ film; (**e**) cross section of films annealed for 120 min in PbCl$_2$ vapors; (**f**) cross section of etched microstructures in CsPbBr$_3$ thin films

The binding energies of Br $3d_{5/2}$ and $3d_{3/2}$ for CsPbBr$_3$ are shown in Fig. 4b. No significant change is observed for ion-exchanged films with either PbCl$_2$; however, the attenuation of Br 3d doublet and the presence in the Cl 2p doublet intensity (Fig. 4c) in PbCl$_2$-treated films demonstrate a successful ion exchange [42]. As expected, the atomic concentration of Cl in films treated with PbCl$_2$ varies linearly with annealing time.

Figure 4d shows the cross section of the as-deposited CsPbBr$_3$ films (8 μm). The film exhibits large grains, with an average size of ~5 μm. Annealing of this film at 425 °C under vacuum in the absence of PbX$_2$ results in the formation of voids without any grain growth. Recrystallization and grain growth are evident when annealed at the same temperature in the presence of PbCl$_2$ vapor (Fig. 4e), with a remarkable change for the films annealed. For films annealed in PbCl$_2$ (Fig. 4e) after 120 min, grain size increased to ~12 μm. This recrystallization is much larger than that obtained by methods such as solution-based treatments [43, 44] and CVD, where the maximum reported grain size was in the order of 0.3 to 0.5 μm [45]. The recrystallization in the presence of PbCl$_2$ vapor not only promoted grain growth but also resulted in a void-free surface with no visible grain boundaries across the film cross section [46]. The CsPb$_{3-x}$Cl$_x$ sample underwent a dry etching process using a combination of HBr and Ar plasma, with which it was possible to obtain the trenches in the perovskite films, as shown in Fig. 4f.

The CsPbBr$_3$ structure was also confirmed by Raman spectroscopy analysis (Fig. 5a). The Raman spectrum corresponds to an orthorhombic phase with bands at 52 cm^{-1} and 74 cm^{-1} assigned to vibrational modes of the [PbBr$_6$]$^{4-}$ octahedron and bands at 127 cm^{-1} and 151 cm^{-1} to Cs$^+$ ion vibrations [47, 48]. The Pb-Br rocking modes in the [PbBr$_6$]$^{4-}$ octahedron for Cs$_4$PbBr$_6$ have two intense bands at ~86 cm^{-1} and ~127 cm^{-1} [9, 49]; however, the absence of the 86 cm^{-1} band and the weak nature of the band at 127 cm^{-1} confirm that no Cs$_4$PbBr$_6$ is present. The weak nature of the 127 cm^{-1} bands is characteristic of CsPbBr$_3$ [9, 48, 49]. From the Raman spectra in Fig. 5a, it is evident that with the introduction of Cl, the degeneracies of Raman modes are lifted, leading to the appearance of additional phonon modes [47].

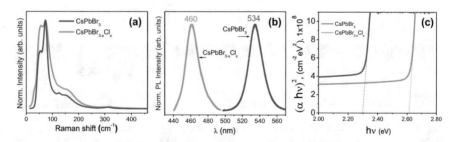

Fig. 5 (a) Raman spectra of the CsPbBr$_3$ and CsPbBr$_{3-x}$Cl$_x$ films; (b) PL bands of the as-deposited CsPbBr$_3$ and CsPbBr$_{3-x}$Cl$_x$ films; (c) estimated bandgap of the CsPbBr$_3$ and ion-exchanged films. The bandgap was determined using the relation $(\alpha E)^2 = A(E - E_g)$; the absorption coefficient (α) was obtained from the UV-Vis spectral measurements

The photoluminescence (PL) spectra of the as-deposited and ion-exchanged $CsPbBr_3$ films are shown in Fig. 5b. As-deposited $CsPbBr_3$ films show a strong emission centered at 534 nm and a weak shoulder at 550 nm. Films annealed in $PbCl_2$ show a strong emission centered at 460 nm. The weak band in the PL spectra of the thin films has been associated with several phenomena, including photon recycling [50, 51], structural differences between surface and bulk leading to slightly different bandgaps [52], bound excitons [4], and defects due to grain size inhomogeneity or traces of precursor [53, 54]. Based on the SEM and XRD analysis ($0.5°$ and $2.0°$ grazing angles), grain size inhomogeneity and precursor traces can be ruled out. Photon recycling can happen in translucent materials. The redshifted PL band can be observed along with the original PL emission when the signal is captured from a wide area [51]. Additionally, the bandgap of the $CsPbBr_3$ film was estimated at 2.32 eV and incremented to 2.62 eV with the ion-exchange treatment that further confirmed the incorporation of Cl into the lattice [46] (Fig. 5c).

6 $Ga_2O_3/CsPbBr_{3-x}Cl_x$ Diode Performance and Neutron Sensing

The performance of $CsPbBr_3$ in a diode configuration $ITO/Ga_2O_3/CsPbBr_3/Au$ (Fig. 2b) has been reported in previous sections. The fully depleted diode had a capacitance as low as 15 pF, a low leakage current of 5×10^{-8} A/mm^2 at -5 V, and rectification of $\sim 10^4$ [24]. Resistivity, mobility, and carrier concentration values of n-Ga_2O_3, determined from Hall measurements, are $\sim 5 \times 10^6$ Ω cm, 1 cm^2 $(V\ s)^{-1}$, and $\sim 10^{15}$ cm^{-3}, respectively [24, 55]. Figure 6a shows the current-voltage plots for the $Ga_2O_3/CsPbBr_3$ devices, both for as-deposited and ion-exchanged in $PbCl_2$ vapor. Lower leakage current density ($\sim 1 \times 10^{-8}$ A/mm^2) at -5 V was obtained with $PbCl_2$ ion-exchanged devices, with a rectification factor of 10^5. This value is $10\times$ higher than the fabricated device with as-deposited $CsPbBr_3$. This performance can be attributed not only to the favorable band alignments and the high bandgap of Ga_2O_3 but also to the recrystallization leading to grain growth, reduction in grain

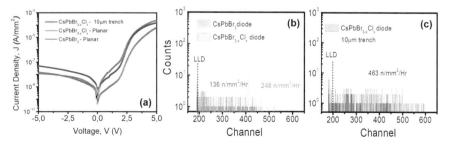

Fig. 6 (a) Effect of $PbCl_2$ vapor and the microstructure in $CsPbBr_3$ thin films on I-V properties of detectors; (b) normalized neutron response of $CsPbBr_3$- and $CsPbBr_{3-x}Cl_x$-fabricated devices; (c) normalized neutron response of $CsPbBr_{3-x}Cl_x$ with trenches

boundaries, and dopant compensation [38] that the $PbCl_2$ ion exchange provides to the $CsPbBr_3$ films. The changes observed between diodes without and with dry etching are mainly due to the generation of surface states on the sidewalls of the microstructures during etching. These surface states can negatively impact the electrical performance of devices.

Figure 6b shows normalized comparison of neutron response between the $CsPbBr_3$ and $CsPbBr_{3-x}Cl_x$ diodes when exposed to ^{252}Cf source. Theoretical efficiency using a ^{10}B conversion layer is ~4.5% (ref). Calculated efficiency shows approx. 2.5% for $CsPbBr_{3-x}Cl_x$ which represents a substantial increase compared to 1% for $CsPbBr_3$. The 2.5% efficiency represents 69.9% of the obtained silicon efficiency (Fig. 3c) [24, 46].

7 Radiation Detection Efficiency of Micro-structured $CsPbBr_{3-x}Cl_x/Ga_2O_3$ Detectors

Calculated efficiency in $CsPbBr_{3-x}Cl_x$ with trenches (10 μm deep) represents a substantial increase compared to planar devices. The device response to the ^{252}Cf source can be seen in Fig. 6c, showing an efficiency of approx. 4.3%. This result represents 130.4% of the alpha efficiency obtained with silicon-based diodes (Fig. 3c).

Figure 7a shows the comparison of experimental thermal neutron detection efficiencies of $CsPbBr_{3-x}Cl_x$-based detectors with the theoretical efficiency calculated by MCNP. The discrepancy in experimental neutron detection efficiency

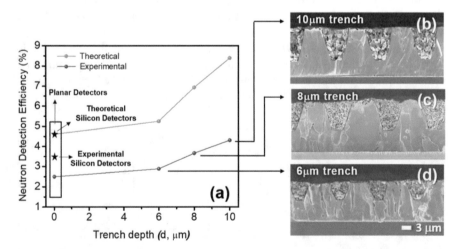

Fig. 7 (**a**) Comparison of neutron detection efficiencies between theoretical simulations and experimental results (the efficiency of planar silicon is represented using star symbol); (**b–d**) cross-section image of etched microstructures in $CsPbBr_{3-x}Cl_x$ thin films filled with ^{10}B conversion film

compared with theoretical efficiency is ~45%. The discrepancy increases with increasing trench depth and can be attributed to two major reasons: the packing density of ^{10}B in the microstructures and poor charge collection efficiency in the detectors. It is known that the discrepancy in the efficiency of up to 25% in micro-structured detectors arises from differences in experimental and theoretical packing density of ^{10}B in the microstructures [28]. In theoretical calculations, the micro-structures were assumed to contain ^{10}B with 100% density, whereas in these experiments, such packing density of ^{10}B powders was not achieved. Figure 7b–d shows the SEM cross-sectional image of microstructures in $CsPbBr_3$ that are backfilled with ^{10}B powders using a sedimentation process designed to preserve the perovskite integrity.

The microstructures were successfully filled without large voids. It is mainly due to the wide opening of trenches that enabled ^{10}B particles to backfill these structures easily. The irregular shape and different sizes of ^{10}B powder particles enable the close-packed filling in the microstructures. However, tiny voids between the powder particles prevent microstructures from attaining 100% fill density. Finally, the discrepancy due to low load collection efficiency in the etched samples may be related to permanent damage to the sidewalls during trench formation. Surface recombination of charges generated by incident alpha particles on these defective surfaces is more pronounced and could negatively impact the charge collection efficiency of the detector [56].

The etched sidewall area increases as the etch depth increases resulting in the formation of more surface defects. This can be seen in the broadening discrepancy of experimental efficiency as a trench depth function where more surface states are present.

8 Conclusions

A solution-free CSS process for depositing high-quality films of $CsPbBr_3$ is dem-onstrated and implemented in the fabrication of solid-state alpha particles and thermal neutron detectors. The CSS process used small crystals as precursors to deposit films with large columnar grain growth across the entire film thickness of 8 to 15 μm. Congruent sublimation of $CsPbBr_3$ resulted in films with a pure orthorhombic phase with no secondary phases. An ion-exchange treatment was developed to control composition and promote crystal growth, positively impacting the overall semiconducting properties of the perovskite film. Good band edge alignment, high resistivity of $CsPbBr_3$ and $CsPbBr_{3-x}Cl_x$, and large bandgap of the n-layer Ga_2O_3 result in diodes with low leakage currents (10^{-8} A/mm^2) with rectification between 10^4 and 10^5. Preliminary studies on the exposure to thermal neutrons resulted in counts ~60 higher than the background noise. The efficiencies of $CsPbBr_3$, $CsPbBr_{3-x}Cl_x$, and $CsPbBr_{3-x}Cl_x$ with trench (10 μm) samples are 1.0%, 2.5%, and 4.3%, respectively, that represent 38.3%, 69.9%, and 130.4% of the alpha capture efficiency seen in planar silicon diodes. These results are highly encouraging and highlight the promise of the CSS process in depositing perovskite films and the dry etching method to obtain micro-structured perovskite films, wallowing large-area fabrication perovskite-based neutron detectors.

References

1. Zhang, Y., Sun, R., Ou, X., et al.: Metal halide perovskite nanosheet for X-ray high-resolution scintillation imaging screens. ACS Nano. **13**, 2520–2525 (2019). https://doi.org/10.1021/acsnano.8b09484
2. Heo, J.H., Shin, D.H., Park, J.K., et al.: High-performance next-generation perovskite nanocrystal scintillator for nondestructive X-Ray imaging. Adv. Mater. **30**, 1801743 (2018). https://doi.org/10.1002/adma.201801743
3. Chen, Q., Wu, J., Ou, X., et al.: All-inorganic perovskite nanocrystal scintillators. Nature. **561**, 88–93 (2018). https://doi.org/10.1038/s41586-018-0451-1
4. Stoumpos, C.C., Malliakas, C.D., Peters, J.A., et al.: Crystal growth of the perovskite semiconductor CsPbBr₃: A new material for high-energy radiation detection. Cryst. Growth Des. **13**, 2722–2727 (2013). https://doi.org/10.1021/cg400645t
5. Li, X., Wu, Y., Zhang, S., et al.: CsPbX3 quantum dots for lighting and displays: room-temperature synthesis, photoluminescence superiorities, underlying origins and white light-emitting diodes. Adv. Funct. Mater. **26**, 2435–2445 (2016). https://doi.org/10.1002/adfm.201600109
6. Yettapu, G.R., Talukdar, D., Sarkar, S., et al.: Terahertz conductivity within colloidal CsPbBr₃ perovskite nanocrystals: remarkably high carrier mobilities and large diffusion lengths. Nano Lett. **16**, 4838–4848 (2016). https://doi.org/10.1021/acs.nanolett.6b01168
7. Shi, Z., Lei, L., Li, Y., et al.: Hole-injection layer-free perovskite light-emitting diodes. ACS Appl. Mater. Interfaces. **10**, 32289–32297 (2018). https://doi.org/10.1021/acsami.8b07048
8. Ding, J., Du, S., Zuo, Z., et al.: High detectivity and rapid response in perovskite CsPbBr₃ single-crystal photodetector. J. Phys. Chem. C. **121**, 4917–4923 (2017). https://doi.org/10.1021/acs.jpcc.7b01171
9. Cha, J.-H., Han, J.H., Yin, W., et al.: Photoresponse of CsPbBr₃ and Cs₄ PbBr₆ perovskite single crystals. J. Phys. Chem. Lett. **8**, 565–570 (2017). https://doi.org/10.1021/acs.jpclett.6b02763
10. Saidaminov, M.I., Haque, M.A., Almutlaq, J., et al.: Inorganic lead halide perovskite single crystals: phase-selective low-temperature growth, carrier transport properties, and self-powered photodetection. Adv. Opt. Mater. **5**, 1600704 (2017). https://doi.org/10.1002/adom.201600704
11. Zhang, H., Liu, X., Dong, J., et al.: Centimeter-sized inorganic lead halide perovskite CsPbBr₃ crystals grown by an improved solution method. Cryst. Growth Des. **17**, 6426–6431 (2017). https://doi.org/10.1021/acs.cgd.7b01086
12. Liu, Z., Peters, J.A., Stoumpos, C.C., et al.: Heavy metal ternary halides for room-temperature x-ray and gamma-ray detection. In: Hard X-Ray, Gamma-Ray, and Neutron Detector Physics XV. International Society for Optics and Photonics (2013), p 88520A
13. Dirin, D.N., Cherniukh, I., Yakunin, S., et al.: Solution-grown CsPbBr₃ perovskite single crystals for photon detection. Chem. Mater. **28**, 8470–8474 (2016). https://doi.org/10.1021/acs.chemmater.6b04298
14. Clark, D.J., Stoumpos, C.C., Saouma, F.O., et al.: Polarization-selective three-photon absorption and subsequent photoluminescence in ${{\mathrm CsPbBr}}_{3}$ single crystal at room temperature. Phys. Rev. B. **93**, 195202 (2016). https://doi.org/10.1103/PhysRevB.93.195202
15. He, Y., Liu, Z., McCall, K.M., et al.: Perovskite CsPbBr₃ single crystal detector for alpha-particle spectroscopy. Nucl. Instrum. Methods Phys. Res., Sect. A. **922**, 217–221 (2019). https://doi.org/10.1016/j.nima.2019.01.008

16. He, Y., Matei, L., Jung, H.J., et al.: High spectral resolution of gamma-rays at room temperature by perovskite CsPbBr$_3$ single crystals. Nat. Commun. **9**, 1609 (2018). https://doi.org/10.1038/s41467-018-04073-3

17. Glodo, J., Wang, Y., Shawgo, R., et al.: New developments in scintillators for security applications. Phys. Procedia. **90**, 285–290 (2017). https://doi.org/10.1016/j.phpro.2017.09.012

18. Cho, H., Wolf, C., Kim, J.S., et al.: High-efficiency solution-processed inorganic metal halide perovskite light-emitting diodes. Adv. Mater. **29**, 1700579 (2017). https://doi.org/10.1002/adma.201700579

19. Duan, J., Zhao, Y., Yang, X., et al.: Lanthanide ions doped CsPbBr$_3$ Halides for HTM-free 10.14%-efficiency inorganic perovskite solar cell with an ultrahigh open-circuit voltage of 1.594 V. Adv. Energy Mater. **8**, 1802346 (2018). https://doi.org/10.1002/aenm.201802346

20. Luo, P., Zhou, Y., Zhou, S., et al.: Fast anion-exchange from CsPbI3 to CsPbBr3 via Br2-vapor-assisted deposition for air-stable all-inorganic perovskite solar cells. Chem. Eng. J. **343**, 146–154 (2018). https://doi.org/10.1016/j.cej.2018.03.009

21. Burwig, T., Fränzel, W., Pistor, P.: Crystal phases and thermal stability of co-evaporated CsPbX3 (X = I, Br) thin films. J. Phys. Chem. Lett. **9**, 4808–4813 (2018). https://doi.org/10.1021/acs.jpclett.8b02059

22. Zhuang, S., Ma, X., Hu, D., et al.: Air-stable all inorganic green perovskite light emitting diodes based on ZnO/CsPbBr3/NiO heterojunction structure. Ceram. Int. **44**, 4685–4688 (2018). https://doi.org/10.1016/j.ceramint.2017.12.048

23. Murphy, J.W., Kunnen, G.R., Mejia, I., et al.: Optimizing diode thickness for thin-film solid state thermal neutron detectors. Appl. Phys. Lett. **101**, 143506 (2012). https://doi.org/10.1063/1.4757292

24. Fernandez-Izquierdo, L., Reyes-Banda, M.G., Mathew, X., et al.: Cesium lead bromide (CsPbBr3) thin-film-based solid-state neutron detector developed by a solution-free sublimation process. Adv. Mater. Technol., 2000534. https://doi.org/10.1002/admt.202000534

25. McGregor, D.S., Klann, R.T., Gersch, H.K., et al.: New surface morphology for low stress thin-film-coated thermal neutron detectors. IEEE Trans. Nucl. Sci. **49**, 1999–2004 (2002). https://doi.org/10.1109/TNS.2002.801697

26. Bellinger, S.L., Fronk, R.G., McNeil, W.J., et al.: Enhanced variant designs and characteristics of the microstructured solid-state neutron detector. Nucl. Instrum. Methods Phys. Res., Sect. A. **652**, 387–391 (2011). https://doi.org/10.1016/j.nima.2010.08.049

27. Fronk, R.G., Bellinger, S.L., Henson, L.C., et al.: Advancements in microstructured semiconductor neutron detector (MSND)-based instruments. In: 2015 IEEE Nuclear Science Symposium and Medical Imaging Conference (NSS/MIC) (2015), pp 1–5

28. Nandagopala Krishnan, S.S., Avila-Avendano, C., Shamsi, Z., et al.: 10B conformal doping for highly efficient thermal neutron detectors. ACS Sens. **5**, 2852–2857 (2020). https://doi.org/10.1021/acssensors.0c01013

29. Fu, Y., Zhu, H., Stoumpos, C.C., et al.: Broad wavelength tunable robust lasing from single-crystal nanowires of cesium lead halide perovskites (CsPbX3, X = Cl, Br, I). ACS Nano. **10**, 7963–7972 (2016). https://doi.org/10.1021/acsnano.6b03916

30. Li, J., Zhang, H., Wang, S., et al.: Inter-conversion between different compounds of ternary Cs-Pb-Br System. Materials (Basel). **11** (2018). https://doi.org/10.3390/ma11050717

31. Chen, H., Zhang, M., Bo, R., et al.: Superior self-powered room-temperature chemical sensing with light-activated inorganic halides perovskites. Small. **14**, 1702571 (2018). https://doi.org/10.1002/smll.201702571

32. Endres, J., Kulbak, M., Zhao, L., et al.: Electronic structure of the CsPbBr3/polytriarylamine (PTAA) system. J. Appl. Phys. **121**, 035304 (2017). https://doi.org/10.1063/1.4974471

33. Shen, Y., Wei, C., Ma, L., et al.: In situ formation of CsPbBr3/ZnO bulk heterojunctions towards photodetectors with ultrahigh responsivity. J. Mater. Chem. C. **6**, 12164–12169 (2018). https://doi.org/10.1039/C8TC04374D

34. Xue, M., Zhou, H., Ma, G., et al.: Investigation of the stability for self-powered CsPbBr3 perovskite photodetector with an all-inorganic structure. Sol. Energy Mater. Sol. Cells. **187**, 69–75 (2018). https://doi.org/10.1016/j.solmat.2018.07.023

35. Li, Y., Shi, Z.-F., Li, S., et al.: High-performance perovskite photodetectors based on solution-processed all-inorganic CsPbBr3 thin films. J. Mater. Chem. C. **5**, 8355–8360 (2017). https://doi.org/10.1039/C7TC02137B

36. McGregor, D.S., Hammig, M.D., Yang, Y.-H., et al.: Design considerations for thin film coated semiconductor thermal neutron detectors—I: basics regarding alpha particle emitting neutron reactive films. Nucl. Instrum. Methods Phys. Res., Sect. A. **500**, 272–308 (2003). https://doi.org/10.1016/S0168-9002(02)02078-8

37. Murphy, J.W., Smith, L., Calkins, J., et al.: Thin film cadmium telluride charged particle sensors for large area neutron detectors. Appl. Phys. Lett. **105**, 112107 (2014). https://doi.org/10.1063/1.4895925

38. Wei, H., DeSantis, D., Wei, W., et al.: Dopant compensation in alloyed CH 3 NH 3 PbBr 3−x Cl x perovskite single crystals for gamma-ray spectroscopy. Nat. Mater. **16**, 826–833 (2017). https://doi.org/10.1038/nmat4927

39. He, X., Qiu, Y., Yang, S.: Fully-inorganic trihalide perovskite nanocrystals: a new research frontier of optoelectronic materials. Adv. Mater. **29**, 1700775 (2017). https://doi.org/10.1002/adma.201700775

40. Hu, Y., Wang, Q., Shi, Y.-L., et al.: Vacuum-evaporated all-inorganic cesium lead bromine perovskites for high-performance light-emitting diodes. J. Mater. Chem. C. **5**, 8144–8149 (2017). https://doi.org/10.1039/C7TC02477K

41. Wu, Y., Li, X., Fu, S., et al.: Efficient methylammonium lead trihalide perovskite solar cells with chloroformamidinium chloride (Cl-FACl) as an additive. J. Mater. Chem. A. **7**, 8078–8084 (2019). https://doi.org/10.1039/C9TA01319A

42. Liashenko, T.G., Cherotchenko, E.D., Pushkarev, A.P., et al.: Electronic structure of CsPbBr3−xClx perovskites: synthesis, experimental characterization, and DFT simulations. Phys. Chem. Chem. Phys. **21**, 18930–18938 (2019). https://doi.org/10.1039/C9CP03656C

43. Li, B., Zhang, Y., Zhang, L., Yin, L.: PbCl2-tuned inorganic cubic CsPbBr3(Cl) perovskite solar cells with enhanced electron lifetime, diffusion length and photovoltaic performance. J. Power Sources. **360**, 11–20 (2017). https://doi.org/10.1016/j.jpowsour.2017.05.050

44. Zheng, L., Zhang, D., Ma, Y., et al.: Morphology control of the perovskite films for efficient solar cells. Dalton Trans. **44**, 10582–10593 (2015). https://doi.org/10.1039/C4DT03869J

45. Zhang, Y., Luo, L., Hua, J., et al.: Moisture assisted CsPbBr3 film growth for high-efficiency, all-inorganic solar cells prepared by a multiple sequential vacuum deposition method. Mater. Sci. Semicond. Process. **98**, 39–43 (2019). https://doi.org/10.1016/j.mssp.2019.03.021

46. Reyes-Banda, M.G., Fernandez-Izquierdo, L., Nandagopala Krishnan, S.S., et al.: Material properties modulation in inorganic perovskite films via solution-free solid-state reactions. ACS Appl. Electron. Mater. (2021). https://doi.org/10.1021/acsaelm.1c00072

47. Calistru, D.M., Mihut, L., Lefrant, S., Baltog, I.: Identification of the symmetry of phonon modes in CsPbCl3 in phase IV by Raman and resonance-Raman scattering. J. Appl. Phys. **82**, 5391–5395 (1997). https://doi.org/10.1063/1.366307

48. Zhang, L., Zeng, Q., Wang, K.: Pressure-induced structural and optical properties of inorganic halide perovskite CsPbBr₃. J. Phys. Chem. Lett. **8**, 3752–3758 (2017). https://doi.org/10.1021/acs.jpclett.7b01577

49. Yin, J., Zhang, Y., Bruno, A., et al.: Intrinsic lead ion emissions in zero-dimensional Cs4PbBr6 nanocrystals. ACS Energy Lett. **2**, 2805–2811 (2017). https://doi.org/10.1021/acsenergylett.7b01026

50. Dursun, I., Zheng, Y., Guo, T., et al.: Efficient photon recycling and radiation trapping in cesium lead halide perovskite waveguides. ACS Energy Lett. **3**, 1492–1498 (2018). https://doi.org/10.1021/acsenergylett.8b00758

51. Fang, Y., Wei, H., Dong, Q., Huang, J.: Quantification of re-absorption and re-emission processes to determine photon recycling efficiency in perovskite single crystals. Nat. Commun. **8**, 1–9 (2017). https://doi.org/10.1038/ncomms14417
52. Wu, B., Nguyen, H.T., Ku, Z., et al.: Discerning the Surface and Bulk Recombination Kinetics of Organic–Inorganic Halide Perovskite Single Crystals. Adv. Energy Mater. **6**, 1600551 (2016). https://doi.org/10.1002/aenm.201600551
53. Priante, D., Dursun, I., Alias, M.S., et al.: The recombination mechanisms leading to amplified spontaneous emission at the true-green wavelength in CH3NH3PbBr3 perovskites. Appl. Phys. Lett. **106**, 081902 (2015). https://doi.org/10.1063/1.4913463
54. Sebastian, M., Peters, J.A., Stoumpos, C.C., et al.: Excitonic emissions and above-band-gap luminescence in the single-crystal perovskite semiconductors $CsPbBr_3$ and $CsPbCl_3$. Giant-magnetoresistance anomaly associated with a magnetization process in UFe4Al8. Phys. Rev. B. **92**, 235210 (2015). https://doi.org/10.1103/PhysRevB.92.235210
55. Pintor-Monroy, M.I., Barrera, D., Murillo-Borjas, B.L., et al.: Tunable electrical and optical properties of nickel oxide (NiOx) thin films for fully transparent NiOx–Ga2O3 p–n junction diodes. ACS Appl. Mater. Interfaces. **10**, 38159–38165 (2018). https://doi.org/10.1021/acsami.8b08095
56. Wright, G., Cui, Y., Roy, U.N., et al.: The effects of chemical etching on the charge collection efficiency of 111 oriented Cd/sub-0.9/Zn/sub-0.1/Te nuclear radiation detectors. IEEE Trans. Nucl. Sci. **49**, 2521–2525 (2002). https://doi.org/10.1109/TNS.2002.803852

Radiation Detection Technologies Enabled by Halide Perovskite Single Crystals

Feng Li, Tiebin Yang, and Rongkun Zheng

Abstract Recently, following the unprecedented rise in the performance of optoelectronic devices based on the halide perovskite polycrystalline or nanostructured films, low-cost halide perovskite single-crystal candidates have attracted much attention due to their excellent optoelectronic properties and improved stability when compared to polycrystalline and nanostructured counterparts for various applications such as photovoltaic cells, photodetectors, light-emitting diodes, and lasers. Their unique optoelectronic properties and improved stability, as well as their low-cost raw materials and growing processes, make halide perovskite single crystals and their various forms greatly suitable for radiation detection technologies. In this book chapter, we summarize various synthesis and growth methods of halide perovskite single crystals and introduce their strong radiation detection performance. The advantages and limitations of halide perovskite single crystals, which are employed as active candidates for various radiation detection applications, are discussed in detail. Finally, outlooks on the remaining challenges and new opportunities in the field of single-crystal perovskite radiation detectors, with particular emphasis on device designs, ion migration, and stability concerns, are highlighted.

1 Introduction

Ionizing radiation, defined as having enough energy to ionize atoms or molecules, can be applied in many areas of our lives [1–3]. In the forms of particles or electromagnetic waves, ionizing radiation can be categorized by being directly ionizing or indirectly ionizing, respectively. Any charged particle with enough

F. Li (✉) · R. Zheng
School of Physics, Australian Centre for Microscopy & Microanalysis, Nano Institute, The University of Sydney, Sydney, NSW, Australia
e-mail: feng.li2@sydney.edu.au; rongkun.zheng@sydney.edu.au

T. Yang
School of Physics, The University of Sydney, Sydney, NSW, Australia
e-mail: tiebin.yang@sydney.edu.au

© The Author(s), under exclusive license to Springer Nature Switzerland AG 2022
K. Iniewski (ed.), *Advanced Materials for Radiation Detection*,
https://doi.org/10.1007/978-3-030-76461-6_5

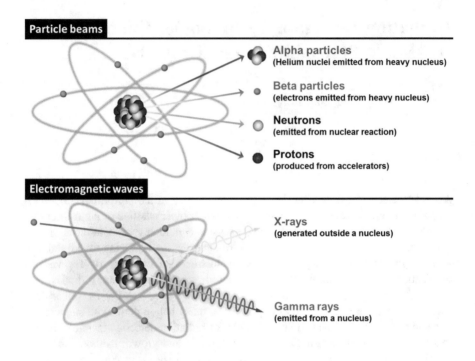

Fig. 1 (**a**) Schematic representation of particle radiation with their formation. (**b**) Schematic representation of the electromagnetic spectrum and the relevant formation

kinetic energy can ionize atoms by Coulomb force, which induces direct ionizing. Directly ionizing radiation particles include atomic nuclei, electrons, muons, charged pions, protons, and energetic charged nuclei stripped of their electrons, of which the most common types are alpha particles and beta particles consisting of helium nuclei and electrons or positrons, respectively. Indirectly ionizing radiation is usually photon radiation or neutron radiation induced by electrically neutral particles, where the atoms are further ionized by beta particles from the interaction between photons or neutrons and other atoms. Photon radiation is typically known as high-energy electromagnetic wave X-rays (0.1–100 keV) and gamma rays (0.1–100 MeV), which will cause the ejection of electrons from atoms and then further ionization by secondary beta particles. Neutrons often cause ionization by interacting with protons in hydrogen via linear energy transfer. Figure 1 represents the formation of common radiation types, including particle radiation and electromagnetic (photo) waves and their related producing/emitting ways. Due to the peculiarity of high penetration and ionization, radiation rays can be widely used in various fields, including nuclear physics, medical imaging, radiotherapy, food industry, security monitoring, crystallography, and so on [4–8].

Generally, semiconductors with an appropriate bandgap can easily detect free charges induced by radiation under applied voltages. Semiconductor-based radiation detectors, such as common photodetectors working in the UV-visible-infrared light ranges, can directly obtain a current signal from free electrons excited by Coulomb force or photoelectric effect. Current radiation detectors are mainly based on silicon (Si), amorphous selenium (a-Se), germanium (Ge), cadmium zinc telluride (CZT), cadmium telluride (CdTe), and mercury iodide (HgI_2) [9–12]. However some prerequisites should be fulfilled before semiconductors can serve as high-performance radiation detectors. Generally, the semiconductor bandgap is required to stay in the range from 1.5 to 2.5 eV, which can ensure a low dark current at room temperature while providing enough of an energy barrier for electron-hole pair production. High average atomic number Z and high mass density lead to the large stopping power required for successfully detecting high-energy radiation. For instance, even though improvements of crystal purity and doping techniques have been applied for the widely used Si semiconductor, the relatively small bandgap (1.12 eV) causes a large dark current at room temperature, and the small atomic number sets a great limitation for fabricating high-performance Si-based radiation detectors. CZT materials, with a relatively large bandgap (1.57 eV) and high atomic number (49.3), have also shown excellent performance in commercial radiation detectors, which can be operated at room temperature [13–15]. The carrier mobility-lifetime product ($\mu\tau$) is another important factor for identifying suitable semiconductors for radiation detection application. A large $\mu\tau$ indicates a long carrier diffusion length, by which the probability of a carrier being captured and recombined is reduced. Radiation can also be detected by indirect detectors consisting of a scintillator and photodetector arrays. Scintillators can convert high-energy particles and photons into visible light via radioluminescence, which is further detected by other sensitive photodetectors such as charge-coupled devices (CCD), photodiode arrays, and complementary metal oxide semiconductors (CMOS). A high-quality scintillator should have the traits of high light yield, long-term stability, and high energy resolution. Most commercially used scintillators are usually based on inorganic crystals such as thallium-doped sodium iodide (NaI(Tl)) and thallium-doped cesium iodide (CsI (Tl)) [16–18]. However, these commercial semiconductors usually require complex growth methods and severe operating conditions, which creates a sharp rise in demand for high-performance and low-cost materials.

As a class of promising photoactive materials, metal halide perovskites $APbX_3$, where the A site can be monovalent organic or inorganic cations such as the methylammonium ($MA^+ = CH_3NH_3^+$) ion or Cs^+, or the mixture thereof, and X is usually a halide component such as Cl^-, Br^-, or I^-, or a mixture thereof, have attracted much attention in energy and optoelectronic applications due to their excellent features, including a long carrier lifetime, large absorption coefficient, high light yield, and cost-effective growth method—encouragingly, halide perovskites, particularly the single crystals, have exhibited many suitable properties for radiation detection applications [19–24]. Ever since halide perovskites were first determined to have the ability to detect radiation [25], many radiation detectors based on this class of perovskites have been successfully fabricated [26–29]. Accordingly, the rapid research progress and great strides made by halide perovskites with different crystal forms in ionizing radiation detection call for a swift and consistent survey into the state of the field. Recently, there have been excellent review papers on the radiation detection devices based on halide perovskites [30, 31]. Each of these corresponding review publications has a different line of focus, depth, and narrative. However, review papers centered on perovskite radiation detection applications that also include the detection of directly ionizing radiation particles are limited. In this regard, this book chapter aims to summarize the recent achievements, ongoing progress, and challenges of halide perovskite single crystals for radiation detection devices with an emphasis on the fundamental detection principles of various radiation photons and charged particles. We begin by offering growth methods for halide perovskite single crystals and their unique physical characteristics for radiation detection. The subsequent section comprehensively presents the mechanism of different types of radiation detectors. Thereafter, the recent achievements of halide perovskite radiation detectors in mainly detecting alpha and beta particles, as well as gamma and X-rays, are summarized and discussed in detail. Finally, we present a brief perspective for the future development of halide perovskites for radiation detection applications and suggest some strategies to enhance their long-term stability and device performance.

2 Halide Perovskite Single Crystals

2.1 Growth of Halide Perovskite Single Crystals

2.1.1 Bulk Halide Perovskite Single Crystals

The solution temperature lowering (STL) method was developed by Tao's group to grow $MAPbI_3$ bulk single crystals (Fig. 2a), since such materials have decreased solubility in HX (X = Cl, Br, and I) solution at low temperatures [32]. Through this method, a large single crystal ($10 \times 10 \times 8$ mm^3) was grown in about 1 month (Fig. 2b). Furthermore, $MAPbBr_{3-x}Cl_x$ and $MAPbI_{3-x}Br_x$ mixed-halide perovskite crystals were grown using this method, during which hydrobromic acid was mixed with hydrochloric acid or hydroiodic acid in different molar ratios into methylamine and lead (II) acetate solution. Because this method has the disadvantage of being time consuming, radically faster perovskite crystal synthesis approaches have thus been developed. In this regard, the inverse temperature crystallization (ITC) method has been widely applied in recent years; it was observed that the exhibited crystals from this method can be shape controlled, can be of higher quality, and can be obtained quicker compared with other growth techniques. Bakr et al. introduced this method to rapidly grow high-quality bulk crystals [33], where an orange $MAPbBr_3$ crystal was grown within 3 h (Fig. 2c). The ITC approach was further modified in order to grow large-scale bulk perovskite crystals. Furthermore, the incorporation of seed crystal growth has been demonstrated to be an effective strategy for the large-scaled single crystals. The growth of various large-sized perovskite crystals via the modified ITC method was also reported by Liu et al., who obtained a number of

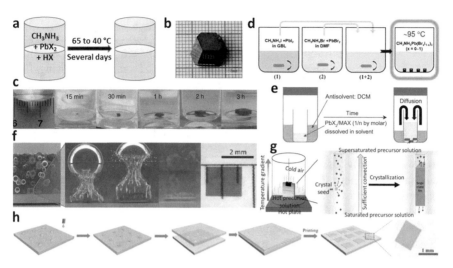

Fig. 2 (a) Schematic of STL method. (b) Image of bulk MAPbI₃ crystal. *Cryst.Eng.Comm* [32], Copyright 2015. (c) MAPbBr₃ crystal growth by ITC method at different time intervals. *Nature Commun* [33], Copyright 2015. (d) Schematic of growth process for the large-sized crystals. *J. Mater. Chem. C* [34], Copyright 2016. (e) Schematic of AVC method. *Science* [35], Copyright 2015. (f) Schematic of cavitation-triggered asymmetrical method. *Adv. Mater* [36], Copyright 2016. (g) Schematic for the growth of perovskite thin crystals. *J. Am. Chem. Soc.* [37], Copyright 2016. (h) Schematic of the scalable growth for perovskite crystal films using an inkjet printing method. *Sci. Adv* [38], Copyright 2018

larger sized crystals (7 mm) by choosing good-quality seed crystals and repeating and carefully controlling the ITC process several times (Fig. 2d) [34]. Taking advantage of the different solubilities of different perovskite single crystals at varying temperatures contributes to the time-saving feature of the ITC method. Another main method to grow perovskite crystals is the anti-solvent vapor-assisted crystallization (AVC) method (Fig. 2e), which was first introduced by Bakr et al. [35]. In this method, two or more solvents are selected, of which one should be a good solvent (solubility) that is less volatile, and the other is a bad solvent that is more volatile. Because of the insolubility of the material in the bad solvent, the proficiency of the perovskite crystal formation increases significantly when the bad solvent slowly diffuses into the precursor solution. Other groups, such as Loi's group and Cao's group, also applied this method to obtain the high-quality crystals [39]. Although the AVC method costs more time than the ITC method, its temperature-independent character is appealing for its widespread use.

2.1.2 Thin Single Crystals

Bulk perovskite single crystals with large thicknesses may cause an increase of charge recombination, thus leading to the degradation of their device performance and impeding practical applications. In this regard, growing thin perovskite crystals represents an effective approach to overcome the above obstacle and thus advance further practical applications. Bakr et al. developed a cavitation-triggered asymmetrical crystallization strategy to grow the thin single crystals. In the processing, they used a very short ultrasonic pulse (≈ 1 s) in the solution to reach a low supersaturation level with anti-solvent vapor diffusion; the μm-sized thin crystals could be grown within hours (Fig. 2f) [36]. Su et al. used a space-limited ITC method and grew a 120 cm^2 single crystal on fluorine-doped tin oxide (FTO)-coated glass [40]. Meanwhile, Wan et al. reported a space-confined solution-processed method to grow the perovskite single-crystal films with adjustable thickness from nanometers to micrometers (Fig. 2g) [37]. Benefitting from the capillary pressure, the perovskite precursor solution filled the whole space between two clean flat substrates, which were clipped together and dipped in the solution.

Currently, more promising approaches have been employed to grow thin single crystals with high quality and large scale. A one-step printing geometrically confined lateral crystal growth method was introduced by Sung et al. to obtain a large-scaled single crystal [41]. During the process, a cylindrical metal roller with a flexible poly-(dimethyl-siloxane) (PDMS) mold was wrapped and then rolled on a preheated SiO$_2$ substrate (180 °C) with an ink supplier filled with the precursor solution. Alternatively, mm-sized single crystals were synthesized by Song et al. by a facile seed-inkjet-printing approach (Fig. 2h) [38]. After injecting the perovskite precursor solution onto the silicon wafer, the ordered seeds could be formed. Another substrate with the perovskite-saturated solution then covered the top, and thin crystals could be grown as the solvent dried at room temperature. Aiming to realize advanced optoelectronic devices, developing more promising growth approaches to synthesize size-controllable perovskite crystals will be rewarding in the future.

2.2 Superb Properties of Perovskite Crystals for Radiation Detection

As mentioned above, the superior photophysical and electronic properties of halide perovskites enable them as promising candidates for next-generation energy and optoelectronic devices [19–24]. As shown in the Introduction, single-crystal perovskites have exhibited many properties that are actually desired for high-performance radiation detectors, particularly for detecting alpha particles, beta particles, gamma rays, and X-rays [25–31, 42]. In this section, the superb features of perovskite single crystals for radiation detection applications are introduced in detail:

2.3 Large Stopping Power

Heavy metal elements (normally Pb or Sn) and halide components (usually Cl, Br, or I) provide this class of perovskite materials with high average atomic numbers, which offer large stopping power and high detection efficiency. For example, all-inorganic halide perovskite $CsPbBr_3$ has an average atomic number of 65.9, which is even larger than that of 49.3 for CZT. Hybrid perovskite $MAPbI_3$ has a density of 4 g/cm^3, corresponding to a linear attenuation coefficient of 10 cm^{-1} at 100 keV, while $CsPbI_3$ holds an even larger linear attenuation coefficient of 14 cm^{-1}. Figure 3a shows the linear attenuation coefficients of halide perovskite materials, including $MAPbI_3$ and $CsPbI_3$, and those of the commonly used Se, CdTe, and TlBr at different photon energies.

2.4 Suitable Bandgap and Large Bulk Resistance

As mentioned above, appropriate bandgaps are crucial for high-performance radiation detectors, especially for room-temperature operation. Encouragingly, halide perovskites hold tuneable bandgaps ranging from 1.3 to 3.2 eV that can be easily realized by adjusting the halide components [33–42, 45–50], which make them perfectly suitable for the radiation detectors (Fig. 3b, c). Meanwhile, the bulk resistance of halide perovskite ranges from 10^7 to 10^{10} Ω cm, which contributes to a low dark current at room temperature and low noise, especially when applied to high voltage bias.

2.5 High Mobility-Lifetime Product

Both the intrinsic and extrinsic defects of the semiconducting materials can capture the charge carriers that are generated by radiation, thus decreasing the signal current of the related devices. The large $\mu\tau$ product of halide perovskites can reduce the recombination of electron-hole pairs at the defects and traps within them, and thus more carriers can be collected by the electrodes, which further increases the detector efficiency. Perovskite single crystals usually exhibit a low trap density of 10^7–10^9 cm^{-3} and thus offer high $\mu\tau$ products [51–53]. For example, the $MAPbI_3$ single crystal holds a high $\mu\tau$ product of 10^{-2} cm^2/V, with an ultralong carrier diffusion length of over 175μm under the illumination of sunlight [54]. Moreover, the internal quantum efficiency of 3 mm thick $MAPbI_3$ single crystal approaches 100% under weak light illumination, which indicates a diffusion length almost exceeding 3 mm.

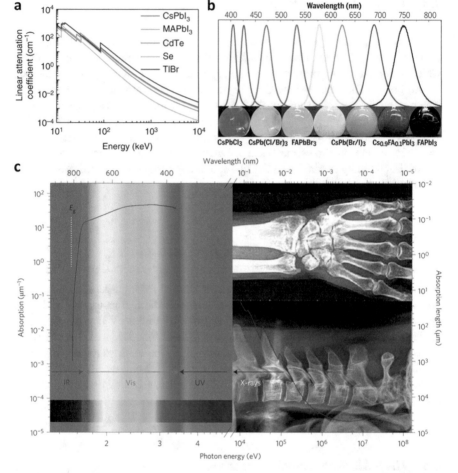

Fig. 3 (**a**) Attenuation coefficients of $CsPbI_3$, $MAPbI_3$, CdTe, Se, and TlBr versus different photon energies. Reproduced with permission [30]. Copyright 2019, Nature Publishing Group. (**b**) PL spectra of tuneable $APbX_3$ nanocrystals. Reproduced with permission [43]. Copyright 2017, AAAS. **c**) Absorption coefficient and length with different photon energy of $MAPbI_3$ crystal. Reproduced with permission [44]. Copyright 2015, Nature Publishing Group

2.6 Facile and Low-Cost Crystal Growth Method

The other advantages of halide perovskites are their cost-effective crystal growth and device fabrication methods. Compared to widely used commercial semiconductors, such as Si, Ge, Se, and CZT, of which the growth processes typically require high temperature, high vacuum, or complex instruments, halide perovskites are usually produced by low-temperature (<150 °C) solution processes such as the abovementioned ITC method [33, 53, 55], AVC method [35, 52, 56], and supersaturated recrystallization [57, 58]. Moreover, the total cost for the growth of a 1 cm³ perovskite single crystal is estimated at around $0.3/cm^3 when scaling up the production, which is three or four times lower than CZT crystals [30]. In addition, the related detection devices can be easily fabricated through one-step spin-coating or spray methods and by directly using the freshly grown perovskite crystals, which offers a large operating range for further integration.

2.7   High Light Yield

The fascinating properties of halide perovskites also enable them to be both direct photoconductors and indirect scintillators for radiation detections. When transformed into different crystal forms, like nanocrystals and quantum dots, halide perovskites also exhibit strong radioluminescence. High-performance scintillators require high light yield (LY) and fast decay time, which are crucial for measuring the time of the initial particle or radiation with high precision and increased timing resolution. When compared to LaBr$_3$-Ce, which exhibited an optimal device performance with a LY value of 70,000 ph/MeV and a decay time of 16 ns, halide perovskites hold a much higher LY value of over 10^6 ph/MeV and a fast decay time of less than 1 ns [59]. In addition, perovskites in their nanocrystal form also permit easy halide ion exchange, from which a tuneable luminescence spectrum across almost the whole visible region can be obtained [60, 61].

Armed with all these advantages, direct radiation detectors based on perovskite single crystals have exhibited excellent device performance by working in photoconductor mode [26, 44, 62–66]. Due to their high light yield under radiation illumination, halide perovskites can serve as scintillators integrated with other semiconductor photodiodes, by which high-performance radiation detectors can be achieved with multiple detection abilities [43, 67–69].

3   Principles of Radiation Detectors

3.1   Direct Radiation Detectors

Suitable semiconductors can directly detect radiation, of which the fabricated detectors are operated in current or voltage modes based on the direct interaction of incident particles or photons with the sensitive semiconductors [70, 71].

3.1.1   Alpha and Beta Particle Detections

The alpha particle was named as such by Rutherford in 1899 as it had the lowest penetration of ordinary objects; the particle was then confirmed to be helium nuclei in 1907 [72]. Emitted from the alpha decay of heavy atoms, alpha particle energy varies with the half-life of the emission process, where higher energies are caused by larger nuclei. Most alpha particles hold energy ranging from 3 to 7 MeV with a speed of 15,000 km/s due to their higher particle mass compared to other radiation rays. Thus, alpha particle penetration is typically lower than beta particles or gamma rays, as it can be blocked by a piece of paper or human skin. However, it is still the most destructive ionizing radiation; alpha-emitting atoms that are inhaled, ingested, or injected into our body will cause serious chromosome damage of DNA that is 20 times greater than that caused by the same dose of gamma rays or beta particles [73]. Developing high-quality detectors for alpha particle radiation is not only for environmental safety concerns but also for the nuclei information carried by alpha particles. An alpha spectrum is commonly used to characterize alpha particles emitting from nuclei. Thus, peak resolution is an important parameter for alpha detectors, which helps distinguish the alpha peak in the alpha spectrum. For high-energy alpha particles, direct radiation detectors usually work in voltage mode, since the particle flux is relatively weak and alpha particles will come into the detector one by one. The electron-hole pairs generated by alpha particles are further collected by the detector, of which the intensity is proportional to the particle energy. Then, a histogram of the energy-resolved spectrum can be obtained. Since the preliminary signal intensity of a histogram is relatively low, a charge-sensitive preamplifier is often integrated with the semiconductor. The amplifier then converts the collected charges into a voltage signal which is later read out by a multiple channel digitizer and finally output as an alpha particle spectrum.

Compared to alpha particles, beta particles usually hold relatively lower energies but have much higher speed, which leads to a different interaction model with atoms. Beta particles can directly interact with other electrons in atoms by inelastic scattering, which excites other electrons to high energy levels or to emit photons. Beta particles can also interact with nuclei through elastic scattering due to their much lower mass, during which only the trajectories of beta particles change. Once electron-holes are generated by beta particles, they can be detected by semiconductor detectors on voltage mode.

3.1.2 Gamma-Ray Detections

Gamma rays, as the form of electromagnetic waves with the shortest wavelength and the highest energy, are usually emitted from atom nuclei. Gamma rays can also be detected by a scintillator or photodetector working in voltage mode that records a gamma-ray spectrum [74]. Based on the different gamma-ray energies, the electron-hole pairs can be generated in different ways. The photoelectric effect often occurs when the gamma photon energy ranges from 10 to 500 keV, where all the energy of gamma photons is used for electron generation. With gamma photon energies from 50 keV to 3 MeV, Compton scattering can occur, where part of the energy of gamma photons is transferred into electrons. If a gamma ray holds energy over the MeV range, electron and positron pairs will be generated in the semiconductor, which is called pair production. The $\mu\tau$ product is an important factor for gamma-ray detectors, which can be derived by fitting the modified Hecht formula [75]:

$$I = \frac{I_0\mu\tau V}{L^2} \frac{1 - \exp\left(-\frac{L^2}{\mu\tau V}\right)}{1 + \frac{Ls}{V\mu}}$$

where I_0 is the saturated photocurrent, L is the material thickness, V is the applied voltage, and s is the surface recombination velocity. Gamma-ray detector performance can be typically characterized by the spectral peak resolution, which is defined by the ratio of the full-width at half-maximum (FWHM) and the incident energy of the radiation source.

3.1.3 Direct X-Ray Detections

X-rays, as electromagnetic waves with high penetration, are some of the most widely used radiations in modern society. Since its discovery by Rontgen in 1895, X-rays were immediately applied to medical applications due to their high sensitivity to different materials, thus enabling imaging applications [76]. In the twentieth century, X-rays have shown their ability in crystallography, imaging, and microscopy, and were even used in World War I. Nowadays, the utility of X-rays has covered medicine, food safety, environmental surveillance, industry, and science research [2, 6, 77]. Similar to gamma rays, X-rays can be detected by either direct photodetectors coupled with semiconductor photodiodes or indirect scintillators. While aiming to get high imaging performance, extra considerations of sensitivity, spatial resolution, and lowest detectable dose rate need to be taken into account for X-ray detectors. Direct X-ray detectors usually work in current mode, where photon flux is strong enough to generate current signals. Compared to high-energy particles or photons, which are collected by the detector separately, multiple X-ray photons come and interact with semiconductors at the same time and are converted into electron-hole pairs by both the photoelectric effect and Compton scattering. Many charges are generated during these two processes, which will further be collected by electrodes under the applied voltage bias to produce the current signal. The key parameter of the direct X-ray detector is sensitivity, which is calculated by

$$S = \frac{\int \left[I_{x-\text{ray}}(t) - I_{\text{dark}}\right]dt}{D \times V}$$

where $I_{X\text{-ray}}$ and I_{dark} are current under the radiation of X-ray and at dark conditions, respectively; D is the X-ray dose rate; and V is the detector volume. Detectors with high sensitivity can generate more current signals under the same irradiation with a higher signal-to-noise ratio, which leads to higher contrast when applied to imaging applications. The linear dynamic range (LDR) describes the range in which the sensitivity remains constant. A long LDR enables detectors to work steadily among a large range of X-rays with different dose rates. Another important parameter of the X-ray detector is the lowest detectable dose rate, which is the detection limit of an X-ray detector and is highly related to X-ray imaging in medical diagnostics. Spatial resolution characterizes the imaging quality obtained by an X-ray detector, which is defined as the number of line pairs that can be distinguished per millimeter. Direct X-ray detectors are often integrated with semiconductor photodiode arrays for imaging and therefore the spatial resolution is affected by both the photodetector and the photodiode arrays.

3.2 Indirect Radiation Detection

Indirect radiation detectors, often consisting of a scintillator and photodiode arrays, are also capable of detecting high-energy particles or photons [78]. As mentioned above, high-energy particles can ionize materials by Coulomb force, while high-energy photons can interact with materials by way of the photoelectrical effect, Compton scattering, and pair production. All these processes will generate many excitons in the scintillator that are transferred to the defect states and further recombine to produce a UV- or visible-light-emitting effect. Then, the UV or visible light is captured by photodiode arrays while the electrical signal can be recorded by an external circuit. The key figures of merit for scintillators are the LY value and decay time. LY is the number of photons that can be converted by the scintillator in units of photon or particle energy, which can be calculated by

$$LY = 10^6 \frac{SQ}{\beta E_g}$$

where S is the efficiency of the transport of electron-hole pairs to the emission center, Q is the radioluminescence efficiency, and β is usually a constant of 2.5. A high LY value means a high number of photons emitted from the scintillator, which leads to high signal output. For a high-performance scintillator, a fast decay time is essential since the time interval between the emission of photons and absorption of radiation is short, which will reduce some side effects such as afterglow.

4   Charged Particle Detection by Halide Perovskites

4.1   Alpha Particle Detectors

As mentioned in the above section, developing high-performance detectors for alpha particle radiation is not only for environmental safety concerns but also for capturing nuclei information carried by alpha particles. Recently, Xu et al. fabricated an alpha particle detector based on a MAPbBr$_3$ single crystal with a metal-semiconductor-

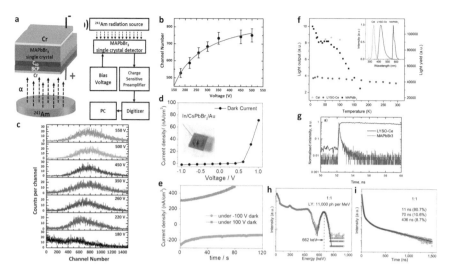

Fig. 4 (**a**) Schematic illustration of the alpha particle detector based on MAPbBr$_3$ crystal. (**b**) Voltage-dependent alpha spectra obtained from the MAPbBr$_3$ single-crystal detector. (**c**) Spectral peak centroids at different voltage biases. Black points represent experimental data, with a red line fitted from the Hecht equation. Reproduced with permission [79]. Copyright 2017, Elsevier. (**d**) *I-V* curve of an In/CsPbBr$_3$/Au alpha particle detector in the dark at biases from −1 V to 1 V. (**e**) Voltage-dependent alpha spectra obtained under different reverse biases. Reproduced with permission [64]. Copyright 2019, Elsevier. (**f**) Scintillator light output measurement as a function of temperature for MAPbBr$_3$ (black), LYSO-Ce (red), and CsI (green) under ^{241}Am source. (**g**) Normalized scintillator decay of MAPbBr$_3$ crystal (red) and LYSO-Ce (black). Reproduced with permission [59]. Copyright 2019, RSC. (**h**) Pulse height spectra of Li-doped (PEA)$_2$PbBr$_4$ with Gaussian fitting to extract light yield. (**i**) Alpha pulse height spectra of (PEA)$_2$PbBr$_4$ scintillator. Reproduced with permission [80]. Copyright 2020, Nature Publishing Group

metal structure [79]. As shown in Fig. 4a, BCP and C$_{60}$ layers were coated for the functions of passivation and electron extraction, respectively, and a preamplifier and digitizer were used for alpha spectrum measurements. A series of alpha energy spectra were obtained under an exposure of a 0.8μCi ^{241}Am source for 900 s at different biases. A voltage-dependent alpha spectral peak with a voltage bias changing from 180 V to 550 V can be seen in Fig. 4b. From the Hecht equation of single-polarity charge transport, the calculated $\mu\tau$ product was $(0.4–1.6) \times 10^{-3}$ cm^2/V (Fig. 4c). All-inorganic perovskite CsPbBr$_3$ also holds great potential for alpha particle detection. He et al. first reported an alpha particle detector with an asymmetric structure of In/CsPbBr$_3$ single crystals/Au [64]. Due to the different work functions between the In ($E_{\mathrm{F,In}} = 4.1$ eV) and Au ($E_{\mathrm{F,Au}} = 5.1$ eV) electrodes, large potential barriers for holes and electrons were formed on the In and Au sides, respectively, which further suppressed the dark current under reverse bias. As shown in Fig. 4d, a low dark current of 2 nA/cm^2 was obtained at −1 V, which was further increased to 100 nA/cm^2 at −100 V. A voltage-dependent alpha spectral peak was also shown for the all-inorganic perovskite alpha detector (Fig. 4e), with a peak resolution of 15%. Notably, this alpha particle detector can be operated at a low bias of −6 V compared to other perovskite-based alpha particle detectors which usually require large operation voltage, exhibiting the competitive device performance of the all-inorganic perovskite working as an alpha particle detector.

Beyond the direct alpha particle detectors, halide perovskites are also suitable for fabricating indirect scintillators for detecting alpha particles. Mykhaylyk et al. reported a high-LY alpha particle scintillator based on MAPbBr$_3$ nanocrystals with an ultrafast response time at low temperature [59]. As depicted in Fig. 4f, MAPbBr$_3$crystals showed increasing light output and light yield with decreasing temperatures that were much higher than those of the commercial LYSO-Ce scintillator at around 150 K and CsI scintillator at around 50 K. Moreover, MAPbBr$_3$ also exhibited a rapid and intense alpha response at 77 K with fast and slow decay times of 0.1 ns and 1 ns, respectively, compared to the relatively longer decay of 1 μs for LYSO-Ce (Fig. 4g). Currently, lithium-doped layer-structured perovskite (PEA)$_2$PbBr$_4$ crystals have also been used for multiple radiation detectors and scintillators [80], which exhibited enhanced performance, including an increased LY value of up to 11,000 ph/MeV and a fast decay time of 11 ns (Fig. 4h). Figure 4i displays the pulse height spectra result of a (PEA)$_2$PbBr$_4$ scintillator under radiation from ^{241}Am and ^{224}Cm sources. Compared to the commercial alpha particle detectors and scintillators, which require high-cost production lines, halide perovskites, taking the obvious advantages of low cost and multiple functional detectability, have already shown impressive potential for next-generation alpha particle detections.

5 Photon-Radiation Detection by Halide Perovskites

5.1 Gamma-Ray Detectors

Yakunin et al. reported a gamma-ray detector based on an FAPbI$_3$ (FA is formamidinium) single crystal at room temperature for the first time [81]. In this work, the large-size (3–12 mm) MAPbI$_3$ and FAPbI$_3$ single crystals were grown via a facile solution-processed method. Figure 5a shows the photocurrent measured as a function of voltage bias under the radiation of a Cu $K\alpha$ X-ray. Through fitting by the Hecht equation, MAPbI$_3$ exhibited a high $\mu\tau$ product of around 10^{-2} cm^2/V. The fabricated gamma-ray detector based on FAPbI$_3$ also showed a good response under radiation from ^{241}Am, as shown in Fig. 5b. Then, He et al. demonstrated a well-resolved gamma-ray spectrum using an MAPbI$_3$ single crystal [82]. Constructed by an asymmetric Ga/MAPbI$_3$/Au structure, the dark current was successfully suppressed at a reverse bias (Fig. 5c). Figure 5d depicts the gamma-ray spectrum measured by a Ga/MAPbI$_3$/Au photodetector, where a superb peak resolution of 6.8% was achieved under 122 keV ^{57}Co radiation. However, the relatively low bandgap and bad thermal stability of MAPbI$_3$ crystals set a huge limit for their usage in high-energy gamma-ray detections. Wei et al. reported a high-performance gamma-ray detector by modifying the halide component ratio in an MAPbBr$_x$Cl$_{3-x}$

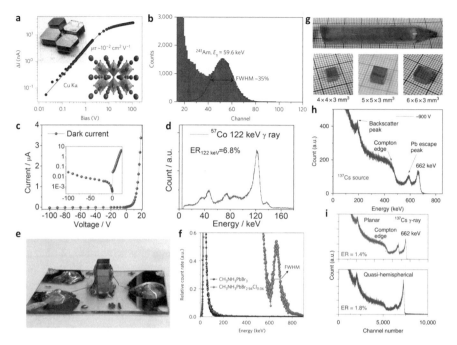

Fig. 5 (**a**) Bias dependence of photocurrent in MAPbI$_3$ single crystal generated by Cu $K\alpha$ X-ray at 8 KeV. The red line indicates a fitting from the Hecht equation with $\mu\tau$ of 10^{-2} cm^2/V. (**b**) Energy-resolved spectrum of ^{241}Am by FAPbI$_3$ single crystal. Reproduced with permission [81]. Copyright 2016, Nature Publishing Group. (**d**) *I-V* characteristic of Ga/MAPbI$_3$/Au photodetector under dark conditions. (**c**) Energy-resolved spectrum of ^{57}Co by MAPbI$_3$ single crystals. Reproduced with permission [82]. Copyright 2017, ACS. (**e**) Side view of a MAPbBr$_{2.94}$Cl$_{0.06}$ single-crystal detector, and electrode sides were encapsulated with epoxy. (**f**) Enlarged view of photopeak region from ^{137}Cs spectrum obtained by MAPbBr$_{2.94}$Cl$_{0.06}$ photodetector (red) and MAPbBr$_3$ photodetector. Reproduced with permission [62]. Copyright 2017, Nature Publishing Group. (**g**) CsPbBr$_3$ single crystal grown by the Bridgman technique. (**h**) Gamma spectrum obtained by CsPbBr$_3$ single crystal under radiation from ^{137}Cs source. Reproduced with permission [26]. Copyright 2018, Nature Publishing Group. (**i**) Gamma spectrum obtained by CsPbBr$_3$ single crystals with different device structures under radiation from a ^{137}Cs source. Reproduced with permission [83]. Copyright 2020, Nature Publishing Group

single crystal [62]. As the halide component varied from totally Br to Cl, the perovskite single crystals showed a transition from *p*-type to *n*-type. Particularly, the mixed hybrid perovskite MAPbBr$_{2.94}$Cl$_{0.06}$ single crystal was almost intrinsic, thus leading to tenfold improved bulk resistivity of 3.6×10^9 Ω cm. The dopant technique also offered an increased $\mu\tau$ product of 1.8×10^{-2} cm^2/V. In addition to crystal quality improvement, efficient device design also greatly contributed to the detecting performance. As shown in Fig. 5e, the surface/edge leakage current was mitigated by a guard-ring electrode. Being capable of detecting higher energy gamma ray, the MAPbBr$_{2.94}$Cl$_{0.06}$ photodetector exhibited a resolution of 6.5% at 665 keV of gamma rays from a ^{137}Cs source (Fig. 5f).

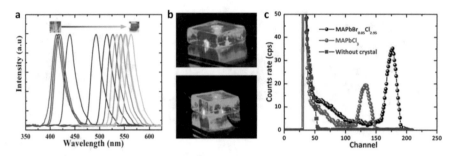

Fig. 6 (**a**) X-ray-induced photoluminescence of $MAPbBr_xCl_{3-x}$ single crystals. (**b**) Optical images of $MAPbBr_xCl_{3-x}$ single crystals excited by 365 nm laser. (**c**) Pulse height spectra acquired by $MAPbCl_3$ and $MAPbBr_{0.05}Cl_{2.95}$ under radiation from a 1.4μCi ^{137}Cs source. Reproduced with permission [84]. Copyright 2019, ACS

All-inorganic perovskite $CsPbBr_3$ with a high average atomic number of 65.9 can lead to a larger attenuation coefficient and higher stopping power, thus holding huge potential for gamma-ray detection. He et al. successfully obtained a high-resolution gamma-ray spectrum by using $CsPbBr_3$ crystals [26]. Grown by the Bridgman technique, the $CsPbBr_3$ single crystal can be easily diced into different shapes (Fig. 5g). An asymmetric device structure was also applied in this work with Ga and Au used as electrodes. As mentioned above, because of the different work functions between the Ga and Au electrodes, the dark current was highly suppressed even under high reverse biases. The $CsPbBr_3$ also maintained a good $\mu\tau$ product of 1.34×10^{-3} cm^2/V. Furthermore, under exposure to a 5μCi ^{137}Cs source, the asymmetric photodetector achieved a peak resolution of 3.8% at 662 keV (Fig. 5h) [26]. Recently, He et al. further improved the low-work-function metal by using eutectic Ga-In alloy (EGaIn) [83], which enabled a record gamma-ray peak resolution of 1.4% by the $EGaIn/CsPbBr_3/Au$ gamma-ray detector with planar structure (Fig. 5i). However, the planar-structured detector suffered a decrease in peak resolution after enlarging the crystal size. With a hemispherical device structure, a unipolar device with electrons being screened was obtained, and larger sized $CsPbBr_3$ single crystals can be applied to use in a gamma-ray detector with a high peak resolution of 1.8%.

Halide perovskites have also been used as indirect gamma-ray scintillators, owing to their high light-emitting property. Recently, Xu et al. reported a gamma-ray scintillator based on an $MAPbBr_{0.05}Cl_{2.95}$ single crystal [84]. In this work, a series of $MAPbBr_xCl_{3-x}$ single crystals were grown via a solution-processed method which were further integrated onto a silicon photomultiplier (SiPM) window as a scintillator. Figure 6a shows the PL spectrum of $MAPbBr_xCl_{3-x}$ single crystals after tuning the ratio of Br and Cl components; optical images excited by 365 nm laser are shown in Fig. 6b. A near-band-edge emission was obtained for the $MAPbBr_{0.05}Cl_{2.95}$ single crystal. The pulse height spectra of the 1.7 μCi ^{137}Cs source were successfully acquired at room temperature (Fig. 6c). In addition, layered perovskite $(C_6H_5C_2H_4NH_3)_2PbBr_4$ has also shown good scintillation when excited by gamma rays [85]. In the referenced work, the large-sized ($27 \times 13 \times 4$ mm^3) crystals were synthesized by a solution-processed method. The bulk sample showed a high light yield of 14,000 ph/MeV and a fast decay time of 11 ns. It should be noticed that even though halide perovskite-based gamma-ray detectors have a relatively lower resolution (about 1%) compared with the state-of-the-art CZT gamma-ray detector, its low-temperature growth method and reasonable production cost indisputably blaze the way for a new generation of gamma-ray detectors that are capable of both direct and indirect detections.

5.2 X-Ray Detectors

As mentioned above, X-rays are some of the most widely used radiations in modern society; their utility has covered medicine, food safety, environmental surveillance, industry, and scientific research [2, 6, 77]. Similar to gamma rays, X-rays can be detected by either direct photodetectors coupled with semiconductor photodiodes or indirect scintillators.

5.2.1 X-Ray Photodetectors

As mentioned before, lead halide perovskites, holding large bulk resistance, strong stopping power, and high mobility-lifetime product are promising for high-energy photon detection, which can also be applied to X-ray detection applications. Recently, Wei et al. reported a highly sensitive X-ray detector based on an MAPbBr$_3$ single crystal [29]. A nonstoichiometric precursor ratio MABr/PbBr$_2$ of 0.8 was used for the growth of high-quality single crystals, where the obtained MAPbBr$_3$ crystal showed a record $\mu\tau$ product of 1.2×10^{-2} cm^2/V (Fig. 7a). As shown in Fig. 7b, the fabricated X-ray detector exhibited a great linear relationship between photocurrent and X-ray dose. A high sensitivity of 80μC/Gy$_{air}$ cm^2 was achieved, which was ten times higher than the CZT detector under the same bias. The lowest detectable X-ray dose rate was observed as 0.5μGy$_{air}$/s, which is enough to satisfy the medical diagnostics (5.5μGy$_{air}$/s) requirement. Lately, a quasi-monocrystalline CsPbBr$_3$-based X-ray detector was fabricated by Pan et al. [65]. Through a simple hot-press method, CsPbBr$_3$ film with the thickness of hundreds of micrometers was obtained, as shown in Fig. 7c. Notably, the X-ray detector held a record sensitivity of 55,684 μC/Gy$_{air}$·cm^2 under a 5.0 V/mm electric field (Fig. 7d).

5.2.2 X-Ray Imaging or Scintillators

The first halide perovskite-based X-ray detector based on polycrystalline MAPbI$_3$ film for imaging applications was reported by Yakunin et al. in 2015 [44]. Wei et al. used molecular (3-aminopropyl) triethoxysilane to connect the native oxide Si and

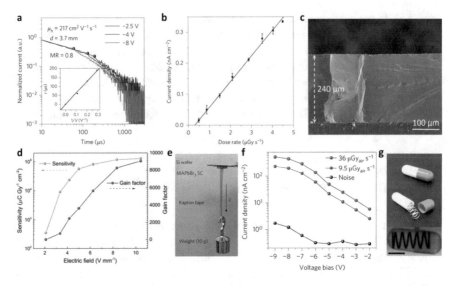

Fig. 7 (**a**) Normalized transient current curves of MAPbBr₃ single-crystal devices with the MABr/PbBr₂ molar ratio of 0.8 under various biases. Inset shows the charge transit time versus the reciprocal of bias. (**b**) X-ray-induced photocurrent at different dose rates. Reproduced with permission [29]. Copyright 2016, Nature Publishing Group. (**c**) SEM image of CsPbBr₃ thick quasi-monocrystalline film grown by hot-press method. (**d**) Sensitivity and gain factor of X-ray detector at various electric field strengths. Reproduced with permission [65]. Copyright 2019, WILEY-VCH. (**e**) Solid connection between MAPbBr₃ single crystal and Si wafer integrated by a small molecule. (**f**) X-ray-induced current density of a 150μm thin crystal device under different X-ray intensities. (**g**) Optical and X-ray images of a capsule with a steel spring. Reproduced with permission [86]. Copyright 2017, Nature Publishing Group

perovskite single crystal by the NH_3Br terminal of the molecule [86]. As shown in Fig. 7e, a solid connection between the Si substrate and MAPbBr₃ single crystals was obtained, which enabled direct X-ray imaging for the device. The dark current was also suppressed because of the molecule interlayer between Si and MAPbBr₃, which led to a high sensitivity of $2.1 \times 10^4 \mu C/Gy_{air}.cm^{-2}$ and the minimum detectable dose rate of 36 nGy_{air}/s (Fig. 7f). High-resolution X-ray images can be obtained with a spatial resolution of around 10 lp/mm (Fig. 7g).

6   Outlook and Perspectives

Halide perovskite single crystals have attracted tremendous research attention for various energy and optoelectronic applications in the past decade due to their promising features. So far, more and more single-crystal perovskite radiation detectors have also been reported, due to their appealingly fast and excellent response to charged particles and photon radiation. The outstanding photophysical and transport properties of halide perovskites allow them to be used as direct radiation detectors with high sensitivity and spectral resolution. Halide perovskites can also be fabricated into scintillators due to their efficient light emission under all types of radiation. Moreover, single-crystal perovskites are usually synthesized via low-cost and facile solution-processed methods compared to complicated requirements of other materials to produce commercial radiation detectors.

Despite the advantages that perovskite single crystals hold for radiation detection, there are still a few challenges that stand in the way of the commercial use of halide perovskites. Firstly, the large-scale production of perovskite crystals is still a daunting challenge and thus limits industrial level radiation detector fabrication. Then, there is still a vast difference in detection performance between single-crystal perovskite detectors and the existing commercial radiation detectors. As for alpha particle detection, the state-of-the-art detector offers the highest spectral resolution of 0.25%, while attempts to detect alpha particles by halide perovskites are still in early stages with a spectral resolution of only 15%. Then, the resistance of halide perovskites (10^7–10^9 Ω cm) is still not high enough to reduce the dark current to appropriate levels, which may reduce the signal quality at high voltage biases. Furthermore, the stability of halide perovskite crystals needs to be taken into consideration because most of them suffer from decomposition in air, especially hybrid perovskite crystals where their organic parts are chemically active. Even though the all-inorganic perovskite crystals show better long-term stability in the air, a thorough encapsulation method is still in high demand. Environmental safety is another concern for commercial applications because of the heavy metal component—Pb. A totally environmentally friendly assembly line is required for perovskite radiation detectors to minimize Pb contamination. Furthermore, the reduction of ion migration in halide perovskite crystals is essential for detector performance, which may necessitate a deep understanding of the physical mechanism inside halide perovskites.

Regarding photon radiation detections, the highest gamma-ray spectral resolution is 0.5%, which was obtained by the commercial CZT crystal-based detector at room temperature. Although impressive progress on single-crystal perovskite gamma-ray detectors has been made, the spectral peak resolution is still far too low, which can be attributed to the relatively higher dark current, especially for Pb types. The long-term stability of hybrid perovskites under the radiation of gamma rays remains elusive, where the organic components in hybrid perovskites may suffer degradation. Single-crystal perovskite X-ray detectors share roughly the same challenges as gamma-ray detectors. Due to imaging requirements, the lowest detectable limit and spatial resolution should be taken into consideration as other factors to consider. The lowest detectable limit of single-crystal perovskite X-ray detectors is continually improving and is more than enough for medical diagnostics at the present time. However, the spatial resolution of direct X-ray detectors is still far behind the first-class HPGe detector, which may be limited by both the perovskite crystal quality and device fabrication method. Despite being confronted by these challenges, a bright future for single-crystal perovskite radiation detection applications is readily apparent considering their tremendous advantages, including the low cost of raw materials, facile production, excellent compatibility, and superb physical properties.

Acknowledgment The author would like to acknowledge financial support from the Discovery Early Career Researcher Award (DECRA) scheme (DE180100167) from the Australian Research Council (ARC).

References

1. Knoll, G.F.: Radiation detection and measurement. John Wiley & Sons, Hoboken (2010)
2. Chapman, D., Thomlinson, W., Johnston, R.E., Washburn, D., Pisano, E., Gmür, N., Zhong, Z., Menk, R., Arfelli, F., Sayers, D.: Phys. Med. Biol. **42**, 2015 (1997)
3. El Ghissassi, F., Baan, R., Straif, K., Grosse, Y., Secretan, B., Bouvard, V., Benbrahim-Tallaa, L., Guha, N., Freeman, C., Galichet, L., Cogliano, V.: Lancet Oncol. **10**, 751 (2009)
4. Chapman, H.N., Fromme, P., Barty, A., White, T.A., et al.: Nature. **470**, 73 (2011)
5. Haff, R.P., Toyofuku, N.: Sens. Instrumen. Food Qual. **2**, 262 (2008)
6. Arfelli, F., Assante, M., Bonvicini, V., Bravin, A., et al.: Phys. Med. Biol. **43**, 2845 (1998)
7. Eisen, Y., Shor, A., Mardor, I.: Nucl. Instrum. Methods Phys. Res., Sect. A. **428**, 158 (1999)
8. Yaffe, M.J., Rowlands, J.A.: Phys. Med. Biol. **42**, 1 (1997)
9. Owens, A.P.: Nucl. Instrum. Methods Phys. Res., Sect. A. **531**, 18 (2004)
10. Kasap, S., Frey, J.B., Belev, G., Tousignant, O., Mani, H., Greenspan, J., Laperriere, L., Bubon, O., Reznik, A., DeCrescenzo, G., Karim, K.S., Rowlands, J.A.: Sensors. **11**, 5112 (2011)
11. Guerra, M., Manso, M., Longelin, S., Pessanha, S., Carvalho, M.L.: J. Instrum. **7**, C10004 (2012)
12. Kasap, S.O.: J. Phys. D. Appl. Phys. **33**, 2853 (2000)
13. Sellin, P.J.: Nucl. Instrum. Methods Phys. Res., Sect. A. **513**, 332 (2003)
14. C: Szeles Phys. Status Solidi (b). **241**, 783 (2004)
15. Schlesinger, T.E., Toney, J.E., Yoon, H., Lee, E.Y., Brunett, B.A., Franks, L., James, R.B.: Mater. Sci. Eng. R Rep. **32**, 103 (2001)
16. Moszyński, M., Nassalski, A., Syntfeld-Każuch, A., Szczęśniak, T., Czarnacki, W., Wolski, D., Pausch, G., Stein, J.: Nucl. Instrum. Methods Phys. Res., Sect. A. **568**, 739 (2006)
17. Bizarri, G., de Haas, J., Dorenbos, P., Eijk, C.W.E.: Nucl. Sci. IEEE Trans. **53**, 615 (2006)
18. Mikhailik, V.B., Kapustyanyk, V., Tsybulskyi, V., Rudyk, V., Kraus, H.: Phys. Status Solidi (b). **252**, 804 (2015)
19. Al-Ashouri, E., Köhnen, B., Li, A.M., et al.: Science. **370**, 1300 (2020)
20. Tan, Z.-K., Moghaddam, R.S., Lai, M.L., Docampo, P., Higler, R., Deschler, F., Price, M., Sadhanala, A., Pazos, L.M., Credgington, D., Hanusch, F., Bein, T., Snaith, H.J., Friend, R.H.: Nat. Nanotech. **9**, 687 (2014)
21. Sutherland, B.R., Sargent, E.H.: Nat. Photonics. **10**, 295 (2016)
22. Li, F., Ma, C., Wang, H., Hu, W., Yu, W., Sheikh, A.D., Wu, T.: Nat. Commun. **6**, 8238 (2015)
23. Zou, Y., Li, F., Zhao, C., Xing, J., Yu, Z., Yu, W., Guo, C.: Adv. Opt. Mater. **7**, 1900676 (2019)
24. de Arquer, F.P.G., Armin, A., Meredith, P., Sargent, E.H.: Nat. Rev. Mater. **2**, 16100 (2017)
25. Stoumpos, C.C., Malliakas, C.D., Peters, J.A., Liu, Z., Sebastian, M., Im, J., Chasapis, T.C., Wibowo, A.C., Chung, D.Y., Freeman, A.J., Wessels, B.W., Kanatzidis, M.G.: Cryst. Growth Des. **13**, 2722 (2013)
26. He, Y., Matei, L., Jung, H.J., McCall, K.M., Chen, M., Stoumpos, C.C., Liu, Z., Peters, J.A., Chung, D.Y., Wessels, B.W., Wasielewski, M.R., Dravid, V.P., Burger, A., Kanatzidis, M.G.: Nat. Commun. **9**, 1609 (2018)
27. Kim, Y.C., Kim, K.H., Son, D.-Y., Jeong, D.-N., Seo, J.-Y., Choi, Y.S., Han, I.T., Lee, S.Y., Park, N.-G.: Nature. **550**, 87 (2017)
28. Yu, D., Wang, P., Cao, F., Gu, Y., Liu, J., Han, Z., Huang, B., Zou, Y., Xu, X., Zeng, H.: Nat. Commun. **11**, 3395 (2020)

29. Wei, H., Fang, Y., Mulligan, P., Chuirazzi, W., Fang, H.-H., Wang, C., Ecker, B.R., Gao, Y., Loi, M.A., Cao, L., Huang, J.: Nat. Photonics. **10**, 333 (2016)
30. Wei, H., Huang, J.: Nat. Commun. **10**, 1066 (2019)
31. Kakavelakis, G., Gedda, M., Panagiotopoulos, A., Kymakis, E., Anthopoulos, T.D., Petridis, K.: Adv. Sci. **7**, 2002098 (2020)
32. Dang, Y., Liu, Y., et al.: CrystEngComm. **17**, 665–670 (2015)
33. Saidaminov, M.I., Abdelhady, A.L., et al.: Nat. Commun. **6**, 1–6 (2015)
34. Zhang, Y., Liu, Y., et al.: J. Mater. Chem. C. **4**, 9172–9178 (2016)
35. Shi, D., Adinolfi, V., et al.: Science. **347**, 519–522 (2015)
36. Peng, W., Wang, L., et al.: Adv. Mater. **28**, 3383–3390 (2016)
37. Chen, Y.-X., Ge, Q.-Q., et al.: J. Am. Chem. Soc. **138**, 16196–16199 (2016)
38. Gu, Z., Huang, Z., et al.: Sci. Adv. **4**, eaat2390 (2018)
39. Wei, H., Fang, Y., et al.: Nat. Photonics. **10**, 333 (2016)
40. Rao, H.S., Li, W.G., et al.: Adv. Mater. **29**, 1602639 (2017)
41. Lee, L., Baek, J., et al.: Nat. Commun. **8**, 1–8 (2017)
42. Pan, W., Wei, H., Yang, B.: Front. Chem. **8**, 268 (2020)
43. Kovalenko, M.V., Protesescu, L., Bodnarchuk, M.I.: Science. **358**, 745 (2017)
44. Yakunin, S., Sytnyk, M., Kriegner, D., Shrestha, S., Richter, M., Matt, G.J., Azimi, H., Brabec, C.J., Stangl, J., Kovalenko, M.V., Heiss, W.: Nat. Photonics. **9**, 444 (2015)
45. Fang, Y., Dong, Q., et al.: Nat. Photonics. **9**, 679 (2015)
46. Xiao, J.-W., Liu, L., Zhang, D., Marco, N.D., Lee, J.-W., Lin, O., Chen, Q., Yang, Y.: Adv. Energy Mater. **7**, 1700491 (2017)
47. Akkerman, Q.A., Motti, S.G., Srimath Kandada, A.R., Mosconi, E., D'Innocenzo, V., Bertoni, G., Marras, S., Kamino, B.A., Miranda, L., De Angelis, F., Petrozza, A., Prato, M., Manna, L., Am, J.: Chem. Soc. **138**, 1010 (2016)
48. Gao, H., Feng, J., Pi, Y., Zhou, Z., Zhang, B., Wu, Y., Wang, X., Jiang, X., Jiang, L.: Adv. Funct. Mater. **28**, 1804349 (2018)
49. Dong, D., Deng, H., Hu, C., Song, H., Qiao, K., Yang, X., Zhang, J., Cai, F., Tang, J., Song, H.: Nanoscale. **9**, 1567 (2017)
50. Zhang, D., Yang, Y., Bekenstein, Y., Yu, Y., Gibson, N.A., Wong, A.B., Eaton, S.W., Kornienko, N., Kong, Q., Lai, M., Alivisatos, A.P., Leone, S.R., Yang, P.: J. Am. Chem. Soc. **138**, 7236 (2016)
51. Lian, Z., Yan, Q., Gao, T., Ding, J., Lv, Q., Ning, C., Li, Q., Sun, J.: J. Am. Chem. Soc. **138**, 9409 (2016)
52. Shi, D., Adinolfi, V., Comin, R., Yuan, M., Alarousu, E., Buin, A., Chen, Y., Hoogland, S., Rothenberger, A., Katsiev, K., Losovyj, Y., Zhang, X., Dowben, P.A., Mohammed, O.F., Sargent, E.H., Bakr, O.M.: Science. **347**, 519 (2015)
53. Maculan, G., Sheikh, A.D., Abdelhady, A.L., Saidaminov, M.I., Haque, M.A., Murali, B., Alarousu, E., Mohammed, O.F., Wu, T., Bakr, O.M.: J. Phys. Chem. Lett. **6**, 3781 (2015)
54. Dong, Q., Fang, Y., Shao, Y., Mulligan, P., Qiu, J., Cao, L., Huang, J.: Science. **347**, 967 (2015)
55. Saidaminov, M.I., Adinolfi, V., Comin, R., Abdelhady, A.L., Peng, W., Dursun, I., Yuan, M., Hoogland, S., Sargent, E.H., Bakr, O.M.: Nat. Commun. **6**, 8724 (2015)
56. Rakita, Y., Kedem, N., Gupta, S., Sadhanala, A., Kalchenko, V., Böhm, M.L., Kulbak, M., Friend, R.H., Cahen, D., Hodes, G.: Cryst. Growth Des. **16**, 5717 (2016)
57. Song, J., Li, J., Li, X., Xu, L., Dong, Y., Zeng, H.: Adv. Mater. **27**, 7162 (2015)
58. Yang, D., Li, X., Zhou, W., Zhang, S., Meng, C., Wu, Y., Wang, Y., Zeng, H.: Adv. Mater. **31**, 1900767 (2019)
59. Mykhaylyk, V.B., Kraus, H., Saliba, M.: Mater. Horiz. **6**, 1740 (2019)
60. Akkerman, Q.A., D'Innocenzo, V., Accornero, S., Scarpellini, A., Petrozza, A., Prato, M., Manna, L.: J. Am. Chem. Soc. **137**, 10276 (2015)
61. Schmidt, L.C., Pertegás, A., González-Carrero, S., Malinkiewicz, O., Agouram, S., Mínguez Espallargas, G., Bolink, H.J., Galian, R.E., Pérez-Prieto, J.: J. Am. Chem. Soc. **136**, 850 (2014)

62. Wei, H., DeSantis, D., Wei, W., Deng, Y., Guo, D., Savenije, T.J., Cao, L., Huang, J.: Nat. Mater. **16**, 826 (2017)
63. Xu, Q., Wang, J., Shao, W., Ouyang, X., Wang, X., Zhang, X., Guo, Y., Ouyang, X.: Nanoscale. **12**, 9727 (2020)
64. He, Y., Liu, Z., McCall, K.M., Lin, W., Chung, D.Y., Wessels, B.W., Kanatzidis, M.G.: Nucl. Instrum. Methods Phys. Res., Sect. A. **922**, 217 (2019)
65. Pan, W., Yang, B., Niu, G., Xue, K.-H., Du, X., Yin, L., Zhang, M., Wu, H., Miao, X.-S., Tang, J.: Adv. Mater. **31**, 1904405 (2019)
66. Wang, W., Meng, H., Qi, H., Xu, H., Du, W., Yang, Y., Yi, Y., Jing, S., Xu, S., Hong, F., Qin, J., Huang, J., Xu, Z., Zhu, Y., Xu, R., Lai, J., Xu, F., Wang, L., Zhu, J.: Adv. Mater. **32**, 2001540 (2020)
67. Zhu, W., Ma, W., Su, Y., Chen, Z., Chen, X., Ma, Y., Bai, L., Xiao, W., Liu, T., Zhu, H., Liu, X., Liu, H., Liu, X., (Michael) Yang, Y.: Light Sci. Appl. **9**, 112 (2020)
68. Zhou, F., Li, Z., Lan, W., Wang, Q., Ding, L., Jin, Z.: Small Methods. **4**, 2000506 (2020)
69. Li, Y., Shao, W., Ouyang, X., Zhu, Z., Zhang, H., Ouyang, X., Liu, B., Xu, Q.: J. Phys. Chem. C. **123**, 17449 (2019)
70. Del Sordo, S., Abbene, L., Caroli, E., Mancini, A.M., Zappettini, A., Ubertini, P.: Sensors. **9**, 3491 (2009)
71. Eskin, J.D., Barrett, H.H., Barber, H.B.: J. Appl. Phys. **85**, 647 (1998)
72. Rutherford, E., Royds, T.: Nature. **78**, 220 (1908)
73. Grellier, J., Atkinson, W., Bérard, P., Bingham, D., Birchall, A., Blanchardon, E., Bull, R., Guseva Canu, I., Challeton-de Vathaire, C., Cockerill, R., Do, M.T., Engels, H., Figuerola, J., Foster, A., Holmstock, L., Hurtgen, C., Laurier, D., Puncher, M., Riddell, A.E., Samson, E., Thierry-Chef, I., Tirmarche, M., Vrijheid, M., Cardis, E.: Epidemiology. **28**, 675 (2017)
74. Richter, M., Siffert, P.: Nucl. Instrum. Methods Phys. Res., Sect. A. **322**, 529 (1992)
75. Androulakis, J., Peter, S.C., Li, H., Malliakas, C.D., Peters, J.A., Liu, Z., Wessels, B.W., Song, J.-H., Jin, H., Freeman, A.J., Kanatzidis, M.G.: Adv. Mater. **23**, 4163 (2011)
76. Spiegel, P.K.: Am. J. Roentgenol. **164**, 241 (1995)
77. Miao, J., Charalambous, P., Kirz, J., Sayre, D.: Nature. **400**, 342 (1999)
78. Büchele, P., Richter, M., Tedde, S.F., Matt, G.J., Ankah, G.N., Fischer, R., Biele, M., Metzger, W., Lilliu, S., Bikondoa, O., Macdonald, J.E., Brabec, C.J., Kraus, T., Lemmer, U., Schmidt, O.: Nat. Photonics. **9**, 843 (2015)
79. Xu, Q., Wei, H., Wei, W., Chuirazzi, W., DeSantis, D., Huang, J., Cao, L.: Nucl. Instrum. Methods Phys. Res., Sect. A. **848**, 106 (2017)
80. Xie, C., Hettiarachchi, F., Maddalena, M.E., Witkowski, M., Makowski, W., Drozdowski, A., Arramel, A.T.S., Wee, S.V., Springham, P.Q., Vuong, H.J., Kim, C., Dujardin, P., Coquet, M. D., Birowosuto, C.D.: Commun. Mater. **1**, 1 (2020)
81. Yakunin, S., Dirin, D.N., Shynkarenko, Y., Morad, V., Cherniukh, I., Nazarenko, O., Kreil, D., Nauser, T., Kovalenko, M.V.: Nat. Photonics. **10**, 585 (2016)
82. He, Y., Ke, W., Alexander, G.C.B., McCall, K.M., Chica, D.G., Liu, Z., Hadar, I., Stoumpos, C. C., Wessels, B.W., Kanatzidis, M.G.: ACS Photonics. **5**, 4132 (2018)
83. He, Y., Petryk, M., Liu, Z., Chica, D.G., Hadar, I., Leak, C., Ke, W., Spanopoulos, I., Lin, W., Chung, D.Y., Wessels, B.W., He, Z., Kanatzidis, M.G.: Nat. Photonics. **15**, 36 (2021)
84. Xu, Q., Shao, W., Liu, J., Zhu, Z., Ouyang, X., Cai, J., Liu, B., Liang, B., Wu, Z., Ouyang, X.: ACS Appl. Mater. Interfaces. **11**, 47485 (2019)
85. Kawano, N., Koshimizu, M., Okada, G., Fujimoto, Y., Kawaguchi, N., Yanagida, T., Asai, K.: Sci. Rep. **7**, 14754 (2017)
86. Wei, W., Zhang, Y., Xu, Q., Wei, H., Fang, Y., Wang, Q., Deng, Y., Li, T., Gruverman, A., Cao, L., Huang, J.: Nat. Photonics. **11**, 315 (2017)

Metal Halide Perovskites for High-Energy Radiation Detection

Murali Gedda, Hendrik Faber, Konstantinos Petridis, and Thomas D. Anthopoulos

Abstract In this chapter, the fundamental principles of high-energy radiation detection and recent progress in the emerging field of metal halide perovskite (MHP)-based direct and indirect X-ray and γ-ray detectors are discussed. The chapter first introduces the underlying principles of high-energy radiation detection, with emphasis on the key performance metrics. This is followed by a comprehensive summary of the recent progress made in the field of perovskite-based radiation detector technologies. Finally, the chapter ends with an overview of current issues and future perspectives on MHP-based direct and indirect (scintillators) radiation detector technologies.

1 Introduction

High-energy radiation (e.g., X-rays, γ-rays; Fig. 1a) detection plays a vital role in many areas, including scientific research, medical applications, nondestructive evaluation, and national security. Current trends in detector technology offer substantial reductions in size, weight, and power as well as reduced exposure times, broadening the scope for emerging opportunities and applications. Recent advancements in the field of high-energy photon detection, which combine enhanced performance and cost-effective, large-area, and high-throughput manufacturing, could pave the way to new technologies beneficial to our society. Metal halide perovskites (MHPs) represent a promising family of materials for radiation detection. Their attractive physical and chemical properties combine a strong stopping power with a large $\mu_{h/e} \times \tau$

M. Gedda · H. Faber · T. D. Anthopoulos (✉)
Kaust Solar Center, King Abdullah University of Science and Technology (KAUST), Thuwal, Saudi Arabia
e-mail: murali.gedda@kaust.edu.sa; hendrik.faber@kaust.edu.sa; thomas.anthopoulos@kaust.edu.sa

K. Petridis
Department of Electronic Engineering, Hellenic Mediterranean University, Chania, Crete, Greece
e-mail: cpetridis@hmu.gr

© The Author(s), under exclusive license to Springer Nature Switzerland AG 2022
K. Iniewski (ed.), *Advanced Materials for Radiation Detection*,
https://doi.org/10.1007/978-3-030-76461-6_6

119

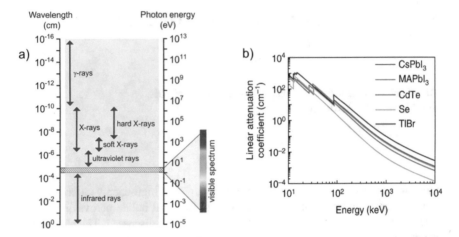

Fig. 1 (**a**) The electromagnetic spectrum ranging from infrared (IR) to γ-rays along with their respective wavelengths and photon energies. The visible wavelength range is expanded for clarity. (**b**) The linear attenuation coefficient of MAPbBr$_3$, CsPbI$_3$, CdTe, Se, and TlBr versus photon energy. Figure 1a is reprinted from [3] with the permission of Wiley-VCH GmbH. Figure 1b is reprinted from [4] with the permission of Springer Nature

[mobility ($\mu_{h/e}$) and lifetime (τ)] product, large linear attenuation coefficients (Fig. 1b), absence of deep traps, and controllable crystallization via scalable solution-based deposition techniques. These features make MHPs ideal for deployment in next-generation high-energy radiation detection technologies. To date, the highest sensitivity reported for MHP-based radiation detector is ~700 mC Gy$_{air}^{-1}$ cm^{-2} and the lowest detection attained is 0.62 nGys^{-1} [1, 2].

2 Key Parameters of High-Energy Radiation Detectors

Depending on the detection principle, high-energy radiation detectors are classified into two types: (I) direct radiation detectors and (II) indirect (scintillation) detectors. Direct radiation detectors rely on photoconductive materials that are sensitive to specific high-energy radiation, while indirect detectors work with scintillator materials that convert high-energy X-rays (0.1–100 keV) or γ-rays (0.1–10 MeV) to ultraviolet light (UV) or visible (Vis) (Fig. 1a), which is subsequently detected by a standard photodiode/array.

2.1 Direct Detectors

The functioning principle of direct high-energy radiation detectors is based on the direct interaction of the incoming photons with the sensing material, characteristically a semiconductor, and the immediate generation of an electrical signal (voltage or current).

2.1.1 X-Ray Detectors

The interaction between soft X-ray photons (Fig. 1a) and sensing material depends on the photoelectric absorption of the sensing material. On the other hand, the detection of hard X-ray photons is due to Compton scattering (photon-electron interaction). Important figures of merit (FOM) that regulate the performance of a direct X-ray detector are discussed in the following.

Stopping Power

The stopping power is defined as the rate of energy lost per unit of path length (x) by a charged particle with kinetic energy (T_E) in a medium of atomic number Z ($Z \propto \rho$, where ρ is the density of the material). It is measured in MeV/cm or J/m and is represented as

$$dT_E/\rho \, dx \tag{1}$$

Ionization Energy ($W_\pm$)

The ionization energy ($W_\pm$) of the sensing material is another critical parameter. It is defined as the energy required to release an electron-hole pair in the photoconductor. The ionization energy minimum is proportional to the energy bandgap (E_g) of the absorbing material employed, and it can be represented as [5]

$$W_\pm = 2.2E_g + E_{\text{photon}} \tag{2}$$

or

$$W_\pm = 3E_g \tag{3}$$

$\mu_{h/e} \times \tau$ Product

At any given X-ray dose, high detection sensitivity is needed to generate good-quality images. The sensitivity of X-ray detectors can be enhanced by means of two different methods: (1) increasing the mobility-lifetime ($\mu_{h/e} \times \tau$) product as it relates to the carrier drift length (L_D) given as $L_D = (k_B T \mu \tau / e)^{1/2}$ and (2) higher applied reverse bias. For efficient direct X-ray detectors, high $\mu_{h/e} \times \tau$ product and low trap density are essential parameters. The $\mu_{h/e} \times \tau$ product for a given sensing material can be estimated from modified Hecht formula for photoconductivity (I) [6]:

$$I = \frac{I_0 \mu \tau V}{L^2} \frac{1 - \exp\left(-\frac{L^2}{\mu \tau V}\right)}{1 + \frac{L}{V}\frac{s}{\mu}}$$

(4)

I_o is the saturated photocurrent, L is the material layer thickness, V is the applied bias, and s is the surface recombination velocity.

Sensitivity

The following formula can calculate the detector's sensitivity:

$$S = \frac{\int \left[I_{X-ray}(t) - I_{dark}\right] dt}{D \times V_d}$$

(5)

where I_{X-ray} and I_{dark} are the generated currents with and without X-ray irradiation, respectively. D is the dose, and V_d denotes the detector volume.

The linear dynamic range (LDR) of the detector that represents the range of X-ray dose rate under which the sensitivity remains constant is a crucial parameter to describe a detector's properties. The higher the LDR, the better the resolution.

Energy Resolution

The energy resolution is defined as the detector's ability to accurately determine the incident radiation's energy [7]. It is derived from the ratio between the FWHM and the photopeak centroid H_{max}. Thus, the energy resolution (R) is calculated from the following equation:

$$R\% = \frac{FWHM}{H_{max}}$$

(6)

An additional requirement is that the active layer's thickness should be at least three times larger than the attenuation length. Direct radiation X-ray detectors with semiconductors operate in current mode, while the magnitude of the produced electrical current is proportional to the incident photon energy.

2.1.2 γ-Ray Detectors

The interactions between γ-ray photons and the sensing element can be described in three mechanisms. (1) *Photoelectric process*: Complete transfer of energy of the γ-ray photon (with energy between 10 and 500 keV) to electrons. (2) *Compton scattering*: Part of the γ-ray energy (50 keV to 3 MeV) is lost and is transferred to the electrons. (3) *Pair production*: An incoming γ-ray photon with energy exceeding 1.022 MeV generates a positron and electron.

The semiconductors with similar characteristics used in X-ray detectors can also serve the purpose of detecting γ-ray photons. The following important characteristics are the prerequisites of semiconductors for high-energy γ-ray detectors.

1. Elements with high atomic number (Z): Facilitates large stopping power (see Eq. (1))
2. Large bipolar $\mu_{h/e} \times \tau$ product: Enables efficient detection
3. Large bulk resistivity of $>10^9$ Ω-cm: Enhances the signal-to-noise ratio (SNR)

The photopeak energy resolution is another FOM parameter, and it is vital as it allows the detector to differentiate between γ-ray photons with various energies. The key FOM parameters of high-energy radiation direct detectors and their significance are listed in Table 1.

2.2 Scintillator Detectors

The ability of scintillators to stop high-energy photons and translate them to lower energy visible photons has found various applications in security and medical imaging [5, 8].

Table 1 Some of the crucial direct radiation detectors' FOM parameters and their significance

FOM parameter	Significance
Mass attenuation coefficient	Penetration depth of high-energy photons
Density of the sensing material	Composition and structural properties of the active material
Signal-to-noise ratio	Defines dark current
Spatial resolution of the detector	Determines the image resolution
Response time	Photogenerated charges and transport
Uniformity of the sensing material	Processing versatility of the active material
Operational stability	Detector's performance during operation
Electron-hole mobility ($\mu_{h/e}$)-lifetime (τ) product ($\mu_{h/e} \times \tau$)	Defines the quality of a semiconductor

2.2.1 X-Ray Scintillators

The interaction between X-ray photons and the scintillator material can occur via (1) photoelectric absorption, (2) Compton scattering, and (3) pair production [9, 10]. All processes are characterized by absorption coefficients that are eventually determined by the atomic number (Z) of scintillator material and the photon energy. When the photoelectric effect is dominant, and for photon energies far from the absorption edge, the linear absorption coefficient (μ) is given by

$$\mu_L \approx \rho Z^n / E^{3.5} \tag{7}$$

where ρ is the material density, E is the photon energy, and n is a constant, which typically varies between 3 and 4. At higher energies, Compton scattering generally occurs. Here, depending on the scattering angle, a part of the photon's energy is transferred to the electron. The Compton scattering linear absorption coefficient (μ_C) is given by [9, 11]

$$\mu_C \approx \rho / (E)^{1/2} \tag{8}$$

For higher energies, the interaction of the radiation with the matter is governed by the generation of low-energy excitons (pair production) within the scintillating material that eventually recombine and produce visible light. The emitted light can then be detected via different photodetectors coupled with the scintillator element. Inexpensive production and stability are the main strengths of indirect scintillator detectors compared to direct detectors [9, 11].

Light yield (LY) is the most important figure of merit for scintillators that describes the number of electron-hole pairs generated during the ionization process per unit energy and is given in photons per MeV:

$$LY = 10^6 \, SQ / (\beta E_g) \tag{9}$$

where Q is the luminescence efficiency, S is the efficiency of transport of electron-hole to the optical (emissive) center, and β is a constant with a typical value of 2.5.

2.2.2 γ-Ray Scintillators

Upon interaction of γ-ray photons (with energies higher than 1.02 MeV) with the sensing element of a γ-ray scintillator, pair production occurs. The absorption coefficient (μ_P) is expressed as

$$\mu_P \approx \rho Z \ln \left(2E / (m_e c^2) \right) \tag{10}$$

Table 2 Important FOM parameters and their significance of high-energy scintillation detectors

FOM parameter	Significance
Radiation absorption efficiency	Density of the scintillator material and atomic number
Light yield (LY)	Number of emitted photons per absorbed energy
Decay time	Kinetics of the light response I(t) characterized by τ
Energy resolution	Ability of the material to discriminate different radiation energies
Spatial resolution	Regulates the spatial frequency response of the photodetector
Radiation hardness	Characterizes the chemical and radiation stability
Proportionality	Linearity of the detected signal to the incoming high-energy radiation intensity
Afterglow	Residual light output occurring after the primary decay time of the main luminescent centers
Stopping power	Attenuation coefficient of the absorbed radiation for a given thickness of a material

where ρ is the material's density, c is the speed of light, and m_e is the mass of electrons. The phenomenon of scintillation can be divided into three main sub-processes: conversion, energy transfer, and luminescence. Photon absorption (i.e., the first stage of scintillation, which lasts ≈ 1 ps) is followed by the charge transport and energy transfer steps. The energy of hot electrons and holes is transferred to the luminescence centers, leading to visible light emission via the aforementioned process. Some of the key FOMs of a scintillator are listed in Table 2.

3   Metal Halide Perovskites for Radiation Detectors

Excellent charge carrier mobility, long carrier diffusion length and lifetime, high absorption coefficient, and low exciton binding energies even for polycrystalline films made from solution are some of the properties that forecast the potential of perovskites for optoelectronic applications in general [12, 13]. In addition, the unique physical and chemical properties of strong stopping power, absence of deep traps, large $\mu_{h/e} \times \tau$ product, and easy crystallization from low-cost solution processes make perovskites suitable for next-generation ionization detection materials. This section focuses on the structure and properties of perovskites that allow them to be superior radiation detection materials.

3.1   Crystal Structure

MHPs with the empirical formula of ABX_3 (e.g., $MAPbX_3$) are classified as 3D perovskites in which BX_6 octahedra are corner-shared along all three fourfold octahedral axes (Fig. 2). 2D perovskites with a layered structure, organized from

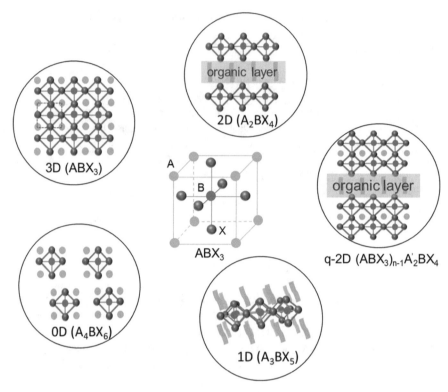

Fig. 2 Schematic representations showing the connectivity of BX_6 octahedra in low-dimensional perovskites and their formation by slicing the 3D structure along crystallographic planes

octahedra connected along two octahedral axes, can be derived from the 3D structure by slicing along specific crystallographic planes. If 2D perovskites are further sliced perpendicular to the inorganic sheets, octahedra remain connected only along one axis, which can be categorized as 1D perovskites. The extreme case is 0D perovskites, derived by further slicing of 1D structures to form non-connected (i.e., isolated) octahedra or octahedral based clusters. Perovskites can also exist in the form of a superposition of two or more classes; for instance, a framework of 3D and 2D is often called quasi-2D perovskites. It is worth mentioning that the perovskites' structural stability of all dimensionalities relies on the cationic organic or inorganic sublattice. For the sake of simplicity, the generalized empirical formulas are used for the perovskites with various dimensionalities in Fig. 2. Even though there are multiple stringent requirements in the material selection for ionizing detectors, which narrow down the choice of effective materials substantially, halide perovskites still maintain many relevant advantages due to their compositional and structural flexibility. The main applications of halide perovskites are focused on optoelectronic devices, and obtaining a high-quality active layer is crucial. In general, the MHP-based radiation detectors consist of either thin films or single crystals (SCs). Solution processes such as drop-casting, spin coating, blade coating, spray coating, and centrifugal casting [14] are widely explored as methods to process these materials, are easy to operate, and are compatible with all kinds of perovskite inks.

3.2 Properties of Metal Halide Perovskites for Radiation Detection

In recent years, research and development efforts towards next-generation materials to detect high-energy radiation have intensified. The majority of ongoing efforts aim to improve the manufacturability and sensitivity of the detection elements and systems [15]. This is why the advancement and/or unearthing of materials that combine key functionalities with economic production has become an intense area of research, with MHPs currently leading the way [4, 15–17]. Due to their high material density ($\approx$4 g cm^{-3}) owing to their ability to integrate atoms with high atomic number (Z), e.g., Pb (Z = 82), Cs (Z = 55), Sn (Z = 50), In (Z = 53), and Br (Z = 35), most of the MHPs are attractive for radiation detectors. Note that the scaling of the X-ray absorption strength, which is derived from Z^4/AE^3 (A: atomic mass; E: energy of the high-energy photons), relies on the Z value. Large absorption cross sections, short penetration depths, large $\mu_{h/e} \times \tau$ products, and short detection times (ns) have propelled MHPs for applications related to high-energy radiation detection.

Tunable and small energy bandgaps of MHPs promote high light yield upon X-ray irradiation (PLQYs: 12,900–250,000 photons/MeV) [18]. The compositional and processing flexibility, high bulk resistivity ($\approx$10^7 Ω-cm), low charge-trap density, and defect-tolerant nature (i.e., $\mu_{h/e} \times \tau$ product) are more beneficial attributes of MHPs [19–22]. Halide perovskites' stopping power (linear attenuation coefficient of 0.09 cm^{-1}) for γ-rays is approximately two times larger than that of commercially deployed CdTe, further emphasizing the advantages of MHP technology [23]. Furthermore, MHPs also offer extraordinary wide-ranging absorption that spans from the visible to hard X-ray (Fig. 1a), making them an exceptional choice for both high-energy direct detectors and scintillators [24]. Apart from their desired radiation detector characteristics, the toxicity of MHPs is the main hurdle for commercial deployment and needs to be addressed.

4 Advances in the Development of Perovskite X-Ray Detectors

Direct detection of X-ray photons using MHPs offers an efficient, simple, and potentially economic technology for various current and evolving applications. Consequently, detection sensitivities of 55,684μC Gy^{-1} cm^{-2} with a low detection

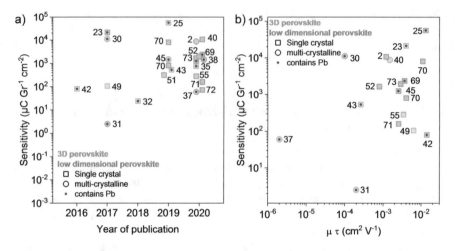

Fig. 3 (**a**) An overview of the achieved sensitivity values for single- and polycrystalline MHPs in recent years and (**b**) combined with their respective mobility-lifetime products. The numbers in the graph indicate the references from which the data have been extracted

limit down to 36 nGys^{-1} have been demonstrated and attributed to the perovskites' superior physical properties [25]. Hence, the application of such a well-performing technology is enormously broad, including flexible and printable large-area X-ray imaging devices as well as futuristic applications such as X-ray photon energy harvesters for powering satellites in space [26, 27]. Two popular forms of MHPs explored for optoelectronic application are polycrystalline thin films and single crystals. For both cases, solution processes have widely been employed to produce these perovskite forms. An overview of the achieved sensitivity values for single- and multi-crystalline MHPs in recent years is given in Fig. 3a and combined with their respective mobility-lifetime products is shown in Fig. 3b.

4.1 Polycrystalline Lead Halide X-Ray Detectors

4.1.1 3D Perovskites

The conventionally employed semiconductors for X-ray detection, amorphous Se, crystalline Si, and CdTe exhibit large photoconduction upon irradiation with X-ray photons [8, 28, 29]. However, uniform film processing onto arbitrary substrates with other device components, e.g., thin-film transistors (TFTs), is highly challenging. This is an area where halide perovskites could provide essential solutions due to their superb processing versatility. For example, the direct X-ray detectors assembled in p-i-n configuration with thick (10–100μm) crystalline MAPbI$_3$ (MAPI) layers processed by the spray-coating method have been used to measure the X-ray-induced charges by monitoring the device's built-in potential (Fig. 4) [24]. Though

Fig. 4 (**a**) Diagrammatic representation of layer stacking of the MAPbI$_3$-based p-i-n photodiode. (**b**) Short-circuit X-ray photocurrent as a function of dose rate. Inset: Sensitivity normalized to the active volume for MAPbI$_3$ layers with different thicknesses. Reprinted from [24] with permission of Springer Nature

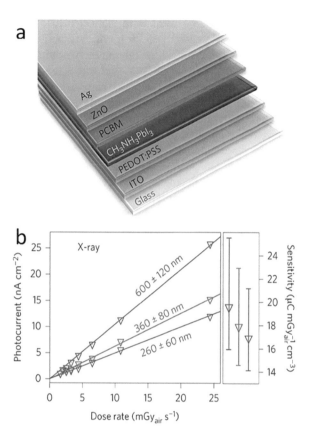

the detectors show relatively long response times, they exhibit high specific sensitivity (25μC mGy$_{air}^{-1}$ cm^{-3}). Apart from the improved response, the devices require a high external bias of 80 V to collect the generated charges efficiently due to the thick perovskite layer. Despite the higher film thickness, the reported strategy demonstrated the potential of spray-coated MAPI for the direct conversion of X-ray photons. For X-ray imaging, such as mammography and digital radiography, thin-film transistor (TFT)-based flat-panel detectors (FPD) are widely used. An 830μm thick polycrystalline MAPI photoconductor was successfully fabricated on a conventional TFT-based backplane and addressed the economic and technical challenges associated with the use of single crystals without adversely affecting the detector performance [30].

For efficient X-ray absorption, the photoconductor layer's thickness should be approximately three times that of the X-ray attenuation length of the material. In MHPs, this characteristic length is on the order of hundreds of micrometers, which represents a significant technical challenge if one considers the required high structural quality of the layers [22]. In an attempt to tackle this challenge, room-temperature mechanically sintered wafers of microcrystalline MAPI (thickness varying from 0.2 to 1 mm) were fabricated [31] to make an X-ray detector. The resulting layers' density ($\approx$3.76 g cm^{-3}) was quite consistent with MAPI single crystals ($\approx$4.15 g cm^{-3}), highlighting the versatility of the sintering process. Upon irradiation of X-rays of 38 keV, the calculated attenuation dept. of the detector (planar-inverted perovskite solar cell architecture) was $\approx$125μm. At an electric field of 0.2 Vμm^{-1}, the device sensitivity was estimated as 2527 μC Gy$_{air}^{-1}$ cm^{-2}. The $\mu_{h/e} \times \tau$ product and ionization energy of the perovskite wafer-based detector were comparable with those of commercial CdTe detectors [32].

Using perovskite nanostructures as the X-ray sensing layer is another approach, which could address the significant challenges of both inorganic and hybrid MHP-based X-ray detectors, such as the limited temporal resolution, nonuniform sensing, and poor stability. For example, an improved response time (1 s) was reported for CsPbBr$_3$ nanoparticles (NPs) decorated with reduced graphene oxide (rGO) as an X-ray sensing element [33]. It was postulated that the generated charge carriers in the NPs were quickly transferred to the rGO nanosheets and ultimately to the electrodes. Besides, the scalable synthesis of CsPbBr$_3$ self-assembled nanosheets (3.1 nm thick) produced by a green synthesis route leads to a high photoluminescence yield, and excellent storage stability is promising for applications as an active material for X-ray imaging screens [34]. The PL quantum yield (PLQY) of cubic phase CsPbBr$_3$ exhibited a higher value of 68% than the orthorhombic counterparts (18.5%). The green synthesis method adopted in this work could potentially provide a route towards a commercially viable perovskite-based X-ray detector technology.

The detector's operational stability, which is quite sensitive to the applied electric field, is another technological bottleneck. To enhance the sensitivity of an X-ray detector, a higher electric field is commonly employed, which leads to increased leakage currents and thus a deteriorated detector performance. To overcome this issue, a Schottky-type sandwiched photodetector comprised of Ag/CsPbBr$_3$/ITO with Ag/perovskite as a rectifying Schottky junction was developed [35]. The device yielded a lower dark current (5 nA/cm^2) and a relatively high sensitivity value of 770μC Gy$_{air}^{-1}$ cm^{-2} upon irradiating with 333.69 nGys^{-1} dose at 8 V. Further, a functional 4$\times$4 X-ray detector array was also demonstrated to show the applicability for more complex sensor layouts.

4.1.2   Low-Dimensional Perovskites

It is evident from the discussions in the previous section that 3D MHPs encounter various challenges related to operational stability and sensitivity. The strategies implemented to obtain a stable and efficient detector performance would add complexity with adverse effects on the economics of manufacturing. Low-dimensional perovskites with intrinsic chemical and moisture stability could be potential alternatives to eliminate such complicated device processes. For example, solution-processed 1D inorganic halide perovskite CsPbI$_3$ crystals for X-ray detection show a maximum sensitivity of 2.37 mC Gy^{-1} cm^{-2} with a lowest

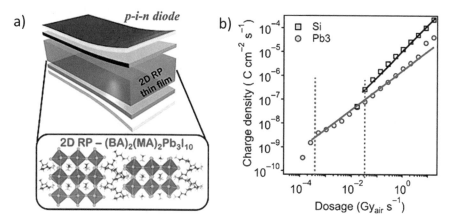

Fig. 5 (**a**) Schematic diagram of the 2D RP-based *p-i-n* thin-film X-ray detector architecture composed of $(BA)_2(MA)_2Pb_3I_{10}$ (named as Pb3) as an absorbing layer. (**b**) Signal-to-noise ratio of X-ray-induced charges for 2D RP and a silicon reference detector. Reprinted from [36] with the permission of American Association for the Advancement of Science

detectable dose rate of $0.219\mu Gy\ s^{-1}$ (minimum signal used in a regular medical diagnostics is $5.5\mu Gy\ s^{-1}$). These values are far superior to the values obtained from their 3D counterparts [69]. Extremely low dark currents (pA), even at 200 V, resulting from an impressive $\mu_{h/e}\times\tau$ product and high bulk resistivity, are the preeminent performance characteristics of the device.

High-quality, thick, single crystals that can endure large applied voltages are frequently used to increase the perovskite layers' resistivity under reverse bias to suppress leakage current. However, developing such single crystals over a large area is quite challenging. A solution-grown 2D Ruddlesden-Popper (RP) phase-layered perovskite film, $(BA)_2(MA)_2Pb_3I_{10}$ (PbI_3), as the X-ray sensing element was proposed to tackle this issue [36]. Evidently, the device shows various exciting features, such as low dark-current (10^{-9} A cm^{-2} at zero bias) and low-voltage operation (self-powered devices). The detector configured in p-i-n ($ITO/PTAA/(PbI_3)/C_{60}$/gold) geometry demonstrated a 10–40-fold higher X-ray absorption coefficient than that of a Si detector (Fig. 5). An excellent X-ray sensitivity of 0.276 CGyair-1 cm^{-3} for 10 keV X-ray photons at zero bias was attributed to the detector's low dark current. The hysteresis-free operation, short response time (1–10μs), and excellent stability are significant outcomes of the reported detectors. However, when it comes to large-area, flexible/conformable medical imaging X-ray detectors, it is a trade-off between the thickness, stopping power, and mechanical flexibility of a sensing element. Inkjet-printed triple-cation perovskites such as $Cs_{0.1}(FA_{0.83}MA_{0.17})_{0.9}Pb$ $(Br_{0.17}I_{0.83})_3$ with 3.7μm thickness were found to provide the required mechanical flexibility without compromising the performance of the X-ray detector [37]. Along with remarkable characteristics, such as good sensitivity of $59.9\mu C$ Gy$_{air}^{-1}$ cm^{-2} and low operating voltage (0.1 V), the detector also showed enhanced stability under

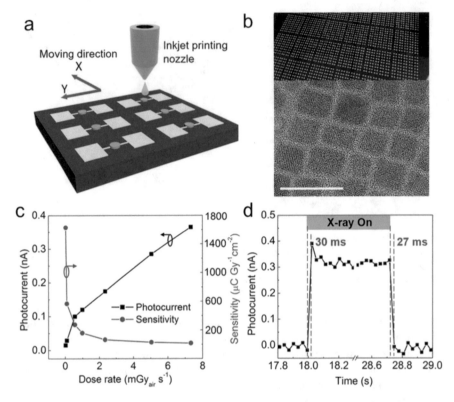

Fig. 6 (**a**) Schematic of perovskite-based device fabrication procedure via inkjet printing. (**b**) Photograph of X-ray detector arrays on a 4-in. wafer (top) and TEM image of CsPbBr$_3$ QDs (bottom). Scale bar: 20 nm. (**c**) X-ray photocurrents and sensitivity as a function of dose rate with 0.1 V bias voltage. (**d**) Temporal response of the device under 7.33 mGy$_{air}$s^{-1} dose rates with 0.1 V bias voltage. Reprinted from [27] with the permission of Wiley-VCH GmbH

X-ray illumination and accumulative exposure of 4 Gy$_{air}$ for 1 h without the need of encapsulation. Notably, the sensor displayed outstanding mechanical flexibility even after 500 bending cycles (bending radius: 3 mm).

When a perovskite film is processed over a large area, developing a homogeneous layer with high X-ray sensitivity is critical. A printable form of 0D CsPbBr$_3$ quantum-dot formulations that were subsequently used to develop X-ray detectors (Fig. 6a) can promote an X-ray sensing element with desired features [27]. Well-controlled crystallinity of the QDs yielded films with a reduced concentration of surface defects (Fig. 6b). The photogenerated current was modulated from 0.1 to 0.36 nA by varying X-ray intensity ranging from 0.55 to 7.33 mGy$_{air}$s^{-1} (Fig. 6c). Prominently, the detector could sense currents down to 9 pA corresponding to an incident X-ray intensity of 0.0172 mGy$_{air}$s^{-1}, with a fast response time of 28 ms (Fig. 6d). The influence of mechanical bending on the performance of the CsPbBr$_3$ QD-based X-ray detector was evaluated through 200 repeated bending cycles. A small change of only 12% confirmed the mechanical robustness of the detector. This study provided an outstanding demonstration of the durability, bendability, and stability of the printed perovskite X-ray detector technology.

The processing versatility of MHPs is further expanded by the melt processing method, which is simple, scalable, and cost effective. $CsPbBr_3$ films deposited onto glass substrates with this method pave the way to large-area, efficient, and low-cost X-ray detectors based on MHPs [38]. Precise control of the cooling rate of the perovskite layer from its melting temperature is the major driving parameter of this process. The subsequent $CsPbBr_3$ layers yield a specific resistance of 8.5×10^9 Ω, which remains similar for films with a thickness ranging from 250µm to 1 mm. Moreover, the estimated sensitivity (1450µC Gy_{air}^{-1} at 300 V) is analogous to conventional Cd(Zn)Te X-ray detectors and superior to α-Si X-ray detectors.

Further enhancement in sensitivity along with reversible, stable, and fast (5 ms) sensing behavior was reported for $CH_3NH_3PbI_2Cl$-based planar-inverted perovskite devices [39]. The demonstrated X-ray detector displayed 550% higher sensitivity than the α-Si reference detector. Along with the high sensitivity and large-area processability of MHPs, flexibility is also another attractive property for next-generation X-ray detectors, which strongly relies on the thickness of the sensing element. The integration of an MHP, $MAPb(I_{0.9}Cl_{0.1})_3$, into a bendable porous nylon membrane is a promising approach to realize real-world flexible X-ray detectors [40]. The device array processed through this method shows the highest $\mu_{h/e} \times \tau$ product, outstanding sensitivity value of 8696 ± 228µCGy$_{air}^{-1}$ cm^{-2}, and excellent operational stability. The proposed high-performing and large-area flexible X-ray detectors offer a solution to some of the existing challenges in medical and industrial X-ray imaging and detection.

4.2 Single-Crystal-Based X-Ray Detectors

Lead halide single crystals are free of grain boundaries and have been demonstrated with lower defect density, better optoelectronic properties, and higher stability than the polycrystalline thin films. The X-ray sensing and harvesting abilities of a single-crystalline MAPI hybrid halide perovskite were first reported in 2015 [41]. The MAPI single crystals' stopping power for X-rays was found to be superior to Si-based detectors. Only 110µm thick MAPI was found sufficient to stop soft X-rays (≈30 keV). In contrast, for detectors with Si, this value is about 1 mm. Absorption of X-ray photons in MHP single crystals occurs in the bulk of the perovskite, where a lower number of defect or trap states exist than those at its surface. Hence, stable and hysteresis-free characteristics were observed. The solution-grown $MAPbBr_3$ single crystals were found to exhibit low defect density, as shown in Fig. 7a. The surface traps on the crystal facets were passivated by UV-O_3 treatment, which resulted in a higher $\mu_{h/e} \times \tau$ product (1.4×10^{-2} cm^2 V^{-1}) and thereby more efficient charge extraction characteristics [42]. A remarkable X-ray photon stopping power, which is higher than that of silicon, CdTe, and $MAPbBr_3$

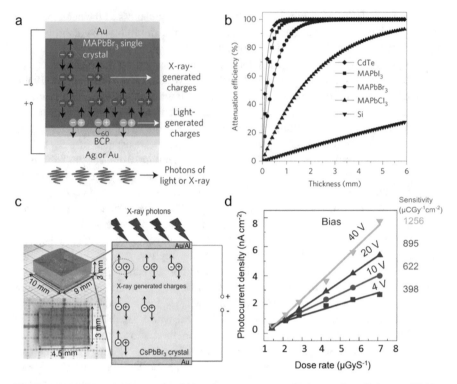

Fig. 7 (**a**) MAPbBr₃ single-crystal radiation detector structure. (**b**) Attenuation efficiency of CdTe, MAPbI₃, MAPbBr₃, MAPbCl₃, and silicon to 50 keV X-ray photons (in terms of the photoelectric effect) versus thickness. (**c**) Photograph of As-grown CsPbBr₃ crystals and detector configuration. (**d**) X-ray response sensitivities of the Al/CsPbBr₃/Au device. Figure 7a and b is reprinted from [42] with permission of Springer Nature, and Fig. 7c and d is reprinted from [45] with the permission of The Royal Society of Chemistry

detectors, was found (Fig. 7b). Additionally, surface engineering techniques were also invented and employed to reduce the surface trap density, thereby achieving high sensitivity and low dark current. A reduced trap density in thermally annealed (AZO)/MAPbBr₃(crystal)/Au detectors compared to non-annealed devices was found, and the effect was attributed to an efficient interface formation between the AZO anode and the perovskite [43]. The interaction of MA^+ and Pb^{+2} dangling bonds with O^{2-} ions was seen as a possible reason for achieving a low-leakage current (nA), thus enhancing sensitivity ($529\mu C\ Gy_{air}^{-1}\ cm^{-2}$ at 50 Vcm^{-1}).

The integration of perovskite crystals with Si-based readout electronics is a step to ensure efficient transport of X-ray-generated charges from the perovskite. A simple technique to monolithically integrate MAPbBr₃ single crystals onto Si substrates via an NH₃Br-terminating molecular interlayer resulted in tremendous figures of merit of X-ray detector (energy range: 8–50 keV) [44]. An outstanding sensitivity ($2.1 \times 10^4 \mu C\ Gy_{air}^{-1}\ cm^{-2}$ under 8 keV X-ray radiation), ≈1000 times higher than commercial a-Se detectors, is one of the remarkable features. Such high-sensitivity detectors are ideal for medical imaging applications.

Single crystals of MAPbBr$_3$ were also explored for energy-sensitive X-ray imaging. The sensitivity of the detector configured in p-i-n diode array was found to vary as a function of the photon energy. This phenomenon was ascribed to carriers being created at different depths of the crystal for dissimilar X-ray photon energies [44]. An increased dark current while applying a high bias voltage to improve the device sensitivity is one of the commonly encountered problems for single- and polycrystalline perovskites. MAPbBr$_3$ crystal X-ray detectors with a Schottky contact were developed that can withstand even high electric fields without increasing the dark current [46]. The X-ray detectors showed a reasonably fast response (76.2µs) and recovery (199.6µs) times, along with good sensitivity (359µC Gy^{-1} cm^{-2} for 50 keV at 200 V). Notably, the higher sensitivity and a threefold faster response were observed for Schottky contact-based detectors than the ohmic device. A similar approach has been employed for inorganic CsPbBr$_3$ single crystals with Al/CsPbBr$_3$/Au configuration, and high sensitivity of 1256µC Gy^{-1} cm^{-2} for 80 keV X-ray photons was reported (Fig. 7c). This is almost 60 times higher than for commercial α-Se detectors (Fig. 7d) [45]. Quasi-monocrystalline CsPbBr$_3$ prepared by the hot-pressed method [25] proved to be highly efficient with the peak sensitivity of 55,684µC Gy$_{air}^{-1}$ cm^{-2}.

Another critical characteristic of MHP-based direct X-ray detectors is their spatial configuration. In 2014 the first single-pixel detector comprising a polycrystalline MAPbI$_3$ that could record 2D X-ray images was introduced [28]. Even though the scanning time was long, the work inspired the scientific community to develop more X-ray detector arrays with MHPs. Soon after, the first linear detector array (LDA) based on MHP-sensing materials was demonstrated featuring 200µm large pixels [44]. The technology was advanced with the development of 2D arrays [30] that were able to deliver faster imaging with improved spatial resolution. Notably, the 2D imaging arrays were fabricated using similar deposition techniques that are used for printable photovoltaics. Since MHPs allow low-temperature processing, their deposition can directly be carried out onto temperature-sensitive readout electronics, further simplifying the overall manufacturing of the detector arrays.

4.3 Lead-Free Perovskites for X-Ray Detectors

It is clear from the preceding sections that lead perovskites have been assiduously studied, as they have established superior properties and proven to be highly promising for various optoelectronic applications. Despite their fascinating properties and astonishing performances, the cytotoxicity of lead and degradation of lead-based MHP devices in the ambient atmosphere have become major obstacles towards their practical use in direct X-ray detectors [47, 48]. The substitution of monovalent Pb by another metal ion with a similar ionic radius, i.e., tin (Sn), can lead to even better optoelectronic properties than for Pb-based materials, such as longer carrier diffusion length, low exciton binding energy, and narrow bandgap. Another approach is to introduce trivalent metals like bismuth (Bi) and antimony (Sb), which are stable in ambient and inherently more environmentally friendly while maintaining a similar electronic structure as that of perovskites with Pb. For example, inorganic $Cs_2AgBiBr_6$ perovskite single-crystal X-ray detectors show four orders of magnitude higher sensitivity (20–105μC Gy_{air}^{-1} cm^{-2}) than commercial a-Se detectors [49, 50]. Along with excellent ambient stability, the detectors also demonstrated high operational stability with the detection limit of 59.7 $nGy_{air}s^{-1}$ along with a low dark current ($\approx 9.55 \times 10^{-16}$ A $Hz^{-1/2}$). Post-deposition treatments, such as thermal annealing and rinsing of $Cs_2AgBiBr_6$ crystals with isopropanol, reduced the number of shallow traps and surface defects [51]. Additionally, the suppression of field-driven ion migration was observed, which helped reduce the vitally important leakage current. With enhanced operational stability, a good sensitivity of 316μ$Gyair^{-1}$ cm^{-2} resulted from these posttreatments.

Another Pb-free single crystal with exceptional optoelectronic properties is $Cs_3Bi_2I_9$. A higher X-ray absorption coefficient (compared to inorganic commercial systems, e.g., CdTe and CsI crystals) and the high responsivity make $Cs_3Bi_2I_9$ single crystals a perfect candidate for X-ray detectors [52]. Some of its characteristic features include a low trap density of 1.4×10^{10} cm^{-3} (significantly lower than commercial inorganic materials (10^{15}–10^{16} cm^{-3}) [53, 54]), the high electrical resistivity of 2.79×10^{10} Ω cm (higher than any Pb-MHPs), and an excellent thermal (up to 550 °C) as well as moisture stability. The X-ray detectors comprised of Au/$Cs_3Bi_2I_9$/Au achieved a sensitivity value of 1652.3μC Gy_{air}^{-1} cm^{-2} and a minimum detection dose of 130 $nGy_{air}s^{-1}$. Another material that could open up an opportunity for broadband detection is $AgBi_2I_7$, with an absorption coefficient for a broad range of X-ray photons, from 0.001 MeV to 10 MeV [55]. The ability of a 0.5 mm thick $AgBi_2I_7$ crystal to stop 100 keV X-rays demonstrates its great potential. The combination of robust operation and low dark current highlights the advantage of these crystals in next-generation direct X-ray detectors. See also Fig. 3b to compare the sensitivity and mobility-lifetime product of recent Pb-containing and Pb-free MHP-based X-ray detectors.

4.4 Scintillators for X-Ray Detection

Most of the commercial large-area X-ray detectors rely on the use of scintillating elements. The primary object of a scintillator is to convert ionizing radiation into high-efficiency visible photons. As conventional inorganic scintillators' fabrication process involves high temperatures (up to 1850 °C), recent efforts have focused on developing scintillators that can be fabricated at reduced temperatures, thus facilitating the technological requirements for large-area production. Even though solution-processable MHPs have tremendous potential for direct radiation detection, they displayed lower responsivities at photon energies above 10 keV. This issue can be addressed to some extent by increasing the thickness of the sensing material and

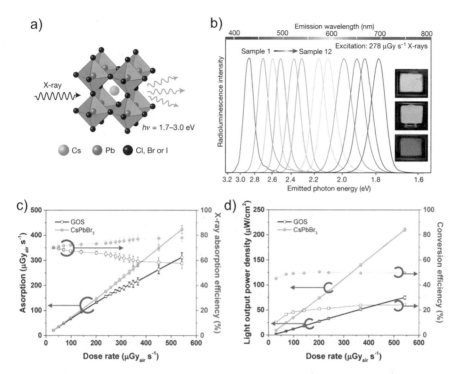

Fig. 8 (**a**) Representation of X-ray-induced luminescence of energy $h\nu$ (where h is the Planck constant and ν is the frequency). (**b**) Tunable luminescence spectra of the perovskite QDs under X-ray illumination with a dose rate of 278μGy s^{-1} at a voltage of 50 kV. (**c**) X-ray absorption and X-ray absorption efficiency of CsPbBr$_3$ and conventional GOS scintillators, (**d**) light output power density and conversion efficiency of CsPbBr$_3$ PNCs and conventional GOS scintillators. Figure 8a and b is reprinted from [57] with the permission of Springer Nature and Fig. 8c and d is reprinted from [59] with the permission of Wiley-VCH GmbH

by further engineering the perovskite structure and composition. Yet, there still exist several technical challenges. This is the area where scintillators would play a dominant role.

The high conversion efficiency of 49% and a decay time as short as 0.7 ns are the best-reported characteristics of perovskite scintillators for X-rays superior to established inorganic systems [56, 57]. Even so, with the versatile chemistry and structural tunability of MHPs, further improvement is quite possible. A MAPbBr$_3$-based X-ray scintillator that works in the temperature range of 50–130 K with LY of 90,000 photons/MeV and 1 ns response time is one example [58]. Ultrasensitive X-ray detectors and flexible, large-area X-ray imaging were further enabled by introducing CsPbBr$_3$ nanocrystals into scintillators [57] (Fig. 8a). Color tunability is an additional advantage of this new class of scintillators (Fig. 8b), including low toxicity, solution synthesis at near-room temperatures, a high emission quantum yield, and fast scintillation response as some of their outstanding attributes. This was followed by exploring CsPbBr$_3$ perovskite nanocrystal-based X-ray detectors, which are cost effective and readily commercialize [59]. Key figures of merit for this study include a high PLQY of 95% at 550 nm, a short PL decay time of about 2.87 ns, a high spatial resolution (9.8 lp mm^{-1}), and high conversion efficiencies (Fig. 8c and d). Even though the origin of nanocrystal scintillation is still not clear, this approach may hold substantial promise for progressing the X-ray sensing and imaging industry.

For the health and safety concerns of the end user, reducing X-ray dose rates by lowering the detector's detection limit is vital. A solution-processed MAPbCl$_3$ single-crystal scintillator was developed to achieve low detection limits [46]. A detection limit of 114.7 nGys^{-1} at 50 keV X-rays, which was comparable to commercial scintillators based on NaI and CsI [60], was attained.

An attractive route for ultrafast and highly efficient scintillation, also known as "quantum scintillation," is pursued by integrating low-dimensional perovskites into scintillators, achieving fast light emission due to quantum confinement effects [61]. Scintillators with 2D perovskites (n-C$_6$H$_{13}$NH$_3$)$_2$PbI$_4$ that consist of multiple natural quantum well structures demonstrated short decay times of 0.7 ns. Along with the fast response, the visible light (558 nm) emission of 2D scintillators is an additional benefit over other technologies that emit UV light. The large exciton binding energies of 2D perovskites can suppress the detected optical signal losses due to thermal quenching. For example, scintillators with the 2D material (EDBE)-PbCl$_4$ showed reduced thermal effects compared to 3D perovskites and an adequate light yield of 9000 photons/MeV even at room temperature [56]. On the other hand, an environmentally friendly 2D perovskite (C$_8$H$_{17}$NH$_3$)$_2$SnBr$_4$ has proven to be efficient, with an excellent absolute PLQY of up to 98% [62]. More importantly, the attained sensitivity was in line with medical safety standards with the emission threshold at 104.23μGys^{-1}. From the discussions so far, it is confirmed that MHPs are very promising scintillator materials in terms of low fabrication costs, nanosecond fast response, low intrinsic trap density, and potentially high light yield.

5   Advance in the Development of Perovskite γ-Ray Detectors

In 2016, the first report appeared on the application of MHPs for the direct detection of γ-rays [63]. Upon exposure with γ-rays (0.96 MeV) from different radioactive sources (^{11}C and ^{137}Cs), the 3D perovskite (MAPbI$_3$, FAPbI$_3$, and MAPbBr) single-crystal devices revealed an overall current to charge efficiency of 19% (Fig. 9a and b). Furthermore, the potential for single γ-photon counting, which is a highly challenging task, was demonstrated. The requirement to develop MHP-based high-resolution energy spectra γ-ray detectors is a high $\mu_{h/e} \times \tau \times E$ product (E: applied external electric field). However, increasing the electrical field enhances both the dark current/noise ratio and ion migration within the perovskite. This issue was addressed by raising the bulk resistivity of the single crystal (3.6 × 10^9 Ω-cm) by

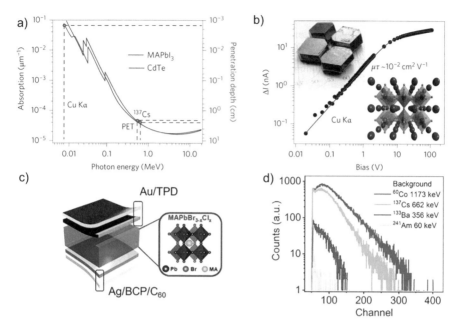

Fig. 9 (**a**) The attenuation coefficient and corresponding penetration depth of MAPbI$_3$ and CdTe as a function of photon energy, from soft X-rays to γ radiation. (**b**) The bias dependence of the photocurrent generated by Cu Kα X-ray radiation (8 keV) in a SC of MAPbI$_3$ perovskite; the red line indicates a fit with the Hecht model showing a high $\mu\tau$ product of ~10^{-2} cm^2 V^{-1}. Top inset: Photograph of typical MAPbI$_3$ perovskite SCs grown from a nonaqueous (retrograde solubility) method, placed on a millimeter ruler. Bottom inset: Schematic of the three-dimensional interconnection of PbI$_6$-octahedra in a perovskite lattice (green, Pb; yellow, I; blue, MA). (**c**) *p-i-n* device structure using MAPbBr$_{3-x}$Cl$_x$ crystal. (**d**) Gamma-ray spectra exposed the device under various radioactive sources collected by a multichannel analyzer (MCA) when the detector was operated at −8 V at room temperature. Figure 9a and b is reprinted from [63] with the permission of Springer Nature, and Fig. 9c and d is reprinted from [66] with the permission of Elsevier

using large-size dopant-compensated CH$_3$NH$_3$PbBr$_{2.94}$Cl$_{0.06}$ crystals that successfully demonstrated a low operation field (1.8 V mm^{-1}) [23]. The crystals showed high hole and electron mobilities of 560 and 320 cm^{2}V^{-1} s^{-1}, respectively, resulting in a significantly higher $\mu_{h/e}\times\tau$ product.

Apart from the performance matrix, the structural stability of MHP single crystals is one of the major pitfalls, in which the phase transformation from cubic to the hexagonal structure within 24 h after the crystal growth was observed [64]. To address this issue, mixed halide perovskite single crystals, Cs$_x$FA$_{1-x}$PbI$_{3-y}$Br$_y$ (x = 0–0.1, y = 0–0.6), with various thicknesses ranging from 0.2 to 15 mm, were fabricated to detect γ-ray within the energy range of 0.02–1 MeV [65]. Evidently, the resulting crystals unveiled good stability up to 20 days, which could be extended further to 60 days by substituting Br in place of I. Despite the encouraging progress, there exist uneven energy detection spectra for MHPs. For example, CsPbBr$_3$ can detect within the 32.3–662 keV energy range while MAPbBr$_{3-x}$Cl$_x$ detects only high-energy γ-photons (0.1–10 MeV) [23]. A unipolar p-i-p device architecture with MAPbBr$_{3-x}$Cl$_x$ γ-ray detector was proposed to resolve this issue (Fig. 9c) [66]. It was found that the use of high-work-function contacts effectively blocks the dark noise originating from thermally activated electron injection from the impurities. Thus, efficient pulse collection could be prompted even at higher electrical fields. Consequently, strong electrical pulses were observed when exposing the detector with different energies of gamma-ray photons emitted from various radioactive sources (Fig. 9d).

5.1   *Scintillator Detectors for γ-Ray Detection*

γ-Ray scintillator detectors are a workhorse for γ-ray spectroscopy that provide high efficiency and, depending on the choice of the scintillator, good energy, and timing resolution. Inorganic γ-ray scintillating materials such as GSO:Ce and LSO:Ge are the most explored systems but often suffer from long decay times (≈40 ns). Hybrid perovskite materials that offer short decay times can be suitable alternatives [4, 66, 67]. However, these are generally unstable with a low material density (1 g/cm^3). Hence, the development of materials that address these difficulties is an immediate requirement for further advancing the various imaging technologies. Hybrid 2D perovskite compound ((C$_6$H$_5$C$_2$H$_4$NH$_3$)$_2$PbX$_4$)-based scintillators with high LY (14,000 photons/MeV), excellent linearity to γ-rays with different energies (122–662 keV), and short decay time (≈11 ns) are showcasing the potential of MHPs as γ-ray scintillators [68], where the natural quantum wells of 2D perovskite promote improved detection characteristics.

6   Summary and Future Perspective

The rise of MHPs as a promising family of materials for application in next-generation X-ray and γ-ray detector technologies has been witnessed from recent reports. Remarkably, the field has skyrocketed since 2017, as proven by the volume of publications and related citations received to date. This global interest stems from the fascinating physical properties that these synthetic perovskites have, which, when combined with unconventional device engineering, can produce detectors with performance features on par with or even better than those of incumbent technologies.

MHPs have been utilized in two types of high-energy radiation detectors, namely direct and scintillator detector technologies. In terms of manufacturability, the simpler one is the direct detector. In contrast, detectors with scintillation systems are bulkier and less portable. Yet, the use of scintillation detectors can address specific difficulties that direct X-ray detectors encountered, so significant ongoing research efforts aim to improve scintillation systems further. Consequently, MHP scintillators have demonstrated the potential to address numerous inadequacies, including manufacturability, sensitivity, response time, and spatial resolution, to the degree that multiple materials are quickly becoming competitive to commercial technologies such as NaI and CsI.

MHPs have also unveiled tremendous potential for direct X-ray detection. The excellent performance of numerous perovskite-based devices reported to date is mainly attributed to the outstanding $\mu_{h/e} \times \tau$ product (1.1×10^{-2} cm^{2}V^{-1}), fast response times, long carrier diffusion lengths (up to 175μm), high conversion efficiencies, high sensitivity, emission in the visible region of the electromagnetic spectrum, quantum confinement (low-dimensional perovskites), and characteristically low trap density of states (10^8 cm^{-3}) even in solution-processed polycrystalline systems. Despite the advantages demonstrated by the perovskite-based high-energy radiation detectors so far, there are still many challenges to be considered before translating this technology to commercialization. Primarily, in direct X-ray detectors, the resistivity of the active layers (films, crystals, etc.) needs to be amplified further in order to decrease the dark current. Larger bias voltages can be applied by using thicker active layers that could improve the charge collection efficiency. But this process induces adverse effects on electrical noise (dark current) and operational stability due to the field-induced ion migration—this is particularly true for polycrystalline films. To overcome these bottlenecks, possible approaches include the application of larger bandgap perovskites and/or the use of higher quality single crystals.

The second challenge is to boost the perovskite X-ray detector's chemical stability towards ambient conditions. While simple device encapsulation could provide a concrete solution for commercial applications, the use of all-inorganic perovskites also shows promise to address this challenge. On the other hand, the simultaneous improvement of LY through enhanced quantum confinement and operational/environmental stability could be attained by developing innovative low-dimension perovskites (e.g., 0D, 1D, 2D). It is clear from the recent reports that the incorporation of such low-dimension perovskite crystals diminishes the ion migration, hence allowing the application of higher bias across the crystals and improving the operational stability of the devices. Increasing the $\mu_{h/e} \times \tau$ product via engineering the perovskite layer/crystal is another challenging part. The use of indirect bandgap perovskites that display extended carrier lifetimes could be exploited as a potential solution. In sandwich-type direct X-ray detectors, incorporating conducting materials, such as graphene or other 2D materials, in the charge transporting layer could improve the conductivity and increase the charge collection efficiency. The next challenge that needs to be addressed is Pb's toxicity in most of the MHPs studied to date. To this end, the exchange of Pb with other high-Z elements that are vital for obtaining high X-ray stopping power has already yielded promising results but with plenty of room for further advancement.

Perovskite-based γ-ray detectors share similar challenges as X-ray detectors. Improving the homogeneity and overall quality of the absorbing material will resolve some of these outstanding issues. Despite other existing hurdles, however, the future of perovskite-based X-ray and γ-ray detectors appears very bright. Only time will tell whether MHP-based high-energy radiation detectors will ultimately make it to the commercial stage in fields ranging from medicine and homeland security to portable radiological identification and energy-harvesting devices for space applications.

References

1. Song, Y., Li, L., Bi, W., Hao, M., Kang, Y., Wang, A., Wang, Z., Li, H., Li, X., Fang, Y., Yang, D., Dong, Q.: Research. **2020**, 1 (2020)
2. Zheng, X., Zhao, W., Wang, P., Tan, H., Saidaminov, M.I., Tie, S., Chen, L., Peng, Y., Long, J., Zhang, W.H.: J. Energy Chem. **49**, 299 (2020)
3. Kakavelakis, G., Gedda, M., Panagiotopoulos, A., Kymakis, E., Anthopoulos, T.D., Petridis, K.: Adv. Sci. **7**, 2002098 (2020)
4. Wei, H., Huang, J.: Nat. Commun. **10**, 1066 (2019)
5. Que, W., Rowlands, J.A.: Med. Phys. **22**, 365 (1995)
6. Zanichelli, M., Santi, A., Pavesi, M., Zappettini, A.: J. Phys. D. Appl. Phys. **46**, 365103 (2013)
7. Gerrish, V.M.: Semicond. Semimetals, pp. 493–530. Elsevier, Amsterdam (1995)
8. Yaffe, M.J., Rowlands, J.A.: Phys. Med. Biol. **42**, 1 (1997)
9. Martz, H.E., Logan, C.M., Schneberk, D.J., Shull, P.J.: X-ray imaging: fundamentals, industrial techniques, and applications. CRC Press, Boca Raton (2017)
10. Rocha, J.G., Lanceros-Mendez, S.: Recent Patents Electron. Eng. **4**, 16 (2011)
11. Guerra, M., Manso, M., Longelin, S., Pessanha, S., Carvalho, M.L.: J. Instrum. **7**, C10004 (2012)
12. Manser, J.S., Christians, J.A., Kamat, P.V.: Chem. Rev. **116**, 12956 (2016)
13. Misra, R.K., El Cohen, B., Iagher, L., Etgar, L.: ChemSusChem. **10**, 3712 (2017)
14. Park, N.G., Zhu, K.: Nat. Rev. Mater. **5**, 333 (2020)
15. He, Y., Kanatzidis, M.: HDIAC J. **6**, 16 (2019)
16. Stoumpos, C.C., Kanatzidis, M.G.: Acc. Chem. Res. **48**, 2791 (2015)
17. Stoumpos, C.C., Malliakas, C.D., Peters, J.A., Liu, Z., Sebastian, M., Im, J., Chasapis, T.C., Wibowo, A.C., Chung, D.Y., Freeman, A.J., Wessels, B.W., Kanatzidis, M.G.: Cryst. Growth Des. **13**, 2722 (2013)
18. Xie, A., Nguyen, T.H., Hettiarachchi, C., Witkowski, M.E., Drozdowski, W., Birowosuto, M. D., Wang, H., Dang, C.: J. Phys. Chem. C. **122**, 16265 (2018)
19. Li, W., Wang, Z., Deschler, F., Gao, S., Friend, R.H., Cheetham, A.K.: Nat. Rev. Mater. **2**, 1 (2017)
20. Ono, L.K., Juarez-Perez, E.J., Qi, Y.: ACS Appl. Mater. Interfaces. **9**, 30197 (2017)
21. Wehrenfennig, C., Eperon, G.E., Johnston, M.B., Snaith, H.J., Herz, L.M.: Adv. Mater. **26**, 1584 (2014)
22. Xing, G., Mathews, N., Sun, S., Lim, S.S., Lam, Y.M., Grätzel, M., Mhaisalkar, S., Sum, T.C.: Science (80-). **342**, 344 (2013)
23. Wei, H., Desantis, D., Wei, W., Deng, Y., Guo, D., Savenije, T.J., Cao, L., Huang, J.: Nat. Mater. **16**, 826 (2017a)
24. Yakunin, S., Sytnyk, M., Kriegner, D., Shrestha, S., Richter, M., Matt, G.J., Azimi, H., Brabec, C.J., Stangl, J., Kovalenko, M.V., Heiss, W.: Nat. Photonics. **9**, 444 (2015)
25. Pan, W., Yang, B., Niu, G., Xue, K.H., Du, X., Yin, L., Zhang, M., Wu, H., Miao, X.S., Tang, J.: Adv. Mater. **31**, 1904405 (2019)
26. Li, H., Shan, X., Neu, J.N., Geske, T., Davis, M., Mao, P., Xiao, K., Siegrist, T., Yu, Z.: J. Mater. Chem. C. **6**, 11961 (2018)

27. Liu, J., Shabbir, B., Wang, C., Wan, T., Ou, Q., Yu, P., Tadich, A., Jiao, X., Chu, D., Qi, D., Li, D., Kan, R., Huang, Y., Dong, Y., Jasieniak, J., Zhang, Y., Bao, Q.: Adv. Mater. **31**, 1901644 (2019)

28. Oh, K.M., Kim, D.K., Shin, J.W., Heo, S.U., Kim, J.S., Park, J.G., Nam, S.H.: J. Instrum. **9**, P01010 (2014)

29. Zhao, W., Rowlands, J.A.: Med. Phys. **22**, 1595 (1995)

30. Kim, Y.C., Kim, K.H., Son, D.-Y., Jeong, D.-N., Seo, J.-Y., Choi, Y.S., Han, I.T., Lee, S.Y., Park, N.-G.: Nature. **550**, 87 (2017)

31. Shrestha, S., Fischer, R., Matt, G.J., Feldner, P., Michel, T., Osvet, A., Levchuk, I., Merle, B., Golkar, S., Chen, H., Tedde, S.F., Schmidt, O., Hock, R., Rührig, M., Göken, M., Heiss, W., Anton, G., Brabec, C.J.: Nat. Photonics. **11**, 436 (2017)

32. Wang, X., Zhao, D., Qiu, Y., Huang, Y., Wu, Y., Li, G., Huang, Q., Khan, Q., Nathan, A., Lei, W., Chen, J.: Phys. Status Solidi Rapid Res Lett. **12**, 1800380 (2018)

33. Liu, X., Xu, T., Li, Y., Zang, Z., Peng, X., Wei, H., Zha, W., Wang, F.: Sol. Energy Mater. Sol. Cell. **187**, 249 (2018)

34. Wang, L., Fu, K., Sun, R., Lian, H., Hu, X., Zhang, Y.: Nano-Micro Lett. **11**, 52 (2019)

35. Xu, Q., Wang, X., Zhang, H., Shao, W., Nie, J., Guo, Y., Wang, J., Ouyang, X.: ACS Appl. Electron. Mater. **2**, 879 (2020b)

36. Tsai, H., Liu, F., Shrestha, S., Fernando, K., Tretiak, S., Scott, B., Vo, D.T., Strzalka, J., Nie, W.: Sci. Adv. **6**, eaay0815 (2020)

37. Mescher, H., Schackmar, F., Eggers, H., Abzieher, T., Zuber, M., Hamann, E., Baumbach, T., Richards, B.S., Hernandez-Sosa, G., Paetzold, U.W., Lemmer, U.: ACS Appl. Mater. Interfaces. **12**, 15774 (2020)

38. Matt, G.J., Levchuk, I., Knüttel, J., Dallmann, J., Osvet, A., Sytnyk, M., Tang, X., Elia, J., Hock, R., Heiss, W., Brabec, C.J.: Adv. Mater. Interfaces. **7**, 1901575 (2020)

39. Gill, H.S., Elshahat, B., Kokil, A., Li, L., Mosurkal, R., Zygmanski, P., Sajo, E., Kumar, J.: Phys. Med. **5**, 20 (2018)

40. Zhao, J., Zhao, L., Deng, Y., Xiao, X., Ni, Z., Xu, S., Huang, J.: Nat. Photonics. **14**, 612 (2020)

41. Náfrádi, B., Náfrádi, G., Forró, L., Horváth, E.: J. Phys. Chem. C. **119**, 25204 (2015)

42. Wei, H., Fang, Y., Mulligan, P., Chuirazzi, W., Fang, H.H., Wang, C., Ecker, B.R., Gao, Y., Loi, M.A., Cao, L., Huang, J.: Nat. Photonics. **10**, 333 (2016)

43. Li, L., Liu, X., Zhang, H., Zhang, B., Jie, W., Sellin, P.J., Hu, C., Zeng, G., Xu, Y.: ACS Appl. Mater. Interfaces. **11**, 7522 (2019)

44. Wei, W., Zhang, Y., Xu, Q., Wei, H., Fang, Y., Wang, Q., Deng, Y., Li, T., Gruverman, A., Cao, L., Huang, J.: Nat. Photonics. **11**, 315 (2017b)

45. Zhang, H., Wang, F., Lu, Y., Sun, Q., Xu, Y., Bin Zhang, B., Jie, W., Kanatzidis, M.G.: J. Mater. Chem. C. **8**, 1248 (2020b)

46. Xu, Q., Shao, W., Li, Y., Zhu, Z., Liu, B., Ouyang, X., Liu, J.: Opt. Lett. **45**, 355 (2020a)

47. Babayigit, A., Ethirajan, A., Muller, M., Conings, B.: Nat. Mater. **15**, 247 (2016)

48. Lyu, M., Yun, J.-H., Chen, P., Hao, M., Wang, L.: Adv. Energy Mater. **7**, 1602512 (2017)

49. Pan, W., Wu, H., Luo, J., Deng, Z., Ge, C., Chen, C., Jiang, X., Yin, W.J., Niu, G., Zhu, L., Yin, L., Zhou, Y., Xie, Q., Ke, X., Sui, M., Tang, J.: Nat. Photonics. **11**, 726 (2017)

50. Zhuge, F., Luo, P., Zhai, T.: Sci. Bull. **62**, 1491 (2017)

51. Zhang, H., Gao, Z., Liang, R., Zheng, X., Geng, X., Zhao, Y., Xie, D., Hong, J., Tian, H., Yang, Y., Wang, X., Ren, T.L.: IEEE Trans. Electron Devices. **66**, 2224 (2019)

52. Zhang, Y., Liu, Y., Xu, Z., Ye, H., Yang, Z., You, J., Liu, M., He, Y., Kanatzidis, M.G., (Frank) Liu, S.: Nat. Commun. **11**, 1 (2020c)

53. Martens, K., Chui, C.O., Brammertz, G., De Jaeger, B., Kuzum, D., Meuris, M., Heyns, M.M., Krishnamohan, T., Saraswat, K., Maes, H.E., Groeseneken, G.: IEEE Trans. Electron. Devices. **55**, 547 (2008)

54. Ni, Z., Bao, C., Liu, Y., Jiang, Q., Wu, W.Q., Chen, S., Dai, X., Chen, B., Hartweg, B., Yu, Z., Holman, Z., Huang, J.: Science (80-). **367**, 1352 (2020)

55. Tie, S., Zhao, W., Huang, W., Xin, D., Zhang, M., Yang, Z., Long, J., Chen, Q., Zheng, X., Zhu, J., Zhang, W.H.: J. Phys. Chem. Lett. **11**, 7939 (2020)
56. Birowosuto, M.D., Cortecchia, D., Drozdowski, W., Brylew, K., Lachmanski, W., Bruno, A., Soci, C.: Sci. Rep. **6**, 37254 (2016)
57. Chen, Q., Wu, J., Ou, X., Huang, B., Almutlaq, J., Zhumekenov, A.A., Guan, X., Han, S., Liang, L., Yi, Z., Li, J., Xie, X., Wang, Y., Li, Y., Fan, D., Teh, D.B.L., All, A.H., Mohammed, O.F., Bakr, O.M., Wu, T., Bettinelli, M., Yang, H., Huang, W., Liu, X.: Nature. **561**, 88 (2018)
58. Mykhaylyk, V.B., Kraus, H., Saliba, M.: Mater. Horiz. **6**, 1740 (2019)
59. Heo, J.H., Shin, D.H., Park, J.K., Kim, D.H., Lee, S.J., Im, S.H.: Adv. Mater. **30**, 1801743 (2018)
60. Gu, Z., Huang, Z., Li, C., Li, M., Song, Y.: Sci. Adv. **4**, eaat2390 (2018)
61. Shibuya, K., Koshimizu, M., Murakami, H., Muroya, Y., Katsumura, Y., Asai, K.: Jpn. J. Appl. Phys. Part 2 Lett. **43**, L1333 (2004)
62. Cao, J., Guo, Z., Zhu, S., Fu, Y., Zhang, H., Wang, Q., Gu, Z.: ACS Appl. Mater. Interfaces. **12**, 19797 (2020)
63. Yakunin, S., Dirin, D.N., Shynkarenko, Y., Morad, V., Cherniukh, I., Nazarenko, O., Kreil, D., Nauser, T., Kovalenko, M.V.: Nat. Photonics. **10**, 585 (2016)
64. Yuan, Y., Huang, J.: Acc. Chem. Res. **49**, 286 (2016)
65. Nazarenko, O., Yakunin, S., Morad, V., Cherniukh, I., Kovalenko, M.V.: NPG Asia Mater. **9**, e373 (2017)
66. Liu, F., Yoho, M., Tsai, H., Fernando, K., Tisdale, J., Shrestha, S., Baldwin, J.K., Mohite, A.D., Tretiak, S., Vo, D.T., Nie, W.: Mater. Today. **37,** 27 (2020)
67. Pan, W., Wei, H., Yang, B.: Front. Chem. **8**, 268 (2020)
68. Kawano, N., Koshimizu, M., Okada, G., Fujimoto, Y., Kawaguchi, N., Yanagida, T., Asai, K.: Sci. Rep. **7**, 14754 (2017)
69. Bin Zhang, B., Liu, X., Xiao, B., Ben Hafsia, A., Gao, K., Xu, Y., Zhou, J., Chen, Y.: J. Phys. Chem. Lett. **11**, 432 (2020)
70. R. Zhuang, X. Wang, W. Ma, Y. Wu, X. Chen, L. Tang, H. Zhu, J. Liu, L. Wu, W. Zhou, X. Liu, Y. (Michael) Yang, Nat. Photonics **13**, 602 (2019)
71. M. Xia, J. Yuan, G. Niu, X. Du, L. Yin, W. Pan, J. Luo, Z. Li, H. Zhao, K. Xue, X. Miao, J. Tang, Adv. Funct. Mater. **30**, 1910648 (2020)
72. L. Yao, G. Niu, L. Yin, X. Du, Y. Lin, X. Den, J. Zhang, J. Tang, J. Mater. Chem. C **8**, 1239 (2020)
73. Y. Liu, Z. Xu, Z. Yang, Y. Zhang, J. Cui, Y. He, H. Ye, K. Zhao, H. Sun, R. Lu, M. Liu, M. G. Kanatzidis, S. (Frank) Liu, Matter **3**, 180 (2020)

Thallium-Based Materials for Radiation Detection

Ge Yang and Ibrahim Hany

Abstract A variety of medical imaging, homeland security, industrial monitoring, environmental survey, and physical science applications can be largely enhanced by employing high-performance room-temperature radiation detectors. To this end, thallium (Tl)-based materials represent an interesting category of detector materials thanks to their unique physical properties. In this regard, TlBr has been intensively studied as a radiation detector material over the past few years. Recently several other Tl-based compounds, e.g., $TlPbI_3$, have also received strong interest in light of their potential for room-temperature radiation detection. In this chapter, we review the development of these Tl-based materials, used as either semiconductor radiation detectors or scintillator radiation detectors, and discuss the performance-determining factors including physical properties, growth processes, and defect control.

Room-temperature radiation detectors are crucial for a wide spectrum of applications including fundamental physical science, medical imaging, homeland security, nuclear safety inspection, environmental survey, nonproliferation, and space exploration [1–3]. For these applications, the information about the type and intensity of the radiation field, as well as the energy-resolving spectrum and the spatial distribution, is commonly desired [4]. There are two major types of solid room-temperature radiation detectors, i.e., semiconductor detectors and scintillator detectors. Semiconductor radiation detectors utilize a "direct" detection mechanism; that is, the electronic signal is directly formed through the bias voltage-driven drift of electrons and holes, which are produced through the interaction of incident ionizing radiation and detection medium. Scintillator radiation detectors involve "indirect" detection processes; that is, the scintillation light will be first produced through the interaction of incident radiation with the detection medium, followed by the conversion of scintillation light signal to the electric signal using the photodetectors, e.g., photomultiplier tubes.

G. Yang (✉) · I. Hany
Department of Nuclear Engineering, North Carolina State University, Raleigh, NC, USA
e-mail: gyang9@ncsu.edu

© The Author(s), under exclusive license to Springer Nature Switzerland AG 2022
K. Iniewski (ed.), *Advanced Materials for Radiation Detection*,
https://doi.org/10.1007/978-3-030-76461-6_7

145

As a unique category of radiation detector materials, Tl-based compounds exhibit many desired physical properties and thus attract strong interest. Over the past few years, Tl-based materials have been successfully developed as semiconductor detectors as well as scintillator detectors. As such, the purpose of this chapter is to offer a timely review of several major Tl-based radiation detector materials, which will help the readers learn the development status and future trend of this promising research area.

1   Tl-Based Semiconductor Detector Materials

1.1   *TlBr*

Among the Tl-based radiation detector materials, TlBr has a long development history. Currently, high-quality TlBr detectors are able to achieve a relatively high energy resolution which is approaching that of the leading room-temperature semiconductor CdZnTe radiation detectors. However, the polarization phenomenon, i.e., the long-term instability issue, is still seriously limiting the deployment of TlBr toward large-scale practical radiation detection applications. To this end, a series of research activities have been actively conducted to address the intrinsic polarization issue of TlBr, aiming to maintain its initial high detection performance over a realistically long period.

TlBr has a high density of 7.56 g/cm^3 and high atomic numbers (Tl $=$ 81, Br $=$ 35), which offer excellent stopping power to high-energy ionizing radiation. Meanwhile, its high bandgap of 2.68 eV could enable high intrinsic resistivity for reducing dark current and achieving low electronic noise detector operation. It should be noted that TlBr also has suitable characteristics for melt growth, i.e., a relatively low melting temperature at 480 °C and a lack of destructive phase transition below the melting point, which enable high-quality growth of TlBr crystals using melt-growth techniques [5]. As a result, TlBr crystals can be grown at a relatively low temperature using melt-growth approaches, which is desired from the points of defect reduction and cost saving.

The initial interest to use TlBr for radiation detection can be tracked back to 1947, when Hofstadter demonstrated the possibility of gamma-ray detection using a prototype TlBr radiation counter [6]. With the employment of raw material purification techniques, especially the zone-refining method, the significantly improved performance of TlBr detectors has been proved to be feasible [5]. Following these early-stage studies, various research groups have shown strong interest to further explore TlBr materials with a focus on identifying optimized crystal purification and growth conditions, achieving a deep understanding of material defects, and adapting different advanced detector designs to enhance energy-resolving capability.

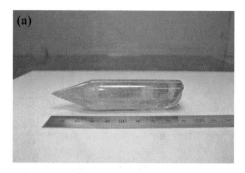

Fig. 1 (**a**) An as-grown TlBr ingot grown by the Bridgman–Stockbarger method [7]; (**b**) an as-grown TlBr ingot grown by the zone-melting method [8]

Different crystal growth techniques have been attempted to grow TlBr materials. These efforts include the Bridgman–Stockbarger method, the traveling molten zone (TMZ) method, and the hydrothermal growth method. Figure 1 shows two as-grown TlBr ingots grown by the Bridgman–Stockbarger method and by the zone-melting method, respectively. As one can see, both techniques can be used to grow large-size TlBr crystals.

The purification plays a key role in determining the performance of TlBr radiation detectors. In this regard, the existence of impurities directly affects the charge transport behavior in as-grown TlBr crystals. As such, different purification strategies, as well as intentional doping efforts, have been developed to address the challenges of impurities. Over the past decades, a variety of purification techniques were employed including the zone-refining method, the vacuum distillation method, the vacuum sublimation method, and the hydrothermal recrystallization method [9]. Among them, the zone-refining method is widely used due to its simple configuration, easy execution, and high purification efficiency. Figures 2 and 3, respectively, show the basic operation principle of zone refining and a representative zone-refining setup in TlBr production. Hitomi et al. demonstrated that the multi-pass zone-refining process can effectively improve the charge transport in TlBr [5]. Their results show a strong correlation between the impurities and the mobility-lifetime product of charge carriers of TlBr. A two-order increase in the mobility-lifetime product was successfully achieved after conducting multi-pass zone refining of TlBr raw materials.

On the other hand, the long-standing polarization phenomenon poses a serious challenge toward the practical deployment of TlBr radiation detectors. In this

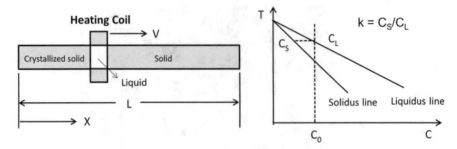

Fig. 2 Basic operation principle of zone-refining method [9]

Fig. 3 A zone-refining
setup used for the
production of TlBr [8]

scenario, TlBr detectors suffer from a slow degradation over time even though their initial performance is excellent. The so-called polarization represents a long-term material instability issue under device operation conditions. Essentially the polarization mechanism is related to the ion migration process under bias voltage across the detector volume. As an ion-type conductor, TlBr can easily exhibit an ion migration behavior once an external electric field is applied across the crystal. In this situation, the positive ion (Tl^+) and the negative ion (Br^-) will diffuse through vacancy-hopping processes. Recent polarization studies show that the faster Br-ion electro-diffusion to the anode has a stronger influence on the stability of TlBr radiation detectors, compared to the slower Tl-ion electro-diffusion toward the cathode [10]. It has been found that Br ions aggressively react with most types of electrode anode materials, thus leading to progressive deterioration of charge collection efficiency and high detector noise [10]. Different approaches have been explored to mitigate the deleterious effects of polarization in TlBr detectors. Operating TlBr detectors at low temperature (~−20 °C) has been effective to control polarization to some extent since both the metal-bromine formation and the diffusivity of ions can be suppressed at low temperature [11]. However, the need for additional cooling apparatus largely increases the complexity of the overall radiation detection system and as such compromises the major motivation to develop TlBr detectors, i.e., utilizing their "room-temperature" detection capability. Several other approaches, e.g., using selected metals as electrodes including Pt, Au, Pd, and Tl or using $TlBr_{1-x}Cl_x$ heterojunction at the metal-TlBr interface, have also been proposed to eliminate the reaction of Bi ion with the electrode metal [10]. While a lot of interesting progress has been made over the years, the corresponding extended detector stability time is still much shorter than the expected detector lifetime for practical operation scenarios. Furthermore, the reproducibility issues of these approaches need to be further addressed to enable the scale-up production of TlBr detectors. Recent exploration of metal oxide electrode materials, e.g., indium tin dioxide, titanium dioxide, and tin dioxide, shows some promising results to mitigate the deleterious reaction between Bi ion and electrode material to provide longer stable detector performance [10]. More systematic experiments are desired to strengthen this effort. The periodic switching of bias voltage was also demonstrated to be effective to help address the polarization issue. A stable operational lifetime of 5 years at 8 h/day was obtained for planar and pixelated detectors [10]. However, frequent change of bias voltage may increase the complexity of the detection system. The long-term reliability of this approach needs to be further studied through a systematic approach.

1.2 TlPbI₃

1.2.1 TlPbI₃ Physical Properties

$TlPbI_3$ is an all-inorganic halide perovskite-like semiconductor that crystallizes in an orthorhombic crystal symmetry that has lattice constant values of 4.625 Å (1), 14.885 Å (2), and 11.857 Å (3). Four molecules are contained in its single unit cell (Z = 4). In this crystal structure, iodine ions form a 3D matrix that hosts thallium and lead cations inside its trigonal prism and octahedral voids, respectively, as shown in Fig. 4. The Pb-I and Th-I coordination bonds are shown in Fig. 5. The two Pb-I bonds are slightly different unlike the four Tl-I bonds, which are largely different. The spatial anisotropy, mainly caused by the coordination between thallium and iodine ions, can lead to anisotropy in the optical and electrical properties similar to the case of thallium iodide (TlI). Lin et al. reported the anisotropy of the static dielectric constant and the electric conductivity [12], and Khyzhun et al. showed experimental evidence of $TlPbI_3$ birefringence using 1540 nm pump laser and 1340 nm probe laser [13]. Moreover, the birefringence varied when changing the pump laser intensity from 0.2 J/cm² to 0.8 J/cm².

$TlPbI_3$ has a low melting point of 346 °C which facilitates its growth using melting methods. The current experimental studies of $TlPbI_3$ report its growth using the vertical Bridgman–Stockbarger (VB) method with thallium iodide (TlI) and lead iodide (PbI_2) as starting material and demonstrate that it undergoes no temperature-dependent structural phase transformation down to room temperature [13–18]. TlI and PbI_2 starting material can be obtained either as purified powders from commercial vendors or as precipitates from the exchange reaction of saturated aqueous

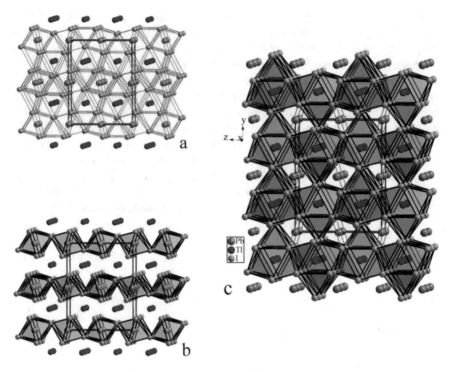

Fig. 4 Framework of (**a**) iodine atoms and the location of the atoms of (**b**) Pb and (**c**) Tl within the framework in the TlPbI₃ structure [13]

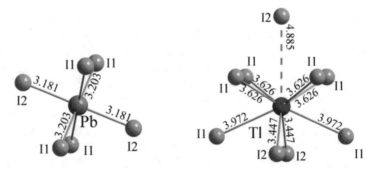

Fig. 5 Coordination surrounding and interatomic distances to iodine atoms in the TlPbI₃ structure [13]

solutions of thallium nitrate and potassium iodide for TlI, or lead nitrate and potassium iodide for PbI₂ since both binary compounds have very low water solubilities (0.006 g/l for TlI and 0.63 g/l for PbI₂). Zone-melting purification can be used to purify TlI and PbI₂ precipitates to mitigate the influence of impurities in the as-grown TlPbI₃ crystals. For further purification, Khyzhun et al. grew transparent yellow TlI single crystals using the vertical Bridgman method to be used for the growth of TlPbI$_3$ [13]. Zone-refining and evaporation techniques have been implemented to purify TlI-PbI$_2$ alloy and polycrystalline TlPbI$_3$ prior to the vertical Bridgman crystal growth as well [12, 15].

TlPbI$_3$ crystals grown at a rate of 0.5 mm/h using the vertical Bridgman method (1) do not have inclusions even in the transition volume connecting the seeding pocket and the main growth chunk, (2) do not suffer from a high density of grain boundaries or large grains, and (3) do not have nucleated grains near the quartz growth ampoules. However, they contain limited elongated crystal grains as shown in Fig. 6 [16].

The theoretical density of TlPbI$_3$ is 6.44 g/cm^3, which can be calculated based on the volume of its unit cell and the number of molecules within the unit cell. Experimentally the density of TlPbI$_3$ is also reported to be 6.60 gm/cm^3 [18]. TlPbI$_3$ has a high average atomic number of 64, which leads to a high attenuation capability for X-rays and gamma rays. Figure 7 shows the photon linear attenuation coefficient comparison of TlPbI$_3$, TlBr, and CdTe. TlPbI$_3$ has a wide

Fig. 6 Picture of TlPbI$_3$ crystal [17]. Infrared transmission (IR) microscopy of TlPbI$_3$, showing the formation of large-volume single crystal even in the transition portion between the conical seeding pocket and the normal growth chunk. Only several small elongated miscellaneous crystal grains are present [16]

Fig. 7 Linear attenuation coefficient for CdTe, TlBr, and TlPbI$_3$ as a function of photon energy [15]

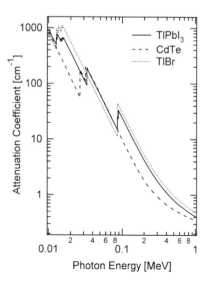

bandgap of 2.17–2.3 eV as can be expected from its red color of the crystal ingot shown in Fig. 3 [14, 15, 17]. It also has a high resistivity of ~10^{11} Ω cm which shows one order of magnitude increase upon cooling to ~10 °C [17]. The lowest conduction bands of TlPbI$_3$ are mainly composed of Pb 6p and Tl 6p states, while the valence bands are composed mostly of the I 5p state [12, 19].

1.2.2 TlPbI$_3$ Radiation Response and Charge Carrier Transport Properties

TlPbI$_3$ exhibits high X-ray-induced conductivity (XRIC) that makes it a potential candidate for X-ray imaging detectors for several important applications, especially medical imaging. Kocsis reported an X-ray-induced current/dark current ratio (~SNR) of 276, and no response delay was observed compared to 500 Hz X-ray shutter for 0.8-mm-thick sample with an X-ray generator operating at 45 mA and 45 kV [14]. The linearity of the XRIC response was also demonstrated in the same work [14].

Hitomi et al. tested TlPbI$_3$ detectors as alpha and gamma spectroscopic detectors [15]. The 0.2-mm-thick detector showed the 5.5 MeV ^{241}Am α-peak upon cathode irradiation (electron signal), while the peak was absent upon anode irradiation (hole signal), and the signal spectrum in both cases was higher than the noise spectrum. However, the same detector did not show a peak for ^{137}Cs gamma-ray spectrum. Instead, it showed the increased counts over the noise spectrum as shown in Fig. 8 [15].

The spectroscopic behavior of TlPbI$_3$ can be explained in terms of its electron and hole transport properties. As shown in Fig. 9, the $\mu\tau$-factor of electrons (**~3.4 × 10^{-5} cm^2/V**) is superior to that of the holes (**~2.3 × 10^{-6} cm^2/V**) by around one order of magnitude [17]. Using these values to calculate the average drift

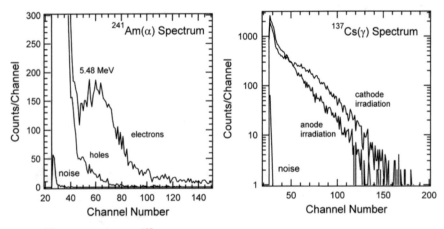

Fig. 8 ^{241}Am α-particle and ^{137}Cs γ-ray spectra obtained from a 0.2-mm-thick TlPbI$_3$ detector with electrodes of 1 mm in diameter at room temperature. The detector was operated at 100 V [15]

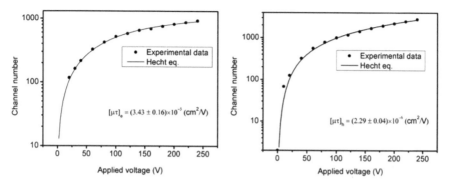

Fig. 9 TlPbI$_3$ electron and hole $\mu\tau$-factors as estimated by Hecht equation fitting upon 405 nm laser pulse irradiation of cathode and anode, respectively [17]

length of electrons and holes for a 0.2-mm-thick sample under 100 V applied voltage results in 1.75 mm and 0.10 mm drift lengths for electrons and holes, respectively [17]. Since the electrons' drift length is longer than the 0.2 mm detector thickness, the cathode irradiation leads to the resolving capability of the 5.5 MeV alpha-peak. On the other hand, the drift length of the holes, which is lower than the thickness of the detector, would not allow resolving the same peak upon anode irradiation. The lack of the 662 keV gamma-ray peak from TlPbI$_3$ detectors can be understood in terms of discrepancy between the drift length values of electrons and holes which leads to the dependence of the collected pulse height on the location inside the detector volume where the energy gets deposited.

1.2.3 TlPbI$_3$ Scintillator

Singh performed an electronic structure calculation study on TlPbI$_3$ to investigate its electronic and optical properties based on the band structure, density of states, and optical spectra [19]. Motivated by the soft lattice characteristics of an iodide, the use of TlPbI$_3$ as a scintillator was considered in that study and it was shown that TlPbI$_3$ may be a promising low-bandgap scintillator if activated by Sn. If that is the case, one of the spectroscopic advantages of TlPbI$_3$ scintillator detectors would be its 2.17–2.3 eV bandgap which leads to low formation energy for e-h pairs (excitons) relative to other inorganic scintillators [20], allowing to reach high energy resolution. Such a relatively low bandgap makes the utilization of compact solid-state SiPMs (silicon photomultipliers) desirable since the energy of the emission photons would be lower than the peak efficiency energies of conventional PMTs (photomultiplier tubes). Moreover, TlPbI$_3$'s relatively long charge carriers' drift length, which allows for resolving α-peak, may consequently be long enough to enable the charge carriers to reach luminescent centers before recombination or trapping at nonradiative defects.

TlPbI$_3$'s luminescence is very weak when grown unintentionally doped. The bandgap of SnPbI$_3$ is smaller than that of TlPbI$_3$; thus Sn^{2+} has the potential to be a successful activator for TlPbI$_3$ [19].

2 Other Emerging Tl-Based Scintillators

Although Tl has been used, in very low concentrations, as an activator for alkali metal halide scintillators for over half a century [21], it was not until the last decade that Tl-based halide compounds, with Tl in the host crystal, attracted strong interest for their applications as inorganic scintillation radiation detectors. One of the motivations for this research effort is the expected high density of Tl-based compounds rendering them suitable for X-ray and gamma-ray detection. The promising scintillation properties (e.g., light yield, decay time, energy resolution) revealed from the initial studies have motivated further research on Tl-based scintillators to cover a large library of candidate compounds. Emerging Tl-based scintillators are classified, based on their chemical structure, into the following four material classes:

- Class A: Tl$_2$ALX$_6$ elpasolites in their pure form and Ce^{3+} doped (activated)
- Class B: Tl$_2$LX$_5$, TlL$_2$X$_7$, and Tl$_3$LX$_6$ (thallium lanthanide halides) in their pure form and Ce^{3+} doped
- Class C: TlM$_2$X$_5$ and TlMX$_3$ (thallium alkaline-earth halides) in their pure form and Eu^{2+} doped, in addition to TlCdCl$_3$ and TlAlF$_4$
- Class D: Tl$_2$BX$_6$ in their pure form

where $\mathbf{A}$ = alkali metal (Li or Na); $\mathbf{B}$ = transition metal (Hf or Zr); $\mathbf{L}$ = rare earth element (Sc, Y, La, Gd, or Lu); $\mathbf{M}$ = alkaline-earth metal (Mg, Ca, or Sr); $\mathbf{X}$ = halogen (F, Br, Cl, or I) [22, 23].

Until now these emerging Tl-based scintillators were grown using the vertical Bridgman melt-growth techniques starting from high-purity materials (>4N). The modified Bridgman techniques were also developed to address specific crystal growth issues (such as impurity segregation and inhomogeneous mixing) [24, 25]. The starting materials (beads or powder form) are commonly handled in a dry environment and dried before the crystal growth process, as most starting materials needed for the growth of this class of materials are highly hygroscopic. Since most of these materials were grown for the first time, their thermodynamic properties, especially their melting temperatures and temperature-dependent phase change, are explored in many studies. Differential scanning calorimetry (DSC) and thermogravimetric analysis (TGA) on Tl-based scintillator materials confirm the absence of secondary-phase changes and reveal moderate melting temperatures, in the range of 400–800 °C, promoting the phase stability of Tl-based scintillators. Moreover, such moderate melting temperatures allow the use of relatively economic melt-growth techniques with no need for expensive high-temperature crucibles.

2.1 Class A: Tl2ALX6 Elpasolites in Their Pure Form and Ce3+ Doped

Ce-doped elpasolites, such as Cs_2LiYCl_6:Ce (CLYC), $Cs_2LiLaBr_6$:Ce (CLLB), and $Cs_2LiLa(Br,Cl)_6$:Ce (CLLBC), have been established as inorganic scintillators that have the capability of detecting and discriminating between neutrons/charged particles and gamma rays (dual-mode operation). However, one of their main drawbacks is the relatively low densities [26–31]. Tl-based elpasolites could address this issue by incorporating Tl ion instead of the Cs ion. This will help increase the density of the elpasolites and improve the intrinsic material efficiency for detecting X-ray and gamma ray.

Table 1 lists all the studied class A materials along with their reported properties. One of the most promising dual-mode scintillator materials in this class is Tl_2LiYCl_6:Ce (TLYC). Not only does it contain Li (^6Li is an efficient thermal

Table 1 Physical and scintillation key parameters of class A Tl-based scintillators

Chemical structure	Doping	M.P. [°C]	Density [g cm⁻³]	E.R.	L.Y. (10³)	λem [nm]	τ [ns]	%	PSD FOM	Ref.
Tl_2LiYCl_6	1% Ce	490	4.58	4.8	30.5	430	69	3	-	[38, 39]
							460	74		
							1400	23		
	5% Ce	-	4.58	3.8	26 (Y) 78 (n)	-	-	-	2.4 (n,Y)	[32–34]
	3% Ce			4.2	-	440	57	3	-	
							431	64		
							1055	33		
$Tl_2LiLaBr_6$	1% Ce	-	6	5.7	53	410			-	[33]
$Tl_2LiGdCl_6$	10% Ce	540	4.75	4.6	58	380	34	81	-	[36]
							191	10		
							1200	9		
$Tl_2LiGdBr_6$	10% Ce	450	5.3	17	17	422	29	92	-	[40]
							197	8		
$Tl_2LiLuCl_6$	1% Ce	400	5.06	5.6	27	428	72	8	-	[41]
							366	30		
							1500	62		
$Tl_2LiScCl_6$	No	550	4.43	8.3	26	385	1147	100	-	[42]
Tl_2NaYCl_6	Ce	-	-	4.1	27.8	430	91	34	-	[37]
							462	52		
							2100	15		
	No	-	-	6.3	23	430	350	80	-	[43]
							2500	20		

M.P.: melting point; E.R.: energy resolution reported as FWHM % at ^{137}Cs 662 keV Y-peak; L.Y.: light yield reported as 10^3 ph/MeV; $λ_{em}$: radioluminescence emission peak; τ: decay time components and their percentages obtained by fitting a single-, double-, or triple-exponential function to X- or Y-ray-induced scintillation; PSD FOM: pulse shape discrimination figure of merit neutron absorber), but it also does not contain La, which has intrinsic radioactivity. The highest energy resolution reported for 5% Ce-doped TLYC is better than 4%. The good pulse shape discrimination figure of merit (PSD FOM of 2.4) shows its high potential for national security and space applications [32–34]. Recently, Watts et al. studied the performance of TLYC in a temperature range from $-20\,^{\circ}C$ to $50\,^{\circ}C$ and demonstrated that the light yield of TLYC stays proportional over that temperature range while the short decay times and the energy resolution degrade with decreasing temperature [35].

$Tl_2LiGdCl_6$:Ce is another less studied candidate material of this class. It has a relatively high light yield and short decay time, which could be promising for many radiation detection applications when the timing characteristics and the intrinsic detection efficiency are desired, e.g., time-of-flight positron-emission tomography (TOF-PET) [36]. Very recently, Tl_2NaYCl_6:Ce was grown and studied, and its reported scintillation results were very similar to those of TLYC. Tl_2NaYCl_6:Ce has the potential to be used as a dual-mode detector for fast neutrons and gamma ray. In this regard, the corresponding fast neutron detection performance based on $^{35}Cl(n, p)^{35}S$ reaction is yet to be studied [37].

2.1.1 Class B: Tl2LX5, TlL2X7, and Tl3LX6 (Thallium Lanthanide Halides) in Their Pure Form and Ce3+-Doped Form

Class B of Tl-based scintillators could be considered as a "new" generation of lanthanide halides [44, 45]. Although $LaBr_3$:Ce broke the energy resolution limit of 3% FWHM at 662 keV, lanthanide halides' densities are moderate (<5 g/cm^3) [45]. To this end, improvements in the stopping power, and thus the intrinsic detector efficiencies of lanthanide halides, could be achieved by the addition of thallium in the newly developed thallium lanthanide halide ternary compounds. A series of related studies on this emerging class of materials and the comparison of their properties are listed in Table 2.

Khan et al. developed a novel melt-rotation crystal growth technique based on the Bridgman method for the growth of Tl_2LaCl_5:Ce (TLC) to overcome the starting materials' inhomogeneous mixing issue due to the incongruent melting [25]. They achieved a high light yield of 82,000 ph/MeV and a high energy resolution of ~3.3% FWHM for 662 keV gamma rays. TLC light yield is highly proportional; moreover, TLC has a fast scintillation decay time of 30–45 ns, depending on the crystal quality and Ce-doping level [25, 46–49]. Studies on Tl_2LaBr_5:Ce (TLB) also show very high energy resolution below 3% FWHM at 662 keV, very high proportionality, and exceptionally fast decay times below 30 ns [46, 48, 50]. These superior scintillation characteristics, in addition to the improved intrinsic efficiency for X- and gamma ray, make TLC and TLB very promising detector materials for applications where high energy resolution, fast timing, and detector size are of high importance, such as nonproliferation, safeguard, medical imaging, and high energy physics applications.

Table 2 Physical and scintillation key parameters of class B Tl-based scintillators

Chemical structure	Doping	M.P. [°C]	Density [g cm^{-3}]	E. R.	L.Y. (10³)	λ_{em} [nm]	τ [ns]	%	Ref.
Tl₂LaCl₅	10% Ce	550	5.16	3.2	68	378	44	98	[46, 48]
	3% Ce	–	–	3.4	76	383	36 / 217 and 1500	89 / 11	[47]
	3% Ce	520	5.17	3.3	82	370 / 384	31 @ 10% Ce	84	[25, 49]
Tl₂LaBr₅	1% Ce	556	5.98	2.8	78	400	28 @ 3% Ce	99	[46, 48]
	5% Ce	548	5.9	6.3	43	375 / 415	25	100	[50]
Tl₂GdCl₅	5% Ce	490	5.1	5	53	388	32 / 271 / 1600	76 / 10 / 14	[53, 54]
TlGd₂Cl₇	5% Ce	560	4.3	6	50	405	89 / 358 / 1780	20 / 21 / 58	[55]
	1% Ce	558	4.49	7	40	412	24 / 87 / 364	28 / 54 / 28	[52]
Tl₃GdCl₆	Ce	Very weak emission (~390 nm) due to poor crystal quality							[52]
Tl₃YCl₆	Ce	Poor crystal quality/hazy polycrystals							[52]
TlY₂Cl₇	5% Ce	550	3.77	9	38	395	95 / 422 / 1200	48 / 36 / 16	[52]

M.P.: melting point; E.R.: energy resolution reported as FWHM % at ^{137}Cs 662 keV Υ-peak; L.Y.: light yield reported as 10³ ph/MeV; λ_{em}: radioluminescence emission peak; τ: decay time components and their percentages obtained by fitting a single-, double-, or triple-exponential function to X- or Υ-ray-induced scintillation

It should be noted that all La-containing scintillators suffer from inherent radioactivity that, although useful for energy calibration, can affect the statistics of energy spectra [51].

van Loef et al. studied Tl₃YCl₆:Ce (TYC₆), Tl₃GdCl₆:Ce (TGC₆), TlY₂Cl₇:Ce (TYC₇), and TlGd₂Cl₇:Ce (TGC₇) and concluded that they are not promising for practical applications and industrial scaling up [52]. This is primarily due to the low material quality of TYC₆ and TGC₆ crystals, and the relatively low densities, poor crystal quality due to the layered crystalline structure, low proportionality, and relatively low energy resolution of TYC₇ and TGC₇ crystals [52].

2.1.2 Class C: TlM2X5 and TlMX3 (Thallium Alkaline-Earth Halides) in Their Pure Form and Eu2+-Doped Form, in Addition to TlCdCl3 and TlAlF4

The rediscovery [56–58] of SrI_2:Eu (SI), enabled by the successful development of economically viable high-quality and high-purity crystals, made it an attractive radiation detector solution for gamma-ray spectroscopy because of its high light yield (>80,000 ph/MeV) and high energy resolution (below 3% FWHM at 662 keV) [59]. Recent research of $TlSr_2I_5$:Eu (TSI) reveals that its scintillation properties are very close to those of SrI_2:Eu, with improved density (~5.3 g/cm^3) compared to that of SrI_2 (~4.5 g/cm^3), and very high proportionality between its light yield and gamma-ray-deposited energy [60]. Table 3 summarizes research efforts and scintillation characterization results on class C of Tl-based scintillators.

Although TSI technology is not as mature as SI, its energy resolution and high intrinsic detector efficiency present it as a very promising scintillator for gamma-ray spectroscopy for national security, space exploration, and high energy physics applications. As a result, it could be a very attractive solution for those scenarios when relatively economic and compact-sized gamma-ray spectrometers are needed, e.g., portable radioisotope identifiers. Although STI does not suffer from inherent

Table 3 Physical and scintillation key parameters of class C Tl-based scintillators

Chemical structure	Doping	M.P. [°C]	Density [g cm^{-3}]	E.R.	L.Y. (10^3)	λ_{em} [nm]	τ [ns]	%	Ref.
TlSr$_2$I$_5$	3% Eu	480	–	4.2	70	463	525 3300	73 27	[61]
	5% Eu		–	2.6	54	420	395 2000	89 11	[62]
	1% Eu	485	5.32	2.8	72	460-470	500	90	[60]
TlSr$_2$Br$_5$	No	560	5.03	4.6	37.6	441	390 1900	66 34	[63]
TlCaCl$_3$	No	680	3.77	5	30.6	425	317 727	44 56	[64]
TlCdCl$_3$	No	–	5.31	–	2.2	450	2.6 45 170	8 47.8 45	[65]
TlMgCl$_3$	No	–	5.26	5	46	405	60 350	25 75	[66]
TlAlF$_4$	No	586	6.1	14	11.8	390	194 992 3500	13.1 84.6 2.3	[67, 68]

M.P.: melting point; E.R.: energy resolution reported as FWHM % at ^{137}Cs 662 keV Y-peak; L.Y.: light yield reported as 10^3 ph/MeV; λ_{em}: radioluminescence emission peak; τ: decay time components and their percentages obtained by fitting a single-, double- or triple-exponential function to X- or Y-ray-induced scintillation radioactivity like $LaBr_3$ high-energy-resolution scintillator, the recent studies of the emission and excitation spectra of TSI reveal that there is an overlap between excitation and emission bands, which possibly leads to self-absorption [61].

2.1.3 Class D: Tl2BX6 in Their Pure Form

Research and development efforts on class D of Tl-based scintillators are listed in Table 4. This class of materials leverage the capabilities of Tl-based scintillators for both fast detection of gamma rays, and dual-mode operation using PSD techniques, without doping/activation. Tl_2HfBr_6 (THB) has a very fast scintillation decay of 22 ns (>99%) and an acceptable energy resolution of >5% FWHM at 662 keV [48]. Given that it also possesses a relatively high effective atomic number (Z_{eff}) of 68, it would be an attractive solution for TOF-PET and other fast gamma-ray detection applications. Its dual-mode operation has not been reported yet. Both Tl_2HfCl_6 (THC) and Tl_2ZrCl_6 (TZC) show good energy resolution of ~4% FWHM at 662 keV and demonstrate good dual-mode operation capabilities. Thus their applications in gamma/neutron or gamma/charged-particle discrimination are promising for national security applications. Especially THC also combines the advantage of high density [24, 69, 70].

Table 4 Physical and scintillation key parameters of class D Tl-based scintillators

Chemical structure	M.P. [°C]	Density [g cm^{-3}]	E. R.	L.Y. (10^3)	λ_{em} [nm]	τ [ns]	%	PSD FOM	Ref.
Tl_2HfBr_6	~664 [a]	–	>5	<20	550	22	+99		[48]
Tl_2HfCl_6	~664 [a]	–	5	20	425	262	34		[48]
	–	5.25	4	32	398	1000		2.6 (α, Υ)	[24]
	–	–	17.7	24.2	465	288 6340			[71]
Tl_2ZrCl_6	<700	4.65	4.3	47	468	2700		3.5 (α, Υ)	[69, 70]
	–	–	5.6	50.8	450-470	696 2360			[71]

M.P.: melting point; E.R.: energy resolution reported as FWHM % at ^{137}Cs 662 keV Υ-peak; L.Y.: light yield reported as 10^3 ph/MeV; λ_{em}: radioluminescence emission peak; τ: decay time components and their percentages obtained by fitting a single-, double-, or triple-exponential function to X- or Υ-ray-induced scintillation; PSD FOM: pulse shape discrimination figure of merit; [a]: assumed values

3   Summary and Outlook

Tl-based materials represent a unique category of detector materials for ionizing radiation detection. These Tl-based materials can be used either in the form of semiconductor detectors or as scintillator detectors. In this chapter, we review the development of classic Tl-based material TlBr and introduce the current status of some emerging Tl-based compounds such as $TlPbI_3$. We note that some Tl-based materials have exhibited great potential as promising scintillators. As such we summarize and compare the properties of these novel Tl-based scintillator materials as well. With further development, these Tl-based compounds could serve as enabling materials for a broader range of radiation detection applications.

References

1. Luke, P.N., Amman, M.: Room-temperature replacement for Ge detectors - are we there yet? In 2006 IEEE Nuclear Science Symposium Conference Record (2007), pp. 3607–3615.
2. Mirzaei, A., Huh, J.S., Kim, S.S., Kim, H.W.: Room temperature hard radiation detectors based on solid state compound semiconductors: an overview. Electron. Mater. Lett. **14**(3), 261–287 (2018) The Korean Institute of Metals and Materials
3. Milbrath, B.D., Peurrung, A.J., Bliss, M., Weber, W.J.: Radiation detector materials: an overview. J. Mater. Res. **23**(10), 2561–2581 (2008) Cambridge University Press
4. Knoll, G.: Radiation detection and measurement. John Wiley & Sons, Hoboken (2010)
5. Churilov, A.V., et al.: Thallium bromide nuclear radiation detector development. IEEE Trans. Nucl. Sci. **56**(4), 1875–1881 (2009)
6. Hofstadter, R.: Thallium halide crystal counter. Physical Review. **72**(11), 1120–1121 (1947) American Physical Society
7. Jin Kim, D., et al.: Characteristics of TlBr single crystals grown using the vertical Bridgman-Stockbarger method for semiconductor-based radiation detector applications. Mater. Sci. Pol. **34**(2), 297–301 (2016)
8. Kim, H., et al.: Thallium bromide gamma-ray spectrometers and pixel arrays. Front. Phys. **8**, 55 (2020)
9. Zhang, X., Friedrich, S., Friedrich, B.: Production of high purity metals: a review on zone refining process. J. Cryst. Process Technol. **08**(01), 33–55 (2018)
10. Datta, A., Becla, P., Motakef, S.: Novel electrodes and engineered interfaces for halide-semiconductor radiation detectors. Sci. Rep. **9**(1) (2019)
11. Dnmez, B., He, Z., Kim, H., Cirignano, L.J., Shah, K.S.: The stability of TlBr detectors at low temperature. Nucl. Instrum. Methods Phys. Res. Sect. A. **623**(3), 1024–1029 (2010)
12. Lin, W., et al.: Inorganic halide perovskitoid TlPbI 3 for ionizing radiation detection. Adv. Funct. Mater. **31**, 2006635 (2021)
13. Khyzhun, O.Y., et al.: Single crystal growth and electronic structure of TlPbI3. Mater. Chem. Phys. **172**, 165–172 (2016)
14. Kocsis, M.: Proposal for a new room temperature x-ray detector-thallium lead iodide. IEEE Trans. Nucl. Sci. **47**(6 I), 1945–1947 (2000)
15. Hitomi, K., Onodera, T., Shoji, T., Hiratate, Y.: Thallium lead iodide radiation detectors. In 2002 IEEE Nuclear Science Symposium Conference Record, vol. 1, pp. 485–488 (2002)
16. Yang, G., Phan, Q.V., Liu, M., Hawari, A., Kim, H.: Material defect study of thallium lead iodide (TlPbI3) crystals for radiation detector applications. Nucl. Instrum. Methods Phys. Res. Sect. A, 954 (2020)

17. Hany, I., Yang, G., Phan, Q.V., Kim, H.J.: Thallium lead iodide (TlPbI3) single crystal inorganic perovskite: Electrical and optical characterization for gamma radiation detection. Mater. Sci. Semicond. Process. **121**, 105392 (2021)
18. Stoeger, W.: Die Kristallstrukturen von TlPbI3 und Tl4PbI6/The Crystal Structures of TlPbI3 and Tl4PbI6. Zeitschrift für Naturforsch B. **32**(9), 978–981 (1977)
19. Singh, D.J.: Electronic structure of TlGeI3, TlSnI3, and TlPbI3: Potential use for spectroscopic radiation detection. J. Appl. Phys. **112**(8), 83509 (2012)
20. Lecoq, P., Annenkov, A., Gektin, A., Korzhik, M., Pedrini, C.: Scintillation and inorganic scintillators. In: Inorganic scintillators for detector systems, pp. 1–34. Springer-Verlag, New York (2006)
21. Hofstadter, R.: Alkali halide scintillation counters. Phys. Rev. **74**(1), 100–101 (1948)
22. Kim, H.J., Rooh, G., Khan, A., Vuong, P.Q., Kim, S.: Discovery, crystal growth, and scintillation properties of novel Tl-based scintillators. Cryst. Res. Technol. **55**(10), 2000074 (2020)
23. Kim, H. J., Khan, A., Rooh, G., Vuong, P., Kim, S., Daniel, J.: Development of Tl-based novel scintillators. In hard X-ray, gamma-ray, and neutron detector physics XXII, vol. 11494, p. 8 (2020)
24. Vuong, P.Q., Tyagi, M., Kim, S.H., Kim, H.J.: Crystal growth of a novel and efficient Tl2HfCl6 scintillator with improved scintillation characteristics. CrystEngComm. **21**(39), 5898–5904 (2019)
25. Khan, A., Vuong, P.Q., Rooh, G., Kim, H.J., Kim, S.: Crystal growth and Ce3+ concentration optimization in Tl2LaCl5: An excellent scintillator for the radiation detection. J. Alloys Compd. **827**, 154366 (2020)
26. Woolf, R.S., Phlips, B.F., Wulf, E.A.: Characterization of the internal background for thermal and fast neutron detection with CLLB. Nucl. Instrum. Methods Phys. Res. Sect. A. **838**, 147–153 (2016)
27. Hull, G., Camera, F., Colombi, G., Josselin, M., Million, B., Blasi, N.: Detection properties and internal activity of newly developed La-containing scintillator crystals. Nucl. Instrum. Methods Phys. Res. Sect. A. **925**, 70–75 (2019)
28. Woolf, R.S., Phlips, B.F., Grove, J.E., Murphy, R.J., Share, G.M.: Novel inorganic scintillators for future space-based solar gamma-ray and neutron research. In Hard x-ray, gamma-ray, and neutron detector physics XXI, vol. 11114, no. 9, p. 7 (2019)
29. Biswas, K., Du, M.H.: Energy transport and scintillation of cerium-doped elpasolite Cs 2LiYCl 6: Hybrid density functional calculations. Phys. Rev. B Condens. Matter Mater. Phys. **86**(1), 014102 (2012)
30. Lee, D.W., Stonehill, L.C., Klimenko, A., Terry, J.R., Tornga, S.R.: Pulse-shape analysis of Cs 2LiYCl 6:Ce scintillator for neutron and gamma-ray discrimination. Nucl. Instrum. Methods Phys. Res. Sect. A. **664**(1), 1–5 (2012)
31. Mesick, K.E., Coupland, D.D.S., Stonehill, L.C.: Pulse-shape discrimination and energy quenching of alpha particles in Cs2LiLaBr6:Ce3+. Nucl. Instrume. Methods Phys. Res. Sect. A. **841**, 139–143 (2017)
32. Hawrami, R., Ariesanti, E., Wei, H., Finkelstein, J., Glodo, J., Shah, K.: Tl2LiYCl6: large diameter, high performing dual mode scintillator. Cryst. Growth Des. **17**(7), 3960–3964 (2017)
33. Hawrami, R., Ariesanti, E., Pandian, L.S., Glodo, J., Shah, K.S.: Tl2LiLaBr6:Ce and Tl2LiYCl6:Ce: new elpasolite scintillators. In 2015 IEEE Nuclear Science Symposium and Medical Imaging Conference, NSS/MIC 2015 (2016)
34. Hawrami, R., Ariesanti, E., Soundara-Pandian, L., Glodo, J., Shah, K.S.: Tl2LiYCl6:Ce: a new elpasolite scintillator. IEEE Trans. Nucl. Sci. **63**(6), 2838–2841 (2016)
35. Watts, M.M., Mesick, K.E., Bartlett, K.D., Coupland, D.D.S.: Thermal characterization of TlLiYCl:Ce (TLYC). IEEE Trans. Nucl. Sci. **67**(3), 525–533 (2020)
36. Kim, H.J., Rooh, G., Park, H., Kim, S.: Luminescence and scintillation properties of the new Ce-doped Tl2LiGdCl6 single crystals. J. Lumin. **164**, 86–89 (2015)
37. Hawrami, R., Ariesanti, E., Burger, A.: Characterization of Tl_2NaYCl_6:Ce scintillation crystals as gamma-ray detectors. J. Cryst. Growth. **565**, 126150 (2021)

38. Kim, H.J., Rooh, G., Park, H., Kim, S.: Tl2LiYCl6 (Ce3+): new Tl-based elpasolite scintillation material. IEEE Trans. Nucl. Sci. **63**(2), 439–442 (2016)
39. Rooh, G., Kim, H.J., Park, H., Kim, S.: Crystal growth and scintillation characterizations of Tl2LiYCl6: Ce3+. J. Cryst. Growth. **459**, 163–166 (2017)
40. Kim, H.J., Rooh, G., Park, H., Kim, S.: Investigations of scintillation characterization of Ce-activated Tl2LiGdBr6 single crystal. Radiat. Meas. **90**, 279–281 (2016)
41. Rooh, G., Kim, H.J., Jang, J., Kim, S.: Scintillation characterizations of Tl2LiLuCl6:Ce 3+ single crystal. J. Lumin. **187**, 347–351 (2017)
42. Kim, M.J., Kim, H.J., Cho, J.Y., Khan, A., Daniel, D.J., Rooh, G.: Characterizations of a New Tl-based Elpasolite Scintillator: Tl2LiScCl6. J. Korean Phys. Soc. **76**(8), 706–709 (2020)
43. Arai, M., et al.: Tl2NaYCl6: a new self-activated scintillator possessing an elpasolite structure. J. Mater. Sci. Mater. Electron., 1–7 (2021)
44. Van Loef, E.V.D., Dorenbos, P., Van Eijk, C.W.E., Krämer, K., Güdel, H.U.: High-energy-resolution scintillator: Ce3+ activated LaBr3. Appl. Phys. Lett. **79**(10), 1573–1575 (2001)
45. Loef, E., Güdel, H., Dorenbos, P., Kraemer, K., Eijk, C.W.: Scintillation properties of LaBr3: Ce3+ crystals: fast, efficient and high-energy-resolution scintillators. Nucl. Instrum. Methods Phys. Res. A. **486**, 254–258 (2002)
46. Shirwadkar, U., et al.: Thallium-based scintillators for high-resolution gamma-ray spectroscopy: Ce 3+-doped Tl 2 LaCl 5 and Tl 2 LaBr 5 ⋆. Nucl. Instrum. Methods Phys. Res. **962**, 163684 (2020)
47. Hawrami, R., Ariesanti, E., Wei, H., Finkelstein, J., Glodo, J., Shah, K.S.: Tl2LaCl5:Ce, high performance scintillator for gamma-ray detectors. Nucl. Instrum. Methods Phys. Res. Sect. A. **869**, 107–109 (2017)
48. van Loef, E.V., Ciampi, G., Shirwadkar, U., Soundara Pandian, L., Shah, K.S.: Crystal growth and scintillation properties of Thallium-based halide scintillators. J. Cryst. Growth. **532**, 125438 (2020)
49. Kim, H.J., Rooh, G., Kim, S.: Tl2LaCl5 (Ce3+): New fast and efficient scintillator for X- and γ-ray detection. J. Lumin. **186**, 219–222 (2017)
50. Kim, H.J., Rooh, G., Khan, A., Kim, S.: New Tl2LaBr5: Ce3+ crystal scintillator for γ-rays detection. Nucl. Instrum. Methods Phys. Res. Sect. A. **849**, 72–75 (2017)
51. Menge, P.R., Gautier, G., Iltis, A., Rozsa, C., Solovyev, V.: Performance of large lanthanum bromide scintillators. Nucl. Instrum. Methods Phys. Res. Sect. A. **579**(1), 6–10 (2007)
52. van Loef, E., et al.: Crystal growth, density functional theory, and scintillation properties of Tl3LnCl6:Ce3+ and TlLn2Cl7:Ce3+ (Ln = Y, Gd). Nucl. Instrum. Methods Phys. Res. Sect. A. **995**, 165047, 2021
53. Khan, A., Rooh, G., Kim, H.J., Kim, S.: Ce3+-activated Tl2GdCl5: Novel halide scintillator for X-ray and γ-ray detection. J. Alloys Compd. **741**, 878–882 (2018)
54. Rooh, G., Khan, A., Kim, H.J., Park, H., Kim, S.: Tl2GdCl5 (Ce3+): a new efficient scintillator for x-and γ -ray detection. IEEE Trans. Nucl. Sci. **65**(8), 2157–2161 (2018)
55. Khan, A., Rooh, G., Kim, H.J., Kim, S.: Scintillation properties of TlGd2Cl7 (Ce3+) single crystal. IEEE Trans. Nucl. Sci. **65**(8), 2152–2156 (2018)
56. Cherepy, N.J., et al.: Scintillators with potential to supersede lanthanum bromide. IEEE Trans. Nucl. Sci. **56**(3), 873–880 (2009)
57. Hawrami, R., et al.: SrI 2: a novel scintillator crystal for nuclear isotope identifiers. In Hard X-ray, gamma-ray, and neutron detector physics X, vol. 7079, p. 70790Y, (2008)
58. H. Robert: Europium activated strontium iodide scintillators. U.S. Patent 3,373,279 A. March 12, 1968.
59. Cherepy, N.J., et al.: History and current status of strontium iodide scintillators, vol. 10392, p. 1 (2017)
60. Soundara Pandian, L., et al.: TlSr2I5:Eu2+- A new high density scintillator for gamma-ray detection. Nucl. Instrum. Methods Phys. Res. Sect. A. **988**, 164876 (2021)
61. Kim, H.J., Rooh, G., Khan, A., Park, H., Kim, S.: Scintillation performance of the TlSr2I5 (Eu2 +) single crystal. Opt. Mater. (Amst). **82**, 7–10 (2018)

62. Hawrami, R., Ariesanti, E., Buliga, V., Burger, A.: Thallium strontium iodide: A high efficiency scintillator for gamma-ray detection. Opt. Mater. (Amst). **100**, 109624 (2020)
63. Rooh, G., Khan, A., Kim, H.J., Park, H., Kim, S.: TlSr2Br5: New intrinsic scintillator for X-ray and γ-ray detection. Opt. Mater. (Amst). **73**, 523–526 (2017)
64. Khan, A., Rooh, G., Kim, H.J., Park, H., Kim, S.: Intrinsically activated TlCaCl3: A new halide scintillator for radiation detection. Radiat. Meas. **107**, 115–118 (2017)
65. Fujimoto, Y., Saeki, K., Yanagida, T., Koshimizu, M., Asai, K.: Luminescence and scintillation properties of TlCdCl3 crystal. Radiat. Meas. **106**, 151–154 (2017)
66. Fujimoto, Y., Koshimizu, M., Yanagida, T., Okada, G., Saeki, K., Asai, K.: Thallium magnesium chloride: A high light yield, large effective atomic number, intrinsically activated crystalline scintillator for X-ray and gamma-ray detection. Jpn. J. Appl. Phys. **55**(9), 090301 (2016)
67. Daniel, D.J., Cho, J., Won, H., Kim, H.J.: TSL kinetic parameters and dosimetric properties of TlAlF4 crystal grown by Bridgman technique. Opt. Mater. (Amst). **1111**, 110636 (2021)
68. Daniel, D.J., Khan, A., Tyagi, M., Kim, H.J.: Scintillation properties of tetrafluoroaluminate crystal. IEEE Trans. Nucl. Sci. **67**(6), 898–903 (2020)
69. Vuong, P.Q., Kim, H.J., Park, H., Rooh, G., Kim, S.H.: Pulse shape discrimination study with Tl2ZrCl6 crystal scintillator. Radiat. Meas. **123**, 83–87 (2019)
70. Phan, Q.V., Kim, H.J., Rooh, G., Kim, S.H.: Tl2ZrCl6 crystal: efficient scintillator for X- and γ-ray spectroscopies. J. Alloys Compd. **766**, 326–330 (2018)
71. Fujimoto, Y., Saeki, K., Nakauchi, D., Yanagida, T., Koshimizu, M., Asai, K.: New intrinsic scintillator with large effective atomic number: Tl2HfCl6 and Tl2ZrCl6 crystals for x-ray and gamma-ray detections. Sens. Mater. **30**(7), 1577–1583 (2018)

CdZnTeSe: A Promising Material for Radiation Detector Applications

Utpal N. Roy and Ralph B. James

Abstract The quest for cost-effective, high-performing radiation detector materials operable at ambient temperature has continued for more than three decades. One key to lower the overall detector cost of production is a high degree of compositional homogeneity. Spatial homogeneity of the charge-transport properties of the detector material is also essential for increasing the yield and active volume of high-performance detectors. Despite its commercial success as a room-temperature radiation detection material, CdZnTe (or CZT) suffers from a lack of compositional homogeneity on both micro- and macroscale and the presence of spatial inhomogeneity in the material's charge-transport properties. The underlying reasons for the spatial inhomogeneity in the charge-transport properties of CZT are the presence of high concentrations of sub-grain boundary (dislocation walls) networks and secondary phases (Te-rich inclusions) on a microscale and the segregation of zinc during growth on a macroscale. This book chapter focuses on the presence of performance-limiting defects in CZT that hinder the yield and cost of high-quality detectors and have restricted widespread deployment for a variety of potential applications, particularly for uses of large-volume detectors where the demands on material perfection are significantly greater. In the recent past, the addition of selenium in CZT matrix was found to be very effective in a drastic reduction of Te-rich secondary phases and dislocation networks, while demonstrating little or no Se segregation during ingot growth. This chapter provides an overview of recent developments on the quaternary CdZnTeSe material as a potential next-generation detector material operable at room temperature.

U. N. Roy (✉) · R. B. James
Savannah River National Laboratory, Aiken, SC, USA
e-mail: Utpal.Roy@srnl.doe.gov; Ralph.James@srnl.doe.gov

© The Author(s), under exclusive license to Springer Nature Switzerland AG 2022
K. Iniewski (ed.), *Advanced Materials for Radiation Detection*,
https://doi.org/10.1007/978-3-030-76461-6_8

1   Introduction

The availability of radiation detection technology with energy-dispersive spectro-scopic capability in the energy range of X- and gamma rays operating at room temperature has opened a wide variety of applications in homeland security, non-proliferation, medical imaging, astrophysics, and high energy physics [1–9]. The enormous potential uses have motivated researchers to develop better materials and processing technology to meet the growing demands worldwide. The properties of candidate materials for radiation detector applications are very stringent and multi-pronged. The basic requirement for the ideal radiation detector material is a high average effective atomic number, hence a high electron density for increased stopping power of the high-energy radiation. In addition, the constituent elements should be nonradioactive. For room-temperature operation the foremost criterion is sufficiently high bandgap energy to achieve an electrical resistivity on the order of $>10^{10}$ Ω-cm. Such high resistivity is required to decrease the dark current in order to enhance the signal-to-noise ratio. Furthermore, as the radiation detector users demand devices with larger thickness to enhance detection efficiency of high-energy gamma rays, the material should have adequately high mobility-lifetime product ($\mu\tau$) to ensure complete charge collection. The operating thickness of the functional detector can be increased with increased $\mu\tau$ value assuming that the internal electric field can be maintained. From the standpoint of the crystal growth, the material should be able to grow from congruent melt at a relatively low temperature and be capable of scale-up to allow for growth of large-volume ingots to keep a lower cost of production. The grown material should be compositionally homogeneous, possess very less defects (intrinsic and extrinsic), and more importantly possess spatial uniformity of the charge-transport characteristics. The chemical and physical stabil-ity of the metal-to-semiconductor junction is also an important aspect for prolonged lifetime of the detector. Over the last three decades, only a handful of semiconductor materials have been discovered with the desired attributes, and CdTe, CdZnTe (CZT), HgI_2, and TlBr are the most prominent ones [10–14]. In the recent years, perovskites have been evolving as a promising material for room-temperature semiconductor detector (RTSD) applications for high-energy radiation [15]. John et al. [16] listed a comprehensive list of possible radiation detector materials in a review article. Each of these materials however do suffer from unique issues associated with difficulty of crystal growth, high concentrations of defects, and device stability.

Despite intense research for the ideal material globally for more than three decades, CZT still remains the gold standard and dominates the commercial sector. It is to be noted that CZT is a by-product of many years of R&D on CdTe. CdTe has multiple uses and has long been known as a substrate material for IR and night-vision applications, and since the 1970s it has been regarded as a potential material for nuclear radiation detector applications [17]. In later years Zn was added to the CdTe matrix for better lattice matching to HgCdTe for night-vision applications, while Zn was added to CdTe to increase the bandgap of the material to achieve higher resistivity and prevent polarization effects associated with Cl-doped CdTe radiation detectors. Most of the CZT material used today for radiation detector applications contains ~10 atomic % of Zn in the CdTe matrix. CZT-based radiation detector applications gained serious momentum from early 1990s after the onset of commercialization. The initial commercial process started with high-pressure Bridgman-grown detector-grade CZT material [18, 19]. Around the same time in 1990, Triboulet et al. [20] demonstrated the growth of CdTe and ZnTe by cold traveling heater method (CTHM), and in 1994 the same group successfully grew CZT by the CTHM technique. In the later years CTHM became commonly known as the traveling heater method (THM). The THM technique uses the growth of the material from a Te-rich solution, which allows the ingots to grow well below its melting point. The THM technique offers several advantages as compared to melt-grown techniques. Because of the lower temperature growth, the ingots possess less defects and thermal stress. The investigators also confirmed that THM-grown ingots show better axial and radial compositional homogeneity with a higher purity as compared to melt-grown ingots [21]. Based on the THM growth technique, Redlen Technologies was founded in 1999 for commercial production of detector-grade CZT. As a consequence of the advantages and successes of the newly invented technique, other companies began using THM to produce detector-grade CZT materials. Over time the THM technique has proven to be the most viable technique to grow CdTe and CZT commercially, and very-large-diameter ingots up to 10 cm can be produced [22]. Despite the intense research for the last three decades, CZT still possesses several shortcomings that severely limit the widespread deployment, especially for large-volume detectors. Although these shortcomings have been very difficult to fully resolve, the quality of CZT material has improved considerably since its early years, and the production cost has steadily decreased. The presence of high concentrations of sub-grain boundary networks and Te inclusions remains as a long-standing issue associated with CZT material. These defects act as the charge trapping and recombination centers and severely affect the uniformity of the charge-transport characteristics of the material resulting in limitations on the thickness of high-performing detectors. Incremental increases in performance have been demon-strated as the size and concentration of Te inclusions have decreased through modifications in the growth and post-processing conditions. In addition to these extended structural defects, CZT suffers heavily from axial and radial compositional inhomogeneity due to the non-unity segregation of Zn in the CdTe matrix [23]. The compositional nonuniformity affects the overall yield of detector-grade material from the grown ingot. In addition the compositional gradient is known to impose considerable strain in the CZT ingot [6, 24]. This chapter discusses the sub-grain boundary networks in CZT and their effect on device performance. One effective way to mitigate such defects is by adding selenium into the CZT matrix, which can increase the yield of high-quality detector-grade material from a grown ingot by reducing the large-scale variations in the alloy composition.

2   Sub-grain Boundaries and Their Networks in CZT

Unlike conventional elemental semiconductors such as Si and Ge, various structural defects are present in CZT, which adversely affect the charge-transport characteristics of the material and the performance of fabricated devices. Sub-grain boundaries and their networks and Te inclusions are the most prominent extended defects in CZT, and they are the main defects responsible for compromising the detector performance and lowering the yield of high-quality detector-grade material from a given ingot. The broadening of photo-peaks in detectors is often attributed to the presence of sub-grain boundaries and their networks and Te inclusions, although point defects and leakage current can be problematic as well. Sub-grain boundaries are in principle dislocation walls, where the dislocations are arranged along the boundary walls [25, 26]. In most cases, these sub-grain boundaries and dislocation walls are arranged in cellular structures because of polygonalization [27] and are highly decorated in CZT material. The configuration of dislocation walls is typically called the sub-grain boundary network. These dislocations are visible after preferential etching of polished surfaces. The sub-grain boundary networks can also be observed by X-ray topographic images, preferably in reflection mode. Figure 1a

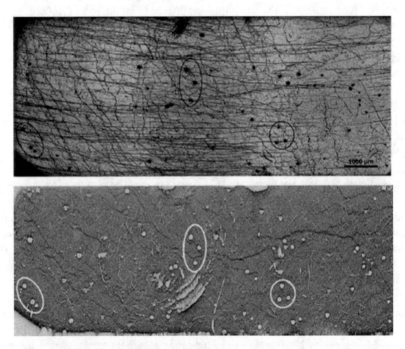

Fig. 1 Sub-grain boundaries and prismatic dislocations, usually generated around Te inclusions after post-growth annealing, are apparent on the surface of a 9×3 mm^2 area of the crystal treated with the Nakagawa etching technique (top) and with X-ray diffraction topography (bottom). The circles denote the prismatic dislocation defects seen in both images (taken from A. Bolotnikov et al., Ref. [25])

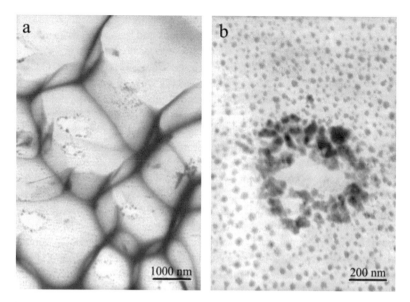

Fig. 2 Sub-grain boundary networks and their Te precipitates. (**a**) Bright-field image of sub-grain boundary networks and (**b**) agglomerated Te precipitates present along sub-grain boundary networks (taken from D. Zeng et al., Ref. [28])

shows a preferentially etched CZT sample surface with dimensions of 9×3 mm^2 and the corresponding X-ray topographic image (Fig. 1b). The dark lines (Fig. 1a), in the cellular structure are arrangements of dislocations along the sub-grain boundary in the CZT sample. The density of sub-grain boundaries is particularly high for this sample. The dark spots in the ellipses denoted in Fig. 1a are commonly known as punching defects, which form on sites of Te inclusions that have been removed by post-growth annealing process. The X-ray topographic image of the same sample also shows the presence of sub-grain boundaries and their network with the appearance of dark and bright lines. The punching defects appear as white spots on the X-ray topographic images as highlighted inside the ellipses. As depicted in Fig. 1, the sub-grain boundary networks are often found to be randomly distributed in the CZT matrix, and the sub-grain boundaries are also known to be decorated with Te inclusions [25, 28]. The sub-grain boundaries are invisible through infrared (IR) transmission microscopy. Since Te inclusions are opaque in the infrared wavelength range, sub-grain boundaries and their networks can often be tracked through the decoration of Te inclusions seen in IR transmission images [25, 28]. The sub-grain boundary network was also observed by bright-field transmission electron microscopy (TEM) by Zeng et al. [28]; they also confirmed the decoration of sub-grain boundaries by Te precipitates as shown in Fig. 2. The sub-grain boundaries and their networks are possibly the most concerning defects in CZT material among those affecting the detector performance. The sub-grain boundaries and their networks were found to be present in CZT samples irrespective of growth techniques. Bolotnikov et al. [25] reported the presence of sub-grain boundaries and their networks in CZT samples by X-ray topographic measurements produced from seven different vendors across the globe. The dislocations and their networks are known to introduce during growth/solidification and subsequent cooling process due to thermal stress related to the inherent poor thermophysical properties of CdTe/CZT. Moreover, the energy of creation of dislocations and stacking faults is inversely proportional to the ionicity [6]. This makes the highly ionic II-VI crystal lattice very sensitive to any strain and prone to introduction of stress-related defects. Thus, the drastic improvement of CdTe/CZT material is limited due to its inherent poor thermophysical properties. The sub-grain boundaries and their network cannot be removed by post-growth annealing; however, eliminating the Te inclusions from the CZT matrix by post-growth annealing is a common practice in CZT detector manufacturing process. Unfortunately, the process is known to produce large starlike defects (punching defects) at the locations of relatively large Te inclusions in post-annealed samples. The starlike defects are invisible in an IR transmission measurement, but they are detectable by charge collection mapping, preferential etching, and X-ray topographic experiments. Depending on their concentration and size, these starlike defects can severely hamper the charge-transport properties [29] and the device performance [30]. Furthermore, Te inclusions are known to be surrounded by a dense field of dislocations, and those remain even after removing the inclusions by thermal annealing [27]. Because of the high concentrations of defects at the dislocation walls, sub-grain boundaries are prone to accumulate impurities [25] in addition to the Te inclusions. In addition to the dislocations itself, all these impurities and the Te inclusions are responsible for creating charge-trapping centers at different binding energies within the bandgap of the material. It has been observed that the dislocations introduced by deformation of CZT produce a trap state at an energy of ~0.27 eV [31]. Te inclusions are reported to introduce a trap level at 1.1 eV above the valence band [32]. Impurities are responsible for shallow and/or deep trap states depending on their position in the energy gap and the Fermi level. For example, the impurities from group III produce shallow levels and the transition elements are accountable for deep states. Most of these defects act as electron-trapping centers and can severely hamper the charge-transport characteristics of the material depending on their concentration and electron capture cross sect [26, 27, 33]. Carini et al. [33] demonstrated direct evidence of the charge loss due to the presence of Te inclusions. The mobility-lifetime product for electrons $[(\mu\tau)_e]$ in CZT can vary over a larger range of $(0.2-20) \times 10^{-3}$ cm^2/V for the regions with Te inclusions compared to the clear regions that are relatively free from large inclusions [33]. This indicates that the random distribution of these localized defects in CZT matrix imposes severe spatial charge-transport inhomogeneity, which eventually broadens the photo-peaks and degrades the detector performance.

The spatial variation of charge-transport characteristics was evaluated by high-spatial-resolution X-ray response mapping with a spatial resolution of better than 10μm [33, 34]. The experiments were conducted using collimated synchrotron light source of beam size ~10 × 10μm^2. Planar CZT detectors were raster scanned across the whole area with a step size of a few microns, and the detector response was registered at each point of interaction for an incident low-energy X-ray beam with an

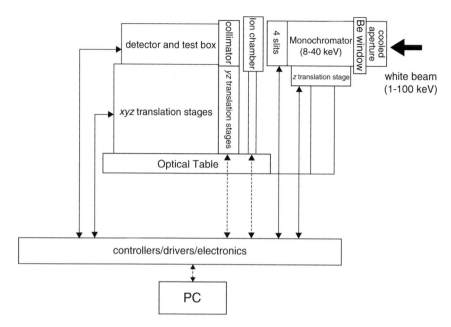

Fig. 3 Block diagram of the X-ray response mapping system (taken from G. A. Carini et al., Ref. [33])

energy of 7–40 keV. The pulse height (i.e., the channel numbers at each point) was then mapped over the whole area of the sample. The schematic of the experimental setup is shown in Fig. 3. The experimental setup and procedure are detailed elsewhere [33, 34] as reported by Carini et al. [33] and Camarda et al. [34]. Charge loss or incomplete charge collection due to the localized defects such as Te inclusions and sub-grain boundaries and their networks were also reported to be dependent on the applied bias, as was demonstrated by Camarda et al. [34]. Figure 4a shows IR image of etched surface of a typical CZT sample revealing different defects such as dislocation walls, sub-grain boundaries, and star-shaped defects. The corresponding X-ray response map of the planar detector fabricated with the same sample of dimensions 12 mm^3 × 5 mm^3 × 5 mm^3 is shown in Fig. 4b under an applied bias of 500 V using a 30 keV energy synchrotron beam. Nonuniform charge collection was observed throughout the entire area. The bright area corresponds to higher charge collection efficiency, while the grey or dark areas are associated with reduced charge collection due to the presence of localized defects. The appearance of dark lines in Fig. 4b shows the presence of sub-grain boundaries (dislocation walls) as a result of incomplete charge collection at the boundary walls. Severe charge loss is also evident for the star-shaped defects as shown in Fig. 4b. Fairly good correlation between the regions of low charge collection and the locations of defects in the crystal displayed in Fig. 4a and b was observed [35].

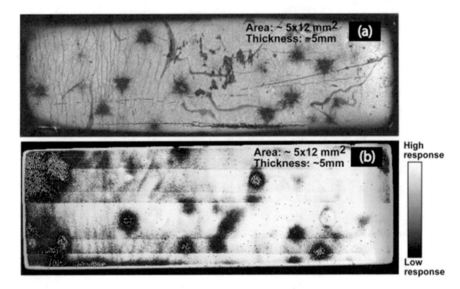

Fig. 4 (a) IR image of etched surface of a CZT sample and (b) raster-scanned image of X-ray response map of the same CZT sample. Sample dimensions: 12 mm^3 × 5 mm^3 × 5 mm^3 (taken from A. Hossain et al., Ref. [35])

Different pulse heights are produced at different locations of the detector sample due to the random distribution of extended defects, which smears the resulting photo-peak of the detector after integrating detector responses over the entire active volume. The performance degradation becomes more prominent with increased thickness. Thus, the spatial homogeneity of charge-transport characteristics in detector material is critically important in achieving high-quality detectors and maximizing their yield. Apart from spatial charge-collection uniformity, high-quality detector material also demands spatial homogeneity of radiation-induced charge generation, and the presence of extended defects and other bandgap variations also adversely affects such uniformity. The effect of sub-grain boundaries on the performance of Frisch grid detectors was very prominent as compared to two different detector samples of the same dimensions [25]. Both the detectors showed the presence of a high density of sub-grain boundary networks, while the sample containing a greater number of major sub-grain boundaries displayed larger broadening of the photo-peaks and low-energy continuum [25]. The presence of sub-grain boundaries/dislocation walls in CdTe family entails a serious issue for medical imaging applications as well. Buis et al. [36] demonstrated that the image of a cowry shell taken with CdTe-pixelated detector contained various lines corresponding to dislocation walls and sub-grain boundaries. The defects had to be eliminated after the flat-field correction. It is thus necessary to develop material free from large Te inclusions and sub-grain boundary networks to avoid the need for charge-loss correction schemes, which impact measurement time, power consumption, and cost-effectiveness. Moreover, material free from the secondary phase and sub-grain boundary networks drastically enhances the yield of high-quality detectors which can potentially lead to a substantial cost reduction.

3 Effects of Se Addition to the CdZnTe Matrix

Most II-VI compounds are prone to strain and extended defects in as-grown ingots due to their high ionicity. Due to the high ionicity of CdTe at 0.55 [6], it is not surprising that CdTe ingots show the presence of sub-grain boundaries and their networks and residual strain. The ionicity is inversely proportional to the energy of creation for dislocations and for stacking faults [6]. Higher ionicity and poor thermophysical properties of CdTe limit the improvement of the material quality and make it difficult to grow large-volume high-quality crystals with less defects. In addition to the growth-related defects, CdTe also suffers from a lower bandgap, which hinders achieving high-resistivity material on the order of $>10^{10}$ Ω-cm. The ternary material CZT came up as the consequence of the desire to improve the properties of CdTe. Adding Zn in CdTe matrix offered several advantages for different applications. For example, adding Zn in CdTe provides better lattice-matching substrates to HgCdTe epilayers for IR or night-vision applications. For detector applications, the bandgap is enhanced to help satisfy the required resistivity of over 10^{10} Ω-cm for making low-noise nuclear detectors with improved performance. The most beneficial reason for adding Zn in CdTe is the higher binding energy of the Zn-Te bond and lower ionicity. Thus, the addition of Zn in CdTe was reported to reduce the dislocation density and sub-grains and enhance the micro-hardness due to the lattice-hardening effect of Zn [6, 37–39]. If we define CdTe as the first-generation radiation detector material, then the ternary compound CZT is the second-generation detector material. Although some lattice hardening was observed after addition of Zn in the CdTe matrix, CZT still comprises a relatively high density of sub-grain boundaries and their networks as discussed in the last section, which affects the yield of high-quality detectors and limits the widespread deployment of CZT especially for large-volume detectors.

Despite incremental improvements over the last three decades, it has been extremely difficult to overcome the inherent poor thermophysical properties of CdTe/CZT and grow large-volume crystals free from sub-grain boundaries and their networks. To improve the crystallinity and increase the volume of defect-free material, one approach has been to pursue isoelectronic doping as an effective means for solid-solution hardening. The substrate community observed a profound reduction of sub-grain boundaries and their networks after doping CdTe by as low as 0.4% (atomic) selenium, and they also showed that solid solution hardening with Se is more effective than with Zn (see, for example, Refs. [7] and [40]). Later, the advantage of Se doping was realized by adding it to the CZT matrix for substrate applications [41]. At the early stage of CZT development in 1994, Fiederle et al. [42] reported a better $\mu\tau$ product for electrons (4.2×10^{-4} cm^2/V) and superior charge-collection efficiency for Bridgman-grown $CdTe_{0.9}Se_{0.1}$ (CTS) crystals compared to

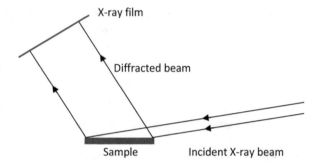

Fig. 5 Schematic of the X-ray diffraction topographic experimental setup in the reflection mode (taken from U.N. Roy et al., Ref. [45])

CZT. In recent years the addition of selenium in the CdTe matrix was also shown to greatly reduce the sub-grain boundaries and produce crystals free from sub-grain boundary networks, plus a reduction in the concentration of Te-rich secondary phases and an improvement in the compositional uniformity [43–45].

White beam X-ray diffraction topography (WBXDT), especially in the reflection mode, is very sensitive in revealing the presence of sub-grain boundaries and their networks. The technique is also sensitive to localized strain present in the crystal. The localized strain results in lattice distortion, which consequently deforms the topographic image. In general, X-ray topographic images in the reflection mode reveal the presence of sub-grain boundaries as the appearance of white and dark lines based on the tilt angle of adjacent sub-grains. The white and dark lines appear due to the separated and overlapping diffracted images of adjacent sub-grains, respectively, subject to the angle of the adjacent sub-grains.

The experimental procedure for WBXDT is rather simple. An intense X-ray beam with an energy range of 6–40 keV from a synchrotron light source reaches the sample's surface, commonly at a grazing incident angle of 4–5 degrees. The diffracted images are recorded on high-resolution X-ray films. The schematic of the setup is shown in Fig. 5.

Significant improvements of the microstructural properties regarding sub-grain boundaries and their networks were demonstrated by adding selenium to the CdTe matrix [6, 40]. They confirmed the resulting CdTeSe to be free from sub-grain boundary networks by using preferential etching to reveal dislocations. Recently, we also confirmed the efficacy of Se addition in CdTe matrix using X-ray topographic studies [45]. Figure 6a shows the results for a lapped wafer along the length of a THM-grown $CdTe_{0.9}Se_{0.1}$ ingot. The arrow indicates the growth direction. The left end is ~1.5 cm from the tip, and the right end is ~1.5 cm from the top of the ingot. The occurrence of a few twins in the material is evident from this figure. The wafer contains two large grains indicated as (a) and (b), as shown in Fig. 6a. The X-ray topographic investigation was carried out for these two large grains. The topographic experiments were carried out on a mirror-finished polished surface followed by etching in a bromine-methanol solution. X-ray topography, especially in the reflection mode, is very sensitive to the surface preparation. Care was taken to remove surface damage from the polishing process by etching the polished samples prior to the experiments. The X-ray diffraction topographic images of the grains "a" and "b"

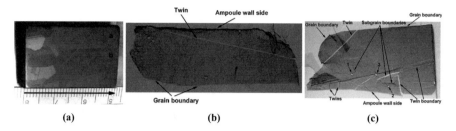

Fig. 6 (**a**) The wafer cut along the length from the central part of the THM-grown $CdTe_{0.9}Se_{0.1}$ ingot. The arrow shows the growth direction. (**b**) X-ray diffraction topographic image of the grains "a" and (**c**) "b" of the CTS sample (taken from U.N. Roy et al., Ref. [45])

as indicated in Fig. 6a are shown in Fig. 6b and c, respectively. As shown in Fig. 6b, the ingot surface touching the ampoule wall shows little or no deformation in the topographic image indicating that the contact is relatively stress free. A slight deformation on the ingot surface touching the ampoule wall is evident from Fig. 6c depicting a small amount of stress-induced lattice deformation. Ampoule walls very commonly introduce strain in melt-grown ingots, especially in CdTe-based compounds [46]. Severe lattice distortions, as observed by the streaky nature of the X-ray topographic images near the periphery of the wafers, were also observed for contactless vapor-grown CdTe and CZT [47, 48]. The distortion near the periphery for the contactless wafer was however reported to be due to possible turbulence in the vapor flow around the periphery of the grown ingot [48]. Due to their inherent structure, twins are very straight in nature. Thus, twins are a good reference frame to observe the presence of any residual stress present in the crystal. Any stress present in the crystal results in localized lattice distortion; hence the twins in the topographic image would be deformed. As shown in Fig. 6b, the topographic image shows the presence of twinning almost diagonally across the entire grain, which is very straight. The appearance of the straight twins in the X-ray topographic image indicates that the entire grain is practically strain free. A similar straight appearance of straight twin lamellae is also evident on the left upper part of the grain "b" as shown in Fig. 6c. No sub-grain boundary was detected in the X-ray topographic image of the grain "b" as shown in Fig. 6b, while few sub-grain boundaries are evident near the lower middle part of the grain "b." These sub-grain boundaries are shown in Fig. 6c as the appearance of white and dark lines in the topographic image. However, no cellular structure, viz. sub-grain boundary networks, was observed in any part of the ingot, which agrees well with earlier results demonstrated back in the mid-1990s [6, 29]. The addition of selenium in the CdTe matrix proved to be an effective solution-hardening element in producing ingots free from sub-grain boundary networks and with greatly reduced residual thermal stress.

Despite profound improvement of the quality, the material does not qualify for radiation detector applications, especially for uses demanding high detection efficiency where lower leakage currents are required. Although the bandgap of CdSe is higher than CdTe, the bandgap of the ternary decreases with Se concentration up to a

Fig. 7 Optical photograph of the grain (left) and the corresponding X-ray topographic image of the grain (taken from U.N. Roy et al., Ref. [50])

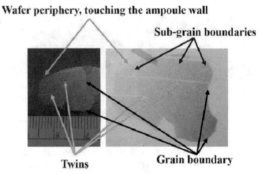

concentration of ~40% [49]. The bowing of the bandgap with Se concentration was ascribed to a compositional disorder effect [49]. Thus, the bandgap of the ternary turns out to be lower than that for CdTe, which restricts the material from attaining the requisite electrical resistivity for room-temperature operation. To achieve the advantages of selenium and also obtain a larger bandgap than CTS, we pursued a quaternary compound by adding selenium to $Cd_{0.9}Zn_{0.1}Te$. Different compositions were evaluated with varying selenium concentrations from 1.5% to 7% (atomic) while keeping the concentration of zinc constant at the value of 10% (atomic). For all the tested concentrations, the resulting CZTS ingots were found to be free from a sub-grain boundary network. Figure 7 shows an optical photograph of a grain (left) from a 2-in. $Cd_{0.9}Zn_{0.1}Te_{0.93}Se_{0.07}$ wafer cut perpendicular to the growth axis for an ingot grown by the THM, and the corresponding X-ray topographic image of the same grain shown on the right side of the photograph [50]. In order to evaluate the effectiveness of selenium as a solution-hardening agent, the ingot was naturally cooled to room temperature after the completion of growth to force introduction of thermal stress. It is to be noted that no lattice distortion was observed near the periphery of the wafer touching the ampoule wall, as indicated by the blue arrow, depicting the absence of any stress introduced from the ampoule wall. This observation agrees well with prior measurements for CTS ingots. However, slight lattice distortion was indeed observed at the wall contact of the CTS ingots, while in case of CZTS, we did not observe any lattice distortion in the X-ray topographic image of the wafer touching the ampoule wall. This observation suggests that the presence of Se and Zn is more effective in arresting the introduction of thermal stress than compared to individually doped Se or Zn in CdTe ingots. Furthermore, the undistorted twins in the X-ray topographic image, as indicated by the green arrows in Fig. 7, provide evidence that the material is free from any thermal stress. The undistorted grain boundaries of the X-ray topographic image also corroborate the effectiveness of selenium as an effective solid solution-hardening element in CZT matrix. The material was observed to be free from sub-grain boundary networks with the presence of very few sub-grain boundaries as illustrated in Fig. 7, albeit the ingot was cooled at a faster rate.

To counter the large thermal stress associated with the higher temperature growth process, similar studies were also carried out with melt-grown CZTS ingots with the

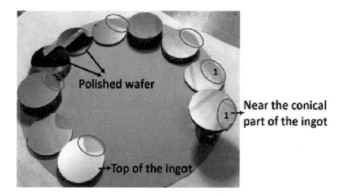

Fig. 8 Optical photograph of the wafers cut perpendicular to the ingot axis (taken from U.N. Roy et al., Ref. [51])

same composition. Very large thermal stress is known to get introduced during the growth and cooling processes of melt-grown ingots by the Bridgman method as compared to the THM. We recently reported an X-ray topographic study of vertical Bridgman-grown $Cd_{0.9}Zn_{0.1}Te_{0.93}Se_{0.07}$ ingot with a diameter of 4 cm [51]. Figure 8 shows the photograph of lapped wafers cut perpendicular to the ~11-cm-long $Cd_{0.9}Zn_{0.1}Te_{0.93}Se_{0.07}$ ingot with a diameter of 4 cm grown by the vertical Bridgman technique [51]. As is evident from Fig. 8, the ingot is composed of two monocrystals with one very large grain except for the conical part of the ingot. A few twins are visible near the conical part of the ingot, and most of the ingot is twin free. As mentioned earlier, all the X-ray topographic diffraction experiments were carried out on mirror-finished surfaces followed by bromine-methanol etching. X-ray topographic analyses were carried out for three different wafers taken from top, middle, and bottom (~1.5 cm above the conical part) parts of the ingot. The very top of the Bridgman-grown ingot undergoes the maximum thermal stress since this part of the ingot remained at the highest temperature for the longest amount of time during growth as well as during the subsequent cooling process.

The ingot was cooled to room temperature at the rate of ~4 °C/h after the completion of the growth process. Figure 9 shows the wafer cut from the top of the ingot and the associated X-ray topographic images. Figure 9a shows the top surface of the ingot. The top of the ingot was barely lapped to a flat surface in order to investigate the X-ray topographic characteristics of the topmost part of the ingot. The X-ray topographic image of the region "a" denoted by the white rectangle in Fig. 9a shows slight lattice distortion near the periphery of the wafer (i.e., the region touching the ampoule wall). The streaky nature of the X-ray topographic image (for grain "b") for the region contacting the ampoule wall illustrates the presence of high lattice distortion. Note that the shape of the grain "b" in the topographic image is undistorted depicting the absence of overall thermal stress in the wafer. Figure 9b shows the opposite face of the top wafer of thickness of ~6.3 mm; the X-ray topographic image of the portion indicated as a rectangle "a" is also shown. One

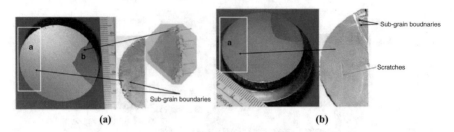

Fig. 9 (**a**) Photograph of the 4-cm-diameter $Cd_{0.9}Zn_{0.1}Te_{0.93}Se_{0.07}$ lapped wafer cut from the very top of the ingot, and the X-ray topographic images of the region denoted by the white rectangle and the grain on the right side of the wafer. (**b**) Photograph of the bottom part of the wafer and the corresponding X-ray topographic image of the region denoted by the white rectangle (the white rectangles and the corresponding X-ray topographic images are not to scale). Thickness of the wafer: 6.3 mm (taken from U.N. Roy et al., Ref. [51])

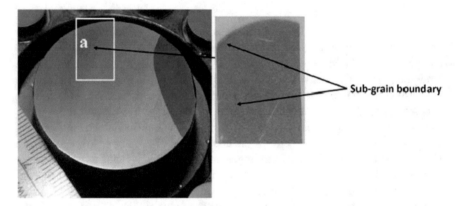

Fig. 10 Photograph of the 4-cm-diameter $Cd_{0.9}Zn_{0.1}Te_{0.93}Se_{0.07}$ lapped wafer cut near the middle of the ingot and the X-ray topographic image of the region denoted by the white rectangle (the white rectangle and the corresponding X-ray topographic images are not to scale) (taken from U.N. Roy et al., Ref. [51])

outstanding feature of the X-ray topographic image is the absence of any lattice distortion on the periphery of the wafer. Even for the portion near the top of the ingot, very few sub-grain boundaries were evident from the topographic image, and no sub-grain boundary network was seen. As expected, the X-ray topographic analyses for the wafers from the middle and the lower parts of the ingot also showed the absence of significant lattice distortion, hence lower thermal stress along the periphery of the wafers for those regions in touch with the ampoule wall. The optical photograph and their corresponding X-ray topographic images are shown in Figs. 10 and 11, respectively. The overall undistorted topographic image also exemplifies the absence of thermal stress in the crystal. Very few sub-grain boundaries were observed for all the wafers from different parts of the ingot. The straight nature of the twin in the topographic image shown in Fig. 11 also indicates that the wafer is

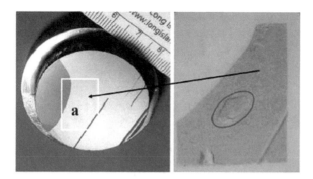

Fig. 11 Photograph of the 4-cm-diameter $Cd_{0.9}Zn_{0.1}Te_{0.93}Se_{0.07}$ lapped wafer cut from the bottom of the ingot and the X-ray topographic image of the region denoted by the white rectangle (the white rectangle and the corresponding X-ray topographic images are not to scale) (taken from U.N. Roy et al., Ref. [51])

stress free. The defect indicated inside the blue ellipse in the topographic image (Fig. 11) is possibly a sub-grain boundary in a closed loop.

The circular nature of the periphery of the wafers, undeformed grain boundaries, and straight nature of the twins in the topographic images imply the absence of overall thermal stress in the wafers as well as the ingots. The features were found to be similar for both THM and Bridgman-grown CZTS ingots. The addition of selenium in the CZT matrix thus proved to be an effective solution-hardening element in arresting the formation of sub-grain boundaries and their networks. It is thus evident that the addition of selenium in CdZnTe successfully resolves persistent unsolved issues associated with difficulties encountered in the CZT crystal growth and provides significant material improvements of importance for radiation-detection applications.

In conclusion selenium was found to play a significant role in improving the material quality of CZT. The microhardness of the new quaternary compound CZTS was observed to be significantly higher [52] as compared to CdTe and CZT. The compositional homogeneity was also greatly enhanced in CTS and CZTS as compared to CZT [43, 50, 53]. The concentrations of the secondary phases, viz. Te inclusions, were also observed to be greatly reduced after adding selenium to the CdTe and the CZT matrix [44, 50, 53, 54]. The reduced performance-limiting defects and enhanced compositional homogeneity ensure higher spatial charge-transport homogeneity in CZTS material resulting in increased yield of high-quality detectors. The achieved energy resolution of Frisch-grid detectors fabricated from as-grown CZTS ingots at 662 keV was in the range of $1 \pm 0.1\%$ [52, 53]. CZTS is thus found to be a very promising material with tremendous potential to replace CZT due to a better yield of high-quality detector material and a reduced cost of production. Due to the multifaceted advantages, CZTS has attracted quick attention from the commercial sector for medical imaging applications. Very recently Yakimov et al. [55] reported superior detector quality as compared to conventional CZT for the intensity of high-flux X-rays expected in many medical imaging applications. Thus, considering the overall superior material properties and homogeneity of CZTS, the material shows the potential to supersede CZT in the future, particularly as the purity of the Se is further increased.

References

1. Schlesinger, T.E., et al.: Cadmium zinc telluride and its use as a nuclear radiation detector material. Mater. Sci. Eng. R. **32**, 103 (2001)
2. Yang, G., James, R.B.: Applications of CdTe, CdZnTe, and CdMnTe radiation detectors. In: Physics, Defects, Hetero- and Nano-structures, Crystal Growth, Surfaces and Applications Part II, (EDAX. Triboulet R. *et al.*) p. 214. Elsevier (2009)
3. Harrison, F.A., et al.: The nuclear spectroscopic telescope array (NuSTAR) high-energy X-ray mission. Astrophys. J. **770**, 103 (2013)
4. Krawczynski, H.S., et al.: X-ray polarimetry with the polarization spectroscopic telescope array (PolSTAR). Astropart. Phys. **75**, 8 (2016)
5. Slomka, P.J., et al.: Solid-state detector SPECT myocardial perfusion imaging. J. Nucl. Med. **60**, 1194 (2019)
6. Triboulet, R.: Fundamentals of the CdTe and CdZnTe bulk growth. Phys. Status. Solidi (c). **5**, 1556 (2005)
7. Rudolph, P.: Fundamental studies on Bridgman growth of CdTe. Prog. Cryst. Growth Charact. **29**, 275 (1994)
8. Jing, W., Chi, L.: Recent advances in cardiac SPECT instrumentation and imaging methods. Phys. Med. Biol. **64**, 06TR01 (2019)
9. Sakamoto, T., et al.: The second swift burst alert telescope gamma-ray burst catalog. Astrophys. J. Suppl. Ser. **195**, 1 (2011)
10. Takahashi, T., et al.: High-resolution CdTe detector and applications to imaging devices. IEEE Trans. Nucl. Sci. **48**, 287 (2001)
11. MacKenzie, J., et al.: Advancements in THM-grown CdZnTe for use as substrates for HgCdTe. J. Electron. Mater. **42**, 3129 (2013)
12. Iniewski, K.: CZT detector technology for medical imaging. J. Instrum. **9**, 1 (2014)
13. Kargar, A., et al.: Charge collection efficiency characterization of a HgI$_2$ Frisch collar spectrometer with collimated high energy gamma rays. Nucl. Instrum. Methods Phys. Res. Sect. A. **652**, 186 (2011)
14. Hitomi, K., et al.: TlBr capacitive Frisch grid detectors. IEEE Trans. Nucl. Sci. **60**, 1156 (2013)
15. He, Y., et al.: High spectral resolution of gamma-rays at room temperature by perovskite CsPbBr$_3$ single crystals. Nat. Commun. **9**, 1609 (2018)
16. Johns, P.M., Nino, J.C.: Room temperature semiconductor detectors for nuclear security. J. Appl. Phys. **126**, 040902 (2019)
17. Takahashi, T., Watanabe, S.: Recent progress in CdTe and CdZnTe detectors. IEEE Trans. Nucl. Sci. **48**, 950 (2001)
18. Szeles, C., Eissler, E.E.: Current issues of high-pressure Bridgman growth of semi-insulating CdZnTe. MRS Symp. Proc. **484**, 309 (1998)
19. Szeles, C., et al.: Development of the high-pressure electro-dynamic gradient crystal-growth technology for semi-insulating CdZnTe growth for radiation detector applications. J. Electron. Mater. **33**, 742 (2004)
20. Triboulet, R., et al.: "Cold travelling heater method", a novel technique of synthesis, purification and growth of CdTe and ZnTe. J. Cryst. Growth. **101**, 216 (1990)
21. El Morki, A., et al.: Growth of large, high purity, low cost, uniform CdZnTe crystals by the cold travelling heater method. J. Cryst. Growth. **138**, 168 (1994)

22. Shiraki, H., et al.: THM growth and characterization of 100 mm diameter CdTe single crystals. IEEE Trans. Nucl. Sci. **56**, 1717 (2009)
23. Zhang, N., et al.: Anomalous segregation during electrodynamic gradient freeze growth of cadmium zinc telluride. J. Cryst. Growth. **325**, 10 (2011)
24. Perfeniuk, C., et al.: Measured critical resolved shear stress and calculated temperature and stress fields during growth of CdZnTe. J. Cryst. Growth. **119**, 261 (1992)
25. Bolotnikov, A.E., et al.: Characterization and evaluation of extended defects in CZT crystals for gamma-ray detectors. J. Cryst. Growth. **379**, 46 (2013)
26. Bolotnikov, A.E., et al.: Extended defects in CdZnTe radiation detectors. IEEE Trans. Nucl. Sci. **56**, 1775 (2009)
27. Szeles, C., et al.: Advances in the crystal growth of semi-insulating CdZnTe for radiation detector applications. IEEE Trans. Nucl. Sci. **49**, 2535 (2002)
28. Zeng, D., et al.: Transmission electron microscopy observations of twin boundaries and sub-boundary networks in bulk CdZnTe crystals. J. Cryst. Growth. **311**, 4414 (2009)
29. Yang, G., et al.: Post-growth thermal annealing study of CdZnTe for developing room-temperature X-ray and gamma-ray detectors. J. Cryst. Growth. **379**, 16 (2013)
30. Bolotnikov, A.E., et al.: CdZnTe position-sensitive drift detectors with thicknesses up to 5 cm. Appl. Phys. Lett. **108**, 093504 (2016)
31. Krsmanovic, N., et al.: Electrical compensation in CdTe and CdZnTe by intrinsic defects. SPIE Proc. **4141**, 219 (2000)
32. Gul, R., et al.: A comparison of point defects in $Cd_{1-x}Zn_xTe_{1-y}Se_y$ crystals grown by Bridgman and traveling heater methods. J. Appl. Phys. **121**, 125705 (2017)
33. Carini, G.A., et al.: High-resolution X-ray mapping of CdZnTe detectors. Nucl. Instrum. Methods Phys. Res. Sect. A. **579**, 120 (2007)
34. Camarda, G.S., et al.: Polarization studies of CdZnTe detectors using synchrotron X-ray radiation. IEEE Trans. Nucl. Sci. **55**, 3725 (2008)
35. Hossain, A., et al.: Extended defects in CdZnTe crystals: effects on device performance. J. Cryst. Growth. **312**, 1795 (2010)
36. Buis, C., et al.: Effects of dislocation walls on image quality when using cadmium telluride X-ray detectors. IEEE Trans. Nucl. Sci. **60**, 199 (2013)
37. Triboulet, R.: Crystal growth technology. In: Scheel, H.J., Fukuda, T. (eds.) CdTe and CdZnTe Growth, p. 373. Wiley, Hoboken (2003)
38. Guergouri, K., et al.: Solution hardening and dislocation density reduction in CdTe crystals by Zn addition. J. Cryst. Growth. **86**, 61 (1988)
39. Imhoff, D., et al.: Zn influence on the plasticity of $Cd_{0.96}Zn_{0.04}Te$. J. Phys. France. **1**, 1841 (1991)
40. Johnson, C.J.: Recent progress in lattice matched substrates for HgCdTe epitaxy. SPIE. **1106**, 56 (1989)
41. Tanaka, A., et al.: Zinc and selenium co-doped CdTe substrates lattice matched to HgCdTe. J. Cryst. Growth. **94**, 166 (1989)
42. Fiederle, M., et al.: Comparison of CdTe, $Cd_{0.9}Zn_{0.1}Te$ and $CdTe_{0.9}Se_{0.1}$ crystals: application for γ- and X-ray detectors. J. Cryst. Growth. **138**, 529 (1994)
43. Roy, U.N., et al.: Growth of $CdTe_xSe_{1-x}$ from a Te-rich solution for applications in radiation detection. J. Cryst. Growth. **386**, 43 (2014)
44. Roy, U.N., et al.: High compositional homogeneity of $CdTe_xSe_{1-x}$ crystals grown by the Bridgman method. Appl. Phys. Lett. Mater. **3**, 026102 (2015)
45. Roy, U.N., et al.: Evaluation of $CdTe_xSe_{1-x}$ crystals grown from a Te-rich solution. J. Cryst. Growth. **389**, 99 (2014)
46. Ching-Hua, S., et al.: Crystal growth and characterization of CdTe grown by vertical gradient freeze. Mater. Sci. Eng. B. **147**, 35 (2008)
47. Palosz, W., et al.: The effect of the wall contact and post-growth cool-down on defects in CdTe crystals grown by 'contactless' physical vapour transport. J. Cryst. Growth. **254**, 316 (2003)

48. Egan, C.K., et al.: Characterization of vapour grown CdZnTe crystals using synchrotron X-ray topography. J. Cryst. Growth. **343**, 1 (2012)
49. Hannachi, L., Bouarissa, N.: Electronic structure and optical properties of $CdSe_xTe_{1-x}$ mixed crystals. Superlattice. Microst. **44**, 794 (2008)
50. Roy, U.N., et al.: Role of selenium addition to CdZnTe matrix for room-temperature radiation detector applications. Sci. Rep. **9**, 1620 (2019)
51. Roy, U.N., et al.: X-ray topographic study of Bridgman-grown CdZnTeSe. J. Cryst. Growth. **546**, 125753 (2020)
52. Franc, J., et al.: Microhardness study of $Cd_{1-x}Zn_xTe_{1-y}Se_y$ crystals for X-ray and gamma ray detectors. Mater. Today Commun. **24**, 101014 (2020)
53. Roy, U.N., et al.: Evaluation of CdZnTeSe as a high-quality gamma-ray spectroscopic material with better compositional homogeneity and reduced defects. Sci. Rep. **9**, 7303 (2019)
54. Roy, U.N., et al.: High-resolution virtual Frisch grid gamma-ray detectors based on as-grown CdZnTeSe with reduced defects. Appl. Phys. Lett. **114**, 232107 (2019)
55. Yakimov, A., et al.: Growth and characterization of detector-grade CdZnTeSe by horizontal Bridgman technique. SPIE Proc. **11114**, 111141N (2019)

Radiation Detection Using n-Type 4H-SiC Epitaxial Layer Surface Barrier Detectors

Sandeep K. Chaudhuri and Krishna C. Mandal

Abstract While CdZnTe (CZT) is one of the best materials for room-temperature radiation detection, they are not quite suitable for high-temperature or harsh-environment applications. This chapter discusses the fabrication and characterization of high-resolution 4H-SiC epitaxial detectors, which are appropriate for use in harsh environments like high temperature, chemically reactive and corrosive environments, and most importantly high-dose nuclear radiation environments. Schottky barrier diodes on thin (≤ 20 μm) 4H-SiC epitaxial layer detectors have been found to be a very promising device for charged particle detection at room and elevated temperatures. While such thin epitaxial layers are sufficient to stop energetic charged particle like alpha particles, highly penetrating radiations such as X- and γ-rays need thicker epitaxial layers to enable substantial photon absorption. This chapter discusses the fabrication of Schottky barrier radiation detectors in n-type 4H-SiC epitaxial layers with different thicknesses (20, 50, and 150 μm); their characterization in terms of alpha, X-rays, and γ-detection; and evaluation of the factors like deep-level defects which regulate their performance as radiation detectors.

1 Introduction

While CZT-based detectors are still the best choice as far as room-temperature detection of energetic gamma rays is concerned, their high-temperature operations are limited. They are reported to operate satisfactorily up to temperatures of 70 °C [1, 2]. CZT detectors also lack the potential to withstand other harsh conditions like mechanical stresses, high radiation field, and chemically reactive and corrosive environments. Silicon carbide (SiC) is a semiconducting material that is known for its wide bandgap, hence high-temperature operability [3], mechanical hardness, and chemical inertness [4–7]. Silicon carbide has an extremely high melting point, which is unachievable under laboratory conditions and directly sublimes at around

S. K. Chaudhuri · K. C. Mandal (✉)
Department of Electrical Engineering, University of South Carolina, Columbia, SC, USA
e-mail: chaudhsk@mailbox.sc.edu; mandalk@cec.sc.edu

© The Author(s), under exclusive license to Springer Nature Switzerland AG 2022
K. Iniewski (ed.), *Advanced Materials for Radiation Detection*,
https://doi.org/10.1007/978-3-030-76461-6_9

183

2700 °C. Among 250 other well-known polytypes of SiC, 4H-SiC is superior as a detector material owing to its wide bandgap of 3.27 eV [5–7], smaller anisotropic conductivity [8], and smaller electron effective mass, or higher mobility [9]. 4H-SiC is known for its radiation hardness which is due to the very high average displacement threshold energy of the constituent elements (19 eV for C sublattice and 42 eV for Si sublattice) [10] in the 4H-SiC matrix and has been reported to withstand radiation doses up to 22 MGy [11]. However, the performance of bulk 4H-SiC-based sensors as radiation detectors is limited due to the presence of defects and impurities in the as-grown crystals. Substantial carrier trapping in deep levels results in polarization which degrades the detector performance and limits its applicability over short period only. Further, recombination or generation of charge carriers in/from the defect centers results in poor efficiency, energy, and timing resolution [12].

The desired crystallinity for high-resolution detector fabrication can nevertheless be obtained using epitaxially grown SiC layers [13–16]. Availability of high-quality 4H-SiC epitaxial layers has further enhanced the possibility of fabrication of nuclear radiation detectors which are compact, operable at room or elevated temperature, physically rugged, and radiation hard. In particular, Schottky barrier diodes (SBD) fabricated on 4H-SiC detectors have demonstrated themselves as high-resolution alpha-ray detectors [13–15]. 4H-SiC SBDs are thus operable at harsh environments like NASA space missions, nuclear core reactors, and accelerator environments, and in laser-generated plasma environments such as facilities for high-energy particle generation [17–27]. SBDs with very high average surface barrier heights (SBH) and diode ideality factors close to unity for large-area detectors can be readily fabricated on n-type 4H-SiC epitaxial layers just by depositing a thin layer of metal like nickel (Ni). Ni metal with a thickness of few nanometers serves as nearly transparent window for charged particles like alpha. Due to low atomic numbers of Si and C, thin 4H-SiC epitaxial layers are fairly transparent to gamma rays and hence are very suitable for neutron detection when coupled with neutron-alpha conversion layers [16, 28]. 4H-SiC has large hole diffusion lengths and due to high achievable built-in potential (greater than 1.15 eV) [17], these SBDs can be used as self-biased detectors [29], which is a sought-after quality for field deployment of standoff detection systems for homeland security purposes.

Apart from charged particle detection, 4H-SiC has been reported to be highly sensitive to soft X-rays of energy up to 50 eV and has achieved high energy resolution of 2.1% for 59.6 keV gamma rays [13, 30, 31]. Bulk semi-insulating 4H-SiC could have been a better choice in terms of absorption of highly penetrating gamma rays but, as has been mentioned previously, the presently available crystal quality is inadequate for radiation detection. Epitaxial 4H-SiC layers, on the other hand, can be grown as highly crystalline detector-grade semiconductor material. Radiation detectors with thicker epitaxial layers are being reported offering better x/γ-photon and neutron absorption [20, 24]. Mere production of SBDs on thicker epitaxial layers will not suffice the purpose of efficient detection of highly penetrating radiation unless the SBDs are fully depleted with widths equaling the epilayer thickness. Full depletion width requires high operating bias conditions where the leakage currents could be too high to maintain low noise conditions. Recently Puglisi and Bertuccio have reported the fabrication of fully depleted position-sensitive microstrip X-ray detectors on 124 µm thick epitaxial 4H-SiC [20]. In this chapter we report alpha detectors fabricated on epitaxially grown 4H-SiC layers with thickness up to 150 µm and electrode area as large as 0.11 cm^2. The 8×8 cm^2 detector wafers were cut from parent wafers of diameter 100 mm.

The performance of the epitaxial layer detectors, like most other semiconductor detectors, is mostly controlled by the extent to which defect-free crystals are grown. The devices that are discussed in this chapter are homoepitaxial devices, meaning 4H-SiC epilayers were grown on 4H-SiC substrates. The substrates, generally grown by sublimation technique, usually contain structural defects in huge concentration. Extrinsic point defects like substitutional nitrogen centers (N(c) and N(h)), substitutional titanium centers (Ti(c) and Ti(h)), and chlorine impurities (Ci1) and intrinsic defects like $Z_{1/2}$, EH_5, EH_6, and EH_7 are prevalent in nitrogen-doped n-type 4H-SiC polytype. The intrinsic defects mentioned above are mostly related to carbon vacancies or carbon clusters or antisites related to carbon and silicon. While the point defects are practically unavoidable and are mostly responsible for the charge carrier trapping, extended structural defects such as micropipes are detrimental to proper functioning of the detector as well. Micropipes [32] are hollow-core threading screw dislocations (TSDs) which are known to induce device failures and reliability issues [33, 34]. TSDs are one of the three structural defects which preexist in the 4H-SiC substrate and can propagate to the epitaxial layers. Micropipes can also dissociate at the substrate-epilayer interface to multiple closed-core TSDs which are believed to be responsible for the premature breakdown in 4H-SiC-based power devices [35]. The other two defect categories which propagate from the substrate to the epitaxial layer are threading edge dislocations (TEDs) and basal plane dislocations (BPDs). TEDs are believed to be benign in comparison to TSDs and BPDs [34]. BPDs normally transform into benign TEDs during epitaxial layer growth but can also propagate through and can cause V_f drift in bipolar devices [36]. With the substantial reduction in the micropipe density (less than 1 cm^{-2}), the performance of the 4H-SiC detectors reported in this work is expected to be mostly limited by the presence of deep-level point defects which can act as potential trapping and recombination centers [37–40].

To summarize this chapter, the primary factors, which have to be taken into consideration for the fabrication of high-energy-resolution and high-efficiency 4H-SiC epitaxial radiation detectors, are discussed at length. The primary factors include the following: (1) minimizing the scattering of the incident radiation at the detector entrance window, (2) ensuring efficient stopping of the incident radiation within the effective volume of the detector (depletion region in the case of Schottky barrier detectors), (3) studying the presence and effect of macro- and microstructural defects in the detector material that limits the charge carrier mobility and lifetime, (4) minimizing the detector leakage current by achieving a high-quality Schottky barrier, and finally (5) appropriate selection of the front-end and filter electronics for minimizing the electronic noise. This chapter discusses how to address the abovementioned aspects to achieve the goal of fabrication of high-resolution radiation detection. The fabrication and characterization of high-resolution n-type 4H-SiC epitaxial layer detectors and the nature of point defects that can affect the performance of these devices have been discussed at length. Deep-level transient spectroscopy (DLTS) in capacitance mode has been used to study the electron-trapping centers by calculating their locations within the bandgap, their concentration, and capture cross sections. Results from 4H-SiC-based SBDs with three different epitaxial layer thicknesses, viz. 20, 50, and 150 μm, have been presented. While 20 μm and 50 μm 4H-SiC SBDs have exhibited high-resolution charged particle detection when fully depleted, the 150 μm thick epilayers, which have the potential of efficient and high X-ray and gamma-ray detection if fully depleted, were found to achieve full-depletion width at very high operating bias conditions.

2   High-Resolution Radiation Detection Using 4H-SiC Epitaxial Layer Detector

The idea to use 4H-SiC as radiation detectors was pioneered by Babcock and Chang [41]. Further works which advanced the detector as high-resolution charge particle detector are as follows. Ruddy et al. [42] reported a percentage energy resolution of 5.8% and 6.6% for a deposited energy of 294 keV and 260 keV alpha particles, respectively. It can be noted that the authors used a collimated ^{238}Pu source and circular diode contacts of 200 and 400 μm. Later on, Ruddy et al. also reported [43] an energy resolution of 5.7% for a deposited energy of 89.5 keV alpha particles from a 100 μm collimated ^{148}Gd source in similar detectors with comparatively larger Schottky contact diameter of 2.5, 3.5, 4.5, 6.0, and 10 μm thick 4H-SiC epitaxial layers. In another work [15], Ruddy et al. reported an energy resolution close to 46 keV (~0.8%) for alpha particles from a ^{238}Pu source and 41.5 keV (~1.3%) for alpha particles from a ^{148}Gd source for devices with aluminum guard ring. Ivanov et al. [29] reported an energy resolution of 20 keV (~0.4%) in the energy range of 5.4–5.5 MeV. In yet another work, Ruddy et al. [44] reported an energy resolution of 20.6 keV (~0.4%) for ^{238}Pu alpha particles. Our group has reported an energy resolution of 2.7% and 0.29% for 5486 keV alpha particles in 50 μm [45] and 20 μm [14] thick n-type Ni/4H-SiC detectors, respectively.

The prospect of 4H-SiC epitaxial layer as a gamma-ray detector was investigated by Bertuccio et al. [30] where they detected well-resolved Np L series X-rays (13.9 and 17.8 keV) from a ^{241}Am radioisotope. Bertuccio et al. [22, 30], in a separate work, has also investigated the possibility of high-resolution X-ray spectroscopy in a wide temperature range, from room temperature up to 100 °C without any cooling system. Bertuccio et al. summarized the advancement in X-ray detection until the end of the last decade using SiC in one of their review works [46]. The beginning of this decade has seen renewed interest of X-ray and low-energy gamma-ray detection with the development of technology of high-quality thicker detector-grade epitaxial layers. Terry et al. [47] have investigated the use of commercial off-the-self (COTS)

4H-SiC photodiodes meant for UV detection as soft X-ray detectors but concluded that the COTS devices were not suitable for X-ray spectrometry purpose. In a follow-up work Mandal et al. [48] showed that Schottky barrier detectors fabricated out of high-quality epitaxially grown 4H-SiC provided much higher sensitivity to soft X-rays compared to that offered by the COTS 4H-SiC UV diode. Mandal et al. in a later work [13] reported the performance of even better detectors with significantly higher X-ray responsivity and an energy resolution of 2.1% for 59.6 keV gamma rays. Radiation detectors with thicker epitaxial layers are being reported which will offer better x/γ-photon absorption. Recently Puglisi and Bertuccio have reported the fabrication of position-sensitive microstrip X-ray detectors on 124 μm thick epitaxial 4H-SiC [20]. We will present in this chapter the recent results from the SBDs fabricated on 150 μm thick epitaxial 4H-SiC.

2.1 Detector Fabrication

4H-SiC epitaxial layers were grown on the (0001) Si face of a 350 μm thick 4H-SiC substrates with a diameter of 100 mm with a thickness of up to 150 μm. The substrates were 8^0 off-cut towards the $\langle 11\bar{2}0 \rangle$ direction to facilitate step-controlled epitaxial growth which in turn helps to reduce the surface defects [49]. A schematic of the crystal faces and the direction convention has been explained in Fig. 1a. Figure 1b illustrates the construction of the epilayer detector with the physical dimensions. The 8 mm^2 × 8 mm^2 wafers were diced out from the parent wafers

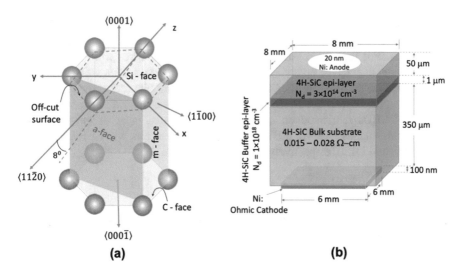

Fig. 1 (**a**) Schematic showing the crystal planes and crystallographic directions in 4H-SiC polytype. The dotted hexagon shows the off-cut plane in the substrates. (**b**) Illustration of the construction of Ni/n-4H-SiC 50μm epitaxial layer detectors

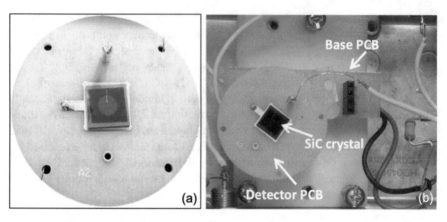

Fig. 2 (a) A 4H-SiC epitaxial detector mounted on a PCB and wire bonded. (b) The detector mounted inside the test box

and cleaned following RCA wafer cleaning procedure [50]. The native oxide layers were removed using concentrated hydrofluoric acid prior to contact deposition. The 1 µm thick buffer layer helps to reduce the propagation of the TSDs, TEDs, and BPDs. The epilayer on the top is the active device layer on which 10–20 nm thick circular (dia ~3.8 mm) nickel contacts were deposited using a quorum Q150T sputter coater. Thin nickel contacts are necessary to reduce energy loss through the incident radiation [51]. Thin Ni contacts formed SBDs with the depletion region extending to the epilayer. A reverse bias condition was obtained when this contact (anode) was at negative potential with respect to the thicker bottom Ni contact. The bottom contacts (cathode) were square-shaped (6 mm × 6 mm) sputter-coated 100 nm thick nickel. For establishing electrical connections with the metal contacts, the detectors were mounted on printed circuit boards (PCB) with metal pad forming the cathode connection. For the anode connection 25 µm thin gold wires were wire bonded using conducting silver epoxy. The other end of the gold wire was attached to another pad on the PCB forming the anode connection. Figure 2 shows the photograph of a detector and the electrical connections. This chapter discusses characterization of SBDs fabricated on n-type 4H-SiC epitaxial layer of three different epilayer thicknesses, viz. 20, 50, and 150 µm. The detectors will be referred to as D20, D50, and D150, respectively.

2.2 Basic Characterization of the Epilayer Schottky Barrier Detectors (SBDs)

The preliminary characterization of the SBDs is current-voltage (*I-V*) measurement which allows to determine crucial interfacial and bulk parameters like the leakage current, Schottky barrier height (SBH), and capacitance-voltage (*C-V*) measurement

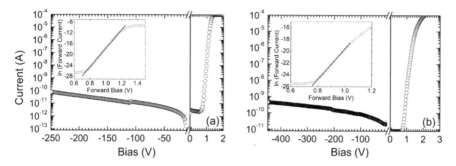

Fig. 3 *I-V* characteristics of the detectors (**a**) D20 and (**b**) D50. The detector D150 has a similar *I-V* behavior as that of D50; hence it has not been shown. Inset shows the fitting of the forward characteristics to a thermionic emission model

which allows to determine the effective doping concentration of the epilayer and the built-in potential of the junction. Figure 3 shows the *I-V* characteristic of D20 and D50. The highly asymmetric characteristic confirms the rectifying behavior of all the devices. The SBDs behaved as reverse biased with a negative polarity bias on the anode (circular nickel contact). The obtained leakage currents are listed in Table 1. It could be noticed from Fig. 3 that leakage currents <1 nA were observed with an applied reverse bias at which the SBDs were fully depleted. A leakage current below 1 nA ensures that the SBDs can be operated as a radiation detector when fully depleted. The device D150 too showed a very low leakage current up to bias voltages as high as 500 V. However, to achieve the full depletion at the given doping concentration of 1.1 cm^{-3} × 10^{14} cm^{-3} (from *C-V* plots) a reverse bias voltage of 2300 V is necessary.

The forward *I-V* characteristic offers a lot of useful information when assumed to follow established theoretical models. Using thermionic emission models of SBD [52, 53] the variation of forward diode current I_F as a function of forward bias V_F can be theoretically modeled as given in Eq. (1) below:

$$I_F = A^{**}AT^2 e^{-q\phi_{b0}/kT} e^{-q\beta V_F/kT} \left\{ e^{qV_F/kT} - 1 \right\}$$ (1)

where A^{**} is the effective Richardson constant (146 A cm^2 K^2 for 4H-SiC), A is the effective contact area participating in current flow, T is the absolute junction temperature, q is the electronic charge, ϕ_{b0} is the zero bias barrier height, k is the Boltzmann's constant, and $\beta = \frac{\partial \phi_b}{\partial V_F}$ gives the variation of the barrier height ϕ_b with bias voltage. The barrier height could vary with the forward bias due to any change in voltage drop across the interfacial layer or due to change in image force with applied bias. The factor β is related to the diode ideality factor n as $1/n = 1 - \beta$. For values of $V_F > 3kT/q$, Eq. 1 can be modified as

$$I_F = A^{**}AT^2 e^{-q\phi_b/kT} e^{-qV_F/nkT}.$$ (2)

Table 1 Surface barrier properties of the 4H-SiC SBDs obtained from I-V and C-V measurements

Detector	Epilayer thickness (μm)	Ideality factor	Φ_b (eV)	$\Phi_{b(C\text{-}V)}$ (eV)	Built-in potential (eV)	Leakage currents (pA) at −200 V	N_{eff} (cm^{-3})	Best energy resolution @ 5486 keV (%)
D20	20	1.2	1.60	1.78	1.5	40.0	2.4×10^{14}	0.29
D50	50	1.5	1.54	2.0	1.7	110.0	8.9×10^{13}	0.80
D150	150	1.2	1.40	2.60	2.3	100.0	1.1×10^{14}	0.45

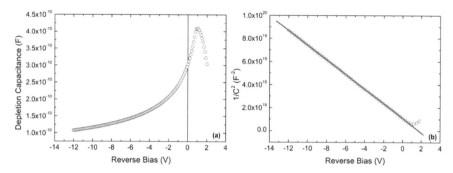

Fig. 4 (**a**) Variation of junction capacitance as a function of reverse bias voltage (*C-V*) of D50 4H-SiC SBD. (**b**) Mott-Schottky plot corresponding to the *C-V* plot. The open circles are the experimental data points and the solid line is the linear regression according to Eq. 3 where N_c is the effective density of states in the conduction band of 4H-SiC and is taken equal to 1.6×10^{19} cm^{-3} [54]

Figure 4a shows the variation of the depletion capacitance of the detector D50 as a function of reverse bias. It could be seen that the depletion capacitance decreased with applied reverse bias which is expected as the depletion width increases with increase in reverse bias. The Mott-Schottky plots (Fig. 4b) were fitted with a straight line following the model provided in Eq. 3 to obtain N_{eff} and V_{bi}. The effective doping concentrations and the built-in potentials for the three detectors are listed in Table 1 and it could be seen that the effective doping concentrations were $\leq 2.4 \times 10^{14}$ cm^{-3}. The barrier heights calculated from the Mott-Schottky plots and Eq. 4 were found to be higher than those obtained from the *I-V* measurements. Barrier height obtained from the forward *I-V* characteristics is dominated by current flow through the low Schottky barrier height locations in a Schottky diode with spatially inhomogeneous barrier height, which results in lower SBH. *C-V* characteristics, on the other hand, give an average value of the barrier height for the whole diode area [55, 56].

The experimentally obtained *I-V* characteristics in the fabricated Ni/n-4H-SiC diode were fitted according to Eq. 2. A straight-line fit (insets of Fig. 3) of the *ln I_F* – *V_F* plot returned the barrier heights and the ideality factors listed in Table 1.

The information from the $C - V$ measurements was obtained as described below. Considering that the metal contact and the substrate layer constitute a parallel plate capacitor with the depletion layer as a dielectric medium, the variation of depletion capacitance C as a function of reverse bias voltage could be written as

$$\frac{1}{C^2} = \frac{2}{qA^2\varepsilon_0\varepsilon_{4H-SiC}N_{eff}}(V + V_{bi}). \tag{3}$$

Here, ε_0 is the permittivity of free space, ε_{4H-SiC} is the dielectric constant of 4H-SiC, N_{eff} is the effective doping concentration in the depletion region, and V_{bi} is the built-in potential. The SBH $\phi_{b(C-V)}$ is calculated using the results obtained from the Mott-Schottky ($\frac{1}{C^2}$ vs V) plots and the equation

$$\phi_{b(C-V)} = V_{bi} + V_n, \tag{4}$$

where V_n is the potential difference between the Fermi-level energy and the conduction band edge in the neutral region of the semiconductor, given by

$$V_n = kT \ln \frac{N_c}{N_{\text{eff}}}, \tag{5}$$

2.3 Crystalline Quality Evaluation of the Epitaxial Layer

The performance of a device obviously depends on the single crystallinity of the epitaxial layer. XRD rocking curve measurement gives very accurate information on the orientation of crystallographic planes. Preferential or defect-delineating etching helps in exposing the sites around defects like dislocations, stacking faults, precipitates, and point defects. The width (FWHM) of the XRD rocking curve peak is a measure of the crystalline quality. The lower the FWHM, the higher the crystalline quality [57]. Molten KOH is well known for its preferential etching on the SiC surface at defect sites [58]. Quality of the 4H-SiC epitaxial layers, used for detector fabrication in this work, has been assessed using defect-delineating chemical etching in molten KOH followed by XRD rocking curve measurements at the exposed defect sites. For the reflection geometry used in our studies, FWHM of the rocking curve can be calculated [58, 59] using the following equation:

$$\text{FWHM} = \frac{2r_e\lambda^2}{\pi V \sin 2\theta_B} \frac{1}{\sqrt{\gamma}} |C| \sqrt{F_{hkl} F_{\bar{h}\,\bar{k}\,\bar{l}}} \tag{6}$$

where r_e is the classical electron radius; λ is the X-ray wavelength; V is the volume of the unit cell; θ_B is the Bragg angle; $\gamma = \cos(\psi_h)/\cos(\psi_0)$ is the asymmetric ratio, where ψ_h and ψ_0 are the angles between the normal to the crystal surface directed inside the crystal and the reflected and incident directions of X-ray waves, respectively; C is the polarization factor ($C = 1$ for σ polarization and $C = \cos 2\theta_B$ for π polarization); and F_{hkl} is the structure factor with the modulus for (000l) reflection in 4H-SiC (back-reflection geometry) given by the equation below [59]:

$$|F_{000l}| = 4\sqrt{f_{Si}^2 + f_C^2 + 2f_{Si}f_C \cos(3\pi l/8)} \tag{7}$$

where f_C and f_{Si} are the atomic scattering factors of C and Si atoms, respectively. The scattering factors were calculated using the nine-parameter equation given below [59–61]:

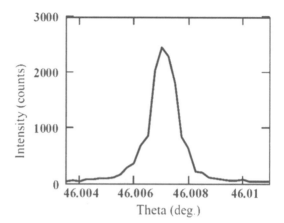

Fig. 5 Rocking curve (0008) reflection of the 50μm 4H-SiC epitaxial layer used for detector fabrication

$$f = c + \sum_{i=1}^{4} a_i \exp\left[-b_i \sin^2(\theta/\lambda)\right] \qquad (8)$$

where a_i, b_i, and c are the atom-specific Cromer-Mann coefficients, which can be found in [60–62]. The FWHM of the (0008) plane reflection was calculated using Eqs. (6–8) and was found to be less than 2.7 arc s. Figure 5 shows the experimentally obtained rocking curve for (0008) reflection on a 50 μm thick 4H-SiC epitaxial layer [13]. The FWHM of the rocking curve peak was found to be ~3.6 arc s, revealing high quality of our epitaxial layer.

2.4 Electron-Hole Pair Creation Energy and Minority Carrier Diffusion Length in 4H-SiC Epitaxial Layers

The experimental studies in the rest of the chapter rely mostly on radiation detection measurements. These measurements were carried out using a standard benchtop radiation detection setup comprising nuclear instrumentation modules. The spectrometer was connected to an EMI-shielded test box equipped with coaxial connectors. The test box was continuously evacuated using a vacuum pump during the alpha particle measurements. The spectrometer comprised of an Amptek CoolFET A250CF charge-sensitive preamplifier which integrates the detector current to produce a voltage output directly proportional to the amount of charge induced in the detector due to the incident ionizing radiation. The preamplifier output was filtered using an ORTEC 572 spectroscopy amplifier. The semi-Gaussian-shaped pulses were digitized using a Canberra Multiport II multichannel analyzer (MCA) which also produced histograms or pulse-height spectra (PHS). A 0.9μCi ^{241}Am radioisotope emitting primarily 5486, 5443, and 5388 keV alpha particles was used to obtain the PHS. The interactions of the alpha particles manifest as peaks in the PHS. To determine the energy resolution of the detection system, the peaks were fitted with Gaussian functions to obtain the width of the peaks (FWHM). The FWHM of the alpha peak expressed as percentage of the actual energy of the detected peak was defined as the percentage energy resolution of the detector at that radiation source energy.

Electron-hole pair (ehp) creation energy ε determines the energy resolution of a radiation detector as the energy resolution is directly linked to the number of *ehp*s created for a single incoming radiation event. The higher the number of *ehp*, the higher the resolution. A method of iterative determination of ε value which involves an absolute calibration using a precision pulser to match the alpha peak energy (5486 keV) observed using a high-resolution 4H-SiC n-type epitaxial Schottky detector has been reported in our earlier publication [51]. The alpha particle spectrometer was calibrated electronically by injecting a pulser signal of known pulse height, V_{pulser} (mV), from a precision pulser through a calibrated feed-through capacitor C_{test}, to the preamplifier input. The peak position of the shaped pulses was recorded in a multichannel analyzer (MCA) for a set of known pulse-height inputs. The SiC equivalent of the MCA peak positions, E_{pulser} in keV, was calculated using Eq. 9 given below:

$$E_{pulser} = \frac{V_{pulser} \times \varepsilon \times C_{test}}{q} \tag{9}$$

where q is the electronic charge. A linear regression of the SiC equivalent peak position as a function of MCA channel number was used to calculate the calibration parameters. The detector D20 was used for this study. The ε value we obtained using the given procedure was 7.28 eV. The value thus calculated differs from the widely accepted value of 7.7 eV as reported earlier [63]. Rogalla et al. calculated an ε value of 8.4 eV for alpha particles in semi-insulating 4H-SiC [64]. An ε value of 8.6 eV for alpha particles in epitaxial n-type 4H-SiC has been reported by Lebedev et al. [65]. Ivanov et al. [29] have reported $\varepsilon = 7.71$ eV for alpha particles in epitaxial n-type 4H-SiC. An ε value of 7.8 eV has been reported by Bertuccio and Casiraghi for 59.5 keV gamma rays [17].

2.5 Minority Carrier Diffusion Length Measurements

Minority carrier diffusion length is the average distance a minority carrier traverses before it recombines. Higher minority carrier diffusion length enhances the detection properties by reducing the effect of ballistic deficit [66]. The minority carrier diffusion length can be indirectly calculated by observing the variation of charge collection efficiency of detectors for ionizing radiations (like alpha particles) with reverse bias voltage. Charge collection efficiency (CCE) is defined as the ratio of charge collected by the collecting electrode at a particular bias to the maximum collected charge, assuming that all the charge carriers have been received by the

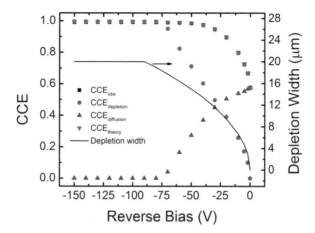

Fig. 6 Variation of CCE$_{obs}$ and CCE$_{theory}$ as a function of reverse bias voltage for SBD D20. The contributions to the total CCE from charge drifts in depletion region (CCE$_{depletion}$) and from hole diffusion in neutral region (CCE$_{diffusion}$) are also shown. The variation of depletion width with bias voltage is also shown

collecting electrode. The collected charge is generally calculated by integrating the current signal received at the input of a charge-sensitive preamplifier. Alternatively, the CCE can also be calculated from alpha spectroscopic measurements. The MCA peak position E_p due to a monoenergetic alpha source can be predicted in a properly calibrated alpha spectrometer, assuming that all the charge carriers created by that particular energy have been received by the collecting electrode. The CCE can then be determined for any MCA peak at E_α by calculating the ratio E_p/E_α. Figure 6 shows the variation of observed charge collection efficiency (CCE$_{obs}$) of a n-type Ni/4H-SiC SBD (D20) as a function of reverse operating bias voltage [14, 67]. By noting that the movement of the charge carriers in rectifying junctions can be due to diffusion as well as drifting, the contribution of each mechanism to CCE can be calculated using a proper model. The contribution of the drift CCE$_{depletion}$- and diffusion CCE$_{diffusion}$-related charge collection to the CCE$_{obs}$ has been calculated using a drift-diffusion model [68] summarized below in Eq. 10:

$$CCE_{theory} = \frac{1}{E_p} \int_0^d \left(\frac{dE}{dx}\right) dx + \frac{1}{E_p} \int_d^{x_r} \left[\left(\frac{dE}{dx}\right) \times \exp\left\{-\frac{(x-d)}{L_d}\right\}\right] dx$$
$$= CCE_{depletion} + CCE_{diffusion} \tag{10}$$

where d is the depletion width at a particular bias, $\frac{dE}{dx}$ is the electronic stopping power of the alpha particles calculated using SRIM [69], and x_r is the projected range of the alpha particles with energy E_p. We have fitted the CCE$_{theory}$ values to the CCE$_{obs}$ values considering L_d, the minority carrier diffusion length, as a free parameter. The L_d value corresponding to the best fit was returned as the average minority carrier diffusion length. For the present SBD, the average value of L_d was found to be ~18.6 μm. From Fig. 6 it was also observed that the CCE$_{diffusion}$ values dominate considerably over those of CCE$_{depletion}$ up to a reverse bias of −30 V. At even higher bias voltages, the depletion width becomes equal or more than the projected range of alpha particles (~18 μm in SiC for 5486 keV alpha particle) and hence charge collection was mainly due to the drifting of charge carriers within the depletion width. Hence, above a bias voltage of -70 V, the $CCE_{depletion}$ matched the CCE_{obs} values.

2.6 Noise Characterization of Detection Electronics for 4H-SiC Radiation Spectrometer

The energy resolution of nuclear spectrometers is dependent on the noise of the detector and associated electronic modules in the spectrometer, the preamplifier in particular. Noise is defined as any statistical fluctuation in currents measured in the detector or associated electronics which constitutes a signal. The most appropriate way to monitor the noise is to capture the pulses from a standard pulser along with the pulses produced by a detector due to the incoming radiation. The pulse-height spectrum then reveals a peak due to the pulser with the actual radiation peaks. The width (FWHM) of the pulser peak then gives the idea of the overall noise of the spectrometer ($FWHM_{total}$).

Figure 7 shows the results of a typical noise-monitoring measurement for D20 [14, 67]. The energy resolution of the detector could be seen to improve with increase in bias voltage up to 100 V reverse bias because of the increase in depletion width (active volume of the detector) and lowering in capacitance. The energy resolution beyond 100 V was seen to deteriorate with increase in bias. The increase in leakage current as a reason behind the deterioration of the resolution was ruled out as it could be seen from the figure that the pulser width barely changed. A possible reason behind the deterioration of the resolution could be the incorporation of the threading dislocation as the depletion width approaches towards the epilayer-substrate interface with the increase in reverse bias. The epilayer-substrate interfacial

Fig. 7 Variation of detector resolution (in keV and percentage) as a function of reverse bias for D20. The variation of the pulser peak width is also shown

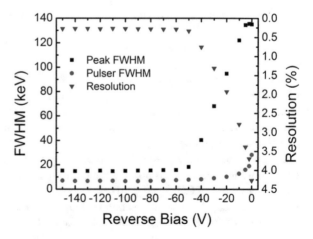

region is prone to have larger threading-type dislocation concentration, which propagates from the substrate to the epilayer.

For the detector D20, the $FWHM_{total}$ was found to be 19.8 keV for 5486 keV alpha particles. The contribution from the noise from the front-end electronics ($FWHM_{elec}$), and the detector leakage current ($FWHM_{leakage}$), can be found from the width of the MCA pulser peak recorded with the detector plugged in and biased. The collective broadening due to $FWHM_{elec}$ and $FWHM_{leakage}$ for this detector was found to be 15.9 keV. The other contributions to the $FWHM_{total}$ are from the statistical fluctuation in the number of *ehps* produced by an ionizing radiation $FWHM_{stat}$, and broadening due to variation of energy due to the entrance window, angle of incidence, self-absorption in the source, etc. ($FWHM_{other}$). All these factors along with the intrinsic detector resolution $FWHM_{det}$ are related to the ultimate peak broadening through the relation given in Eq. 11. $FWHM_{stat}$ and $FWHM_{other}$ were calculated in this detector and found to be 5.3 keV and 0.44 keV, respectively [51]. The intrinsic detector resolution was calculated from Eq. 11 and found to be 10.5 keV:

$$FWHM_{total}^2 = FWHM_{det}^2 + FWHM_{leakage}^2 + FWHM_{stat}^2 + FWHM_{elec}^2$$
$$+ FWHM_{other}^2 \tag{11}$$

The electronic noise has various sources such as detector leakage current and capacitance, and input FET (preamplifier) noise. The contribution of the different sources to the signal-to-noise ratio is dependent on the filtering or shaping operation and in particular the shaping time (except the FET noise). A clear understanding of the electronic noise is thus very essential to the proper tuning of the spectrometer. A pioneering work in this area has been conducted by Bertuccio and Pullia [70]. According to their formalism the electronic noise is expressed in terms of equivalent noise charge (*ENC*) and plotted as a function of the shaping time τ of the shaping amplifier. The plots were then fitted to Eq. 12 below using a least square estimation method:

$$ENC^2 = \left(aC_{tot}^2 A_1\right) \frac{1}{\tau} + \left[\left(2\pi a_f C_{tot}^2 + b_f/_{2\pi}\right)A_2\right] + (bA_3)\tau. \tag{12}$$

Here C_{tot} is the total input capacitance; A_1, A_2, and A_3 are constants depending on the shaping network response; a is the coefficient of white series noise contribution due to the thermal noise of the FET channel; a_f is the coefficient of the FET $1/f$ noise; b_f is the dielectric noise coefficient; and b is the coefficient related to the sum of the white parallel noise due to the shot noise and the detector leakage current.

Figure 8a, b shows the variation of *ENC* with shaping time without and with the detector D20 connected [67]. The contributions from the three different terms were plotted separately. The minimum noise without the detector was observed to correspond to a shaping time value between 1 and 2 μs. The same shifted to higher range of shaping time (between 3 and 6 μs) when the detector was plugged in. As a result, the best energy resolution for 5486 keV alpha particle was found to be at 3 μs from

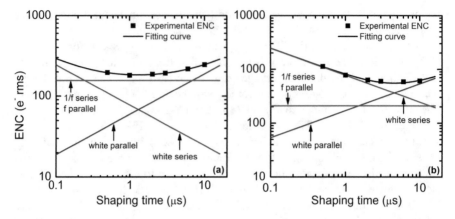

Fig. 8 Variation of equivalent noise charge as a function of shaping time without detector connected (**a**), and with the detector D20 connected (**b**). The separate contributions from white series noise, white parallel noise, and pink noise are also shown

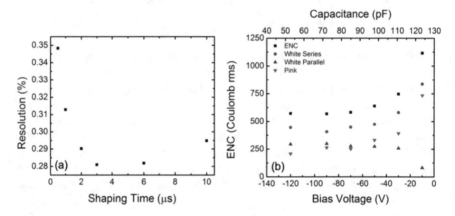

Fig. 9 (**a**) Variation of D20 resolution as a function of amplifier shaping time measured for 5486 keV alpha particles. The detector was reverse biased at −90 V. (**b**) Variation of *ENC*, white series noise, pink noise, and white parallel noise, measured using 3μs shaping time, as a function of different bias voltages/detector capacitances

the shaping-time dependence study of detector performance as shown in Fig. 9a. It can also be seen that the white parallel noise increased almost by a factor of 5 after connecting the detector and the pink noise marginally increased for any given τ after connecting the detector. In contrast, the white series noise increased by an order of magnitude when the detector was connected. The increase in white parallel noise can be attributed to the increase in the leakage current (from the detector). The increase in white series and parallel noise is supposedly due to the increase in input capacitance when the detector is plugged in. In order to study the effect of detector capacitance and leakage current on the electronic noise, a bias-dependent study of the electronic noise of the detection system was carried out. The *ENC*s have been measured at six different reverse bias voltages, viz. -10, -30, -50, -70, -90, and -120 V. Figure 9b shows the variation of the *ENC* and the separate contributions to the electronic noise as a function of applied reverse bias for a shaping time of 3 µs. The increase in reverse bias reduces the detector junction capacitance and simultaneously increases the leakage current too. From Fig. 9b it can be noticed that the contribution of the white series noise dominates towards the overall noise and decreases with the increasing reverse bias or decreasing capacitance which is consistent with Eq. 12. The pink noise follows a similar trend which again is in agreement with Eq. 12. The white parallel noise, which incorporates the detector leakage current, was seen to contribute the least at lower biases and increase steadily with reverse bias due to the increase in leakage current. It can also be noticed that beyond a reverse bias of -50 V, the contribution of the white parallel noise exceeded that of the pink noise.

2.7 Defect Characterization

The epilayers were characterized for defects using capacitance-mode deep-level transient spectroscopy (DLTS) [71]. The DLTS measurements were carried out using a SULA DDS-12 modular DLTS system. The DLTS system comprised of a pulse generator module for applying repetitive bias pulses, a 1 MHz oscillator for capacitance measurements, a sensitive capacitance meter involving self-balancing bridge circuit, and a correlator/preamplifier module which automatically removes DC background from capacitance meter and amplifies the resultant signal change. The correlators were based on a modified double-boxcar signal averaging system. The sample was mounted in a Janis VPF 800 LN$_2$ cryostat for temperature variation, which was controlled by a Lakeshore LS335 temperature controller. The DDS-12 system allows the user to collect four DLTS spectra simultaneously corresponding to four different rate windows in a single temperature scan. The signals were digitized using a National Instruments (NI) digitizer card integrated with the DLTS system for online processing using a personal computer (PC). The entire system including the modules and the temperature controller is controlled using a dedicated LabVIEW interface, which also allows the user to analyze the recorded data.

A steady-state reverse bias voltage of -2 V was used, and the samples were pulsed to 0 V with a pulse width of 1 ms followed by the capacitance transient measurements. A temperature scan in the range of 80–800 K was used to obtain the DLTS spectra. We have used at least four rate windows for the calculation of the activation energies.

The emission rates are given by

$$C(t) = C_0 + \Delta C \exp(-te_n), \tag{16}$$

where C_0 is the steady-state capacitance value, ΔC is the difference in capacitance values between the beginning and the end of the filling pulse, and e_n is the electron emission rate. The emission rates corresponding to the maximum of a trap peak in the DLTS spectra obtained with various rate windows were used along with the temperatures (T) corresponding to the peak maxima to obtain a semilog plot of T^2/e_n vs. $1000/T$ Arrhenius plots. The slope of the linear fit to the Arrhenius plots gives the activation energy of the particular trap. The capture cross section σ_n was calculated from the relation

$$e_n = (\sigma_n \langle v_{th} \rangle N_c / g_1) \exp(-\Delta E/kT), \tag{17}$$

where $\langle v_{th} \rangle$ is the mean electron thermal velocity, N_c is the effective density of states in the conduction band, g_1 is the degeneracy of the trap level which is assumed to be unity, and ΔE is the energy separation between the trap level and the conduction band or the activation energy. The trap concentration N_t was calculated using the following expression:

$$N_t = 2\left(\frac{\Delta C}{C_0}\right)(N_{\text{eff}}). \tag{18}$$

Figure 10 shows the DLTS spectra obtained for the D20, D50, and D150 and the corresponding Arrhenius plots. As can be seen from the plots, fully or partially resolved DLTS peaks were found at various temperatures indicating the presence of multiple types of defects. The negative peaks suggest that the associated defects are majority carrier traps (electron traps in this particular case). Table 2 enlists all the defect parameters extracted from the DLTS spectra for the three detectors.

All the detectors exhibited three DLTS peaks in the entire temperature scan range. The detectors D50 and D150 exhibited similar features where peaks #1, #2, and #3 correspond to Ti(c), $Z_{1/2}$, and $EH_{6/7}$ defect centers, respectively. Trap center Ti (c) has been identified as titanium-related impurity-type defect which has been assigned to ionized titanium acceptor Ti^{3+} residing at cubic Si lattice sites [72–75]. The defect center $Z_{1/2}$ can be related to defect complexes involving equal number of carbon and silicon sites such as silicon and carbon vacancy complexes ($V_{Si} + V_C$) and antisite complex ($Si_C + C_{Si}$) pairs [74]. $Z_{1/2}$ center has also been established as a carrier lifetime-deteriorating defect [37, 73, 76, 77], which is known to play a crucial role in defining the energy resolution of the detectors. Defect center $EH_{6/7}$, situated at around 1.59 eV below the conduction band, is usually identified as carbon vacancies or carbon-silicon di-vacancies [78]. The detector D20 showed the $Z_{1/2}$ and $EH_{6/7}$ defect centers as well. It did not show the presence of Ti(c) center; however, it showed the presence of a peak (peak #2) situated 1.45 eV below the

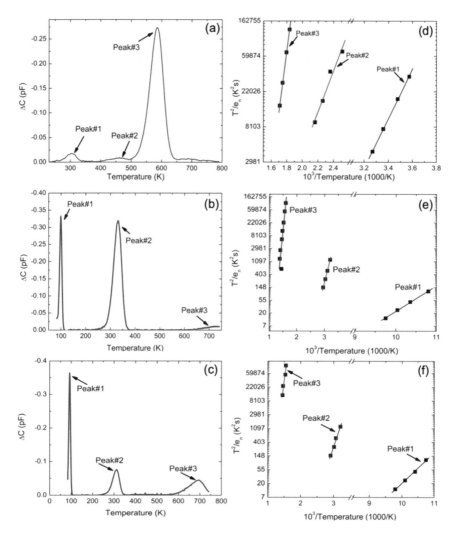

Fig. 10 Capacitance-mode DLTS spectra (one correlator is shown for clarity) for detectors (**a**) D20, (**b**) D50, and (**c**) D150. The corresponding Arrhenius plots obtained using at least four correlator delay settings for (**d**) D20, (**e**) D50, and (**f**) D150

conduction band edge which was not present in the detector D50 or D150. It could be seen from Table 2 that the capture cross section and the concentration of the $Z_{1/2}$ defect center in the detector D20 were way lower than those in the detector D50 or D150. In the next section it has been shown that the energy resolution of the detector D20 was the best among all the detectors presented in this chapter.

Table 2 Defect parameters obtained from the DLTS scans of the SBDsD20, D50, and D150

Detector ID	Peak #1				Peak #2				Peak #3			
	Trap	$\sigma_n \times 10^{-14}$ cm^2	ΔE eV	$N_t \times 10^{12}$ cm^{-3}	Trap	$\sigma_n \times 10^{-13}$ cm^2	ΔE eV	$N_t \times 10^{12}$ cm^{-3}	Trap	$\sigma_n \times 10^{-13}$ cm^2	ΔE eV	$N_t \times 10^{12}$ cm^{-3}
D20	$Z_{1/2}$	5.0×10^{-5}	0.60	0.09	Ci1	0.5	1.45	1.27	EH$_{6/7}$	0.07	1.60	0.1
D50	Ti(c)	5.61	0.17	2.03	$Z_{1/2}$	1.28	0.73	2.27	EH$_{6/7}$	1.88	1.59	1.76
D150	Ti(c)	0.62	0.21	3.36	$Z_{1/2}$	0.02	0.61	0.70	EH$_{6/7}$	3.1×10^{-4}	1.23	0.7

2.8 Radiation Detection Using 4H-SiC Epitaxial Detectors

Figure 11 shows the alpha pulse-height spectraobtained by exposing the detectors to a ^{241}Am alpha particle source. The energy resolution of the detectors at 5486 keV alpha particles is listed in Table 1. It could be seen that the best detector performance resolution-wise was obtained from the detector D20 most probably because of the lowest defect concentration corresponding to the $Z_{1/2}$ defects.

The epilayers were also tested for X-ray and gamma-ray responses. Absolute measurement of photo-responsivity and probing of physical construction of photonic sensors can be very effectively done using synchrotron light sources. The detector D50 was studied at NSLS at BNL for detection of low-energy X-rays. The results were compared to a commercial off-the-shelf (COTS) SiC UV photodiode by IFW, model JEC4, which was known to be the best commercially available for such applications [13]. X-ray spectrometer for such a low-energy spectral range is not available commercially. Figure 12a shows the responsivity of one of our detectors and a IFW JEC4 SiC UV photodiode to soft X-ray energy ranges biased at 250 V and 120 V, respectively [13]. The following results were derived using a statistical analysis of these data based on energy-dependent X-ray attenuation lengths [79].

Responsivity measurements were carried out using the U3C [80] and X8A [47] lines by recording successive measures of photocurrent in response to a high-flux, monoenergetic beam of photons in a well-calibrated silicon sensor (with known responsivity) and in the sensor of interest (Fig. 12a). Dead layers and a limited active volume thickness led to responsivity that varies greatly with photon energy. Further, edges were also apparent in the responsivity curve, arising from discrete atomic transitions. Edges associated with silicon and carbon are clearly observed in the data, providing a quantitative measure of the composition and dimension of the active and dead layers [79]. The general feature of a steep decrease starting at 2–3 keV provides information on active layer thicknesses, which is deduced to be 21 μm in our detector compared to roughly 6 μm for the JEC4 diode [47]. Because of the higher active

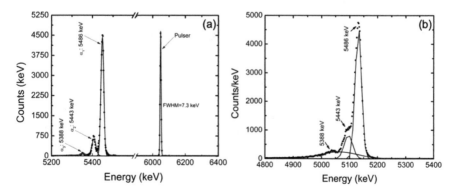

Fig. 11 Pulse-height spectra obtained for the detectors: D20 (**a**) and D150 (**b**). A pulser spectrum was also acquired along with the alpha spectrum to monitor the electronic noise as shown in part (**a**). The detector D50 exhibited a PHS similar to that of D150

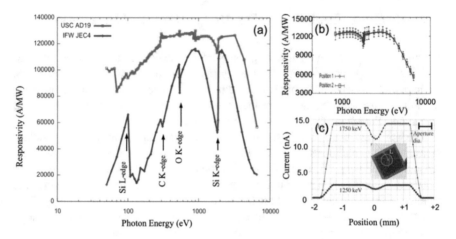

Fig. 12 (**a**) X-ray responsivity measured on a 50μm 4H-SiC epilayer biased at 250 V and an IFW JEC4 photodiode biased at 120 V. (**b**) Responsivity of the detectors on same 4H-SiC n-type epitaxial layer at two different locations and (**c**) surface scan profiles along line L obtained to assess detector's uniformity

layer thickness D50 showed significant improvement of responsivity in the few keV range compared to COTS SiC UV photodiode. Our detector has shown much higher response in the low-energy part of the spectra as well, which could be attributed to a much thinner dead/blocking layer, deduced from the responsivity curve to result solely from the 10 nm thick nickel layer (which leads to the pronounced edge at 70 eV). In comparison, the JEC4 diode has been found to include a significant oxidation and inactive SiC layer on the order of 100 nm each, which limits responsivity at low photon energies [47]. It should be noted that the JEC4 diode is intended for UV detection, for which it is well suited. The significant dead layers are likely due to passivation, which may be preferred over reducing the thickness of dead layers on the active face of the sensor.

Our detectors have also exhibited very good spatial uniformity in measured responsivity. Figure 12b shows responsivity at two different locations and line scan profiles for two different X-ray energies. Note that the decrease of the detector's current at about 0 mm (Fig. 12b) is due to the crossing of the location of wire bonding and not due to detector's imperfection. The SiC detectors were connected to low-noise front-end electronics developed in-house for pulse-height spectroscopy. Pulse-height spectra measurements showed high resolution of our 4H-SiC detectors in detecting 59.6 keV gamma rays from [241]Am. Figure 13 shows a pulse-height spectrum recorded using an [241]Am radiation source with the detector biased at 250 V, with an FWHM of 1.2 keV (2.1%) at 59.5 keV, which is comparable to the resolution achieved using high-quality CdZnTe detectors [81–84].

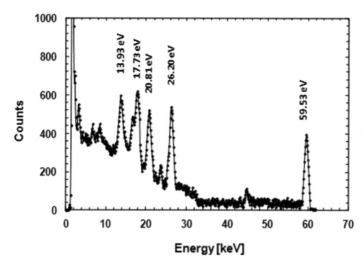

Fig. 13 Pulse-height spectrum obtained from a D50 4H-SiC detector at 300 K, 250 V bias, and 4μs shaping time when exposed to a [241]Am source

Acknowledgments The authors acknowledge financial support provided by the DOE Office of Nuclear Energy's Nuclear Energy University Programs (NEUP), Grant No. DE-NE0008662, and by Los Alamos National Laboratory/DOE (Grant No. 143479). The work was also partially supported by the Advanced Support Program for Innovative Research Excellence-I (ASPIRE-I) of the University of South Carolina (UofSC), Columbia, Grant No. 15530-E404.

References

1. Egarievwe, S.U., Chen, K.-T., Burger, A., James, R.B., Lisse, C.M.: Detection and electrical properties of CdZnTe at elevated temperatures. J. Xray Sci. Technol. **6**(4), 309–315 (1996)
2. Burger, A., Groza, M., Cui, Y., Roy, U.N., Hillman, D., Guo, M., Li, L., Wright, G.W., James, R.B.: Development of portable CdZnTe spectrometers for remote sensing of signatures from nuclear materials. Phys. Status Solidi C. **2**(5), 1586–1591 (2005)
3. Neudeck, P.G., Chen, L.Y., Meredith, R.D., Lukco, D., Spry, D.J., Nakley, M.L., Hunter, G.W.: Operational testing of 4H-SiC JFET ICs for 60 days directly exposed to Venus surface atmospheric conditions. J. Electron Devices Soc. **7**, 100–110 (2019)
4. She, X., Huang, A.Q., Lucia, Ó., Ozpineci, B.: Review of silicon carbide power devices and their applications. IEEE Trans. Ind. Electron. **64**(10), 8193–8205 (2017)
5. Kinoshita, T., Itoh, K.M., Muto, J., Schadt, M., Pensl, G., Takeda, K.: Calculation of the anisotropy of the hall mobility in n-type 4H- and 6H-SiC. Mater. Sci. Forum. **294–268**, 295–298 (1998)
6. Choyke, W.J., Matsunami, H., Pensl, G.: Silicon Carbide—A Review of Fundamental Questions and Applications to Current Device Technology. Springer Verlag, Berlin (2004)
7. Harris, G.L.: Properties of SiC, vol. 13. The Institute of Electrical Engineers, London (1995)
8. Ellis, B., Moss, T.: The conduction bands in 6H and 15R silicon carbide I. Hall effect and infrared Faraday rotation measurements. Proc. R. Soc. London, Ser. A. **299**, 383–392 (1967)

9. Lomakina, G.A.: Silicon carbide - 1973. In: Proceedings of 3rd International Conference on Silicon Carbide, Miami, Florida, 1973

10. Lucas, G., Pizzagalli, L.: Ab initio molecular dynamics calculations of threshold displacement energies in silicon carbide. Phys. Rev. B. **72**(16), 161202 (2005)

11. Nava, F., Bertuccio, G., Cavallini, A., Vittone, E.: Silicon carbide and its use as a radiation detector material. Meas. Sci. Technol. **19**, 102001 (2008)

12. Mandal, K.C., Krishna, R.M., Muzykov, P.G., Das, S., Sudarshan, T.S.: Characterization of semi-insulating 4H silicon carbide for radiation detectors. IEEE Trans. Nucl. Sci. **58**(4), 1992–1999 (2011)

13. Mandal, K.C., Muzykov, P.G., Chaudhuri, S.K., Terry, J.R.: Low energy X-ray and γ-ray detectors fabricated on n-type 4H-SiC epitaxial layer. IEEE Trans. Nucl. Sci. **60**(4), 2888–2893 (2013)

14. Chaudhuri, S.K., Zavalla, K.J., Mandal, K.C.: High resolution alpha particle detection using 4H SiC epitaxial layers: fabrication, characterization, and noise analysis. Nucl. Instrum. Methods Phys. Res. A. **728**, 97–101 (2013)

15. Ruddy, F.H., Seidel, J.G., Chen, H., Dulloo, A.R., Ryu, S.: High-resolution alpha-particle spectrometry using 4H silicon carbide semiconductor detectors. IEEE Trans. Nucl. Sci. **53**, 1713 (2006)

16. Ruddy, F.H., Dulloo, A.R., Seidel, J.G., Das, M.K., Ryu, S., Agarwal, A.K.: The fast neutron response of 4H silicon carbide semiconductor radiation detectors. IEEE Trans. Nucl. Sci. **53**, 1666–1670 (2006)

17. Bertuccio, G., Casiraghi, G.: Study of silicon carbide for X-ray detection and spectroscopy. IEEE Trans. Nucl. Sci. **50**, 175–185 (2003)

18. Mandal, K.C., Kleppinger, J.W., Chaudhuri, S.K.: Advances in high-resolution radiation detection using 4H-SiC epitaxial layer devices. Micromachines. **11**(3), 254 (2020)

19. Schifano, R., Vinattieri, A., Bruzzi, M., Miglio, S., Logomarsino, S., Sciortino, S., Nava, F.: Electrical and optical characterization of 4H-SiC diodes for particle detection. J. Appl. Phys. **97**, 103539 (2005)

20. Puglisi, D., Bertuccio, G.: Silicon carbide microstrip radiation detectors. Micromachines. **10**(12), 835 (2019)

21. Nguyen, K.V., Mannan, M.A., Mandal, K.C.: Improved n-type 4H-SiC epitaxial radiation detectors by edge termination. IEEE Trans. Nucl. Sci. **62**(6), 3199–3206 (2015)

22. Bertuccio, G., Casiraghi, R., Cetronio, A., Lanzieri, C., Nava, F.: Silicon carbide for high resolution X-ray detectors operating up to 100°C. Nucl. Instrum. Methods Phys. Res. Sect. A. **522**(3), 413–419 (2004)

23. Neudeck, P.G., Spry, D.J., Krasowski, M.J., Prokop, N.F., Chen, L.Y.: Demonstration of 4H-SiC JFET digital ICs across 1000°C temperature range without change to input voltages. Mater. Sci. Forum. **963**, 813–817 (2019)

24. Wu, J., Jiang, Y., Li, M., Zeng, L., Gao, H., Zou, D., Bai, Z., Ye, C., Liang, W., Dai, S., Lu, Y., Rong, R., Du, J., Fan, X.: Development of high sensitivity 4H–SiC detectors for fission neutron pulse shape measurements. Rev. Sci. Instrum. **88**, 083301 (2017)

25. Torrisi, L., Foti, G., Guiffrid, L., Puglisi, D., Wolowski, J., Badziak, J., Parys, P., Rosinski, M., Margarone, D., Krasa, J., Velyhan, A., Ullschmied, U.: Single crystal silicon carbide detector of emitted ions and soft X-rays from power laser-generated plasmas. J. Appl. Phys. **105**, 123304 (2009)

26. Bertuccio, G., Puglisi, D., Torrisi, L., Lanzieri, C.: Silicon carbide detector for laser-generated plasma radiation. Appl. Surf. Sci. **272**, 128–131 (2013)

27. Nava, F., Vanni, P., Bruzzi, M., Lagomarsino, S., Sciortino, S., Wagner, G., Lanzieri, C.: Minimum ionizing and alpha particles detectors based on epitaxial semiconductor silicon carbide. IEEE Trans. Nucl. Sci. **51**(1), 238–244 (2004)

28. Seshadri, S., Dulloo, A.R., Ruddy, F.H., Seidel, J.G., Rowland, L.B.: Demonstration of a SiC neutron detector for high-radiation environments. IEEE Trans. Electron Dev. **46**(3), 567–571 (1999)

29. Ivanov, A., Kalinina, E., Kholuyanov, G., Strokan, N., Onushkin, G., Konstantinov, A., Hallen, A., Kuznetsov, A.: High energy resolution detectors based on 4H-SiC. In: Proc. 5th Euro. Conf. Silicon Carbide and Related Materials, Zurich, Switzerland (2005)
30. Bertuccio, G., Casiraghi, R., Nava, F.: Epitaxial silicon carbide for X-ray detection. IEEE Trans. Nucl. Sci. **48**(2), 232–233 (2001)
31. Bertuccio, G., Puglisi, D., Pullia, A., Lanzinieri, C.: X-γ ray spectroscopy with semi-insulating 4H-silicon carbide. IEEE Trans. Nucl. Sci. **60**(2), 1436–1441 (2013)
32. Huang, X.R., Dudley, M., Vetter, W.M., Huang, W., Si, W., Carter, J.C.H.: Superscrew dislocation contrast on synchrotron white-beam topographs: an accurate description of the direct dislocation image. J. Appl. Crystallogr. **32**, 516–524 (1999)
33. Stein, R.A., Lanig, P., Leibenzeder, S.: Influence of surface energy on the growth of 6H- and 4H-SiC polytypes by sublimation. Mat. Sci. Eng. B. **11**, 69–71 (1992)
34. Sumakeris, J.J., Jenny, J.R., Powell, A.R.: Bulk crystal growth, epitaxy, and defect reduction in silicon carbide materials for microwave and power devices. MRS Bull. **30**, 280–286 (2005)
35. Neudeck, P.G.: Electrical impact of SiC structural crystal defects on high electric field devices. Mater. Sci. Forum. **338-342**, 1161–1166 (1999)
36. Dahlquist, F., Johansson, N., Soderholm, R., Nillson, P.A., Bergman, J.P.: Long term operation of 4.5kV PiN and 2.5kV JBS diodes. Mater. Sci. Forum. **353–356**, 727–730 (2001)
37. Kimoto, T., Danno, K., Suda, J.: Lifetime-killing defects in 4H-SiC epilayers and lifetime control by low-energy electron irradiation. Phys. Status Solidi B. **245**, 1327–1336 (2008)
38. Mandal, K.C., Chaudhuri, S.K., Nguyen, K.V., Mannan, M.A.: Correlation of deep levels with detector performance in 4H-SiC epitaxial Schottky barrier alpha detectors. IEEE Trans. Nucl. Sci. **61**(4), 2338–2344 (2014)
39. Mannan, M.A., Chaudhuri, S.K., Nguyen, K.V., Mandal, K.C.: Effect of $Z_{1/2}$, EH_5, and Ci1 deep defects on the performance of n-type 4H-SiC epitaxial layers Schottky detectors: alpha spectroscopy and deep level transient spectroscopy studies. J. Appl. Phys. **115**, 224504 (2014)
40. Li, J., Meng, C., Yu, L., Li, Y., Yan, F., Han, P., Ji, X.: Effect of various defects on 4H-SiC Schottky diode performance and its relation to epitaxial growth conditions. Micromachines. **11**(6), 609 (2020)
41. Babcock, R.V., Chang, H.C.: SiC neutron detectors for high-temperature operation, neutron dosimetry. In: Proceedings of the Symposium on Neutron Detection, Dosimetry and Standardization, Vienna, December 1962, vol. 1, pp. 613–622. International Atomic Energy Agency (IAEA), Vienna (1962)
42. Ruddy, F.H., Dulloo, A.R., Seidel, J.G., Seshadri, S., Rowland, L.B.: Development of a silicon carbide radiation detector. IEEE Trans. Nucl. Sci. **45**, 536–541 (1998)
43. Ruddy, F.H., Dulloo, A.R., Seidel, J.G., Palmour, J.W., Singh, R.: The charged particle response of silicon carbide semiconductor radiation detectors. Nucl. Instrum. Methods Phys. Res. Sect. A. **505**, 159–162 (2003)
44. Ruddy, F.H., Seidel, J.G., Sellin, P. High-resolution alpha spectrometry with a thin-window silicon carbide semiconductor detector. In: Proc. IEEE Nucl. Sci. Symp. Conf. Record (NSS/MIC) (2009)
45. Chaudhuri, S.K., Krishna, R.M., Zavalla, K.J., Mandal, K.C.: Schottky barrier detectors on 4H-SiC n-type epitaxial layer for alpha particles. Nucl. Instrum. Methods Phys. Res. Sect. A. **701**, 214–220 (2013)
46. Bertuccio, G., Caccia, S., Puglisi, D., Macera, D.: Advances in silicon carbide X-ray detectors. Nucl. Instrum. Methods Phys. Res. Sect. A. **652**(1), 193–196 (2011)
47. Terry, J.R., Distel, J.R., Kippen, R.M., Schirato, R., Wallace, M.S.: Evaluation of COTS silicon carbide photodiodes for a radiation-hard, low-energy X-ray spectrometer. In: Proc. IEEE Nucler Science Symp., Valencia, 2011, vol. NP1.M-236, pp. 485–488 (2011)
48. Mandal, K.C., Muzykov, P.G., Terry, J.R.: Highly sensitive X-ray detectors in the low-energy range on n-type 4H-SiC epitaxial layers. Appl. Phys. Lett. **101**, 051111 (2012)
49. Chen, W., Capano, M.A.: Growth and characterization of 4H-SiC epilayers on substrates with different off-cut angles. J. Appl. Phys. **98**, 114907 (2005)

50. Kern, W.: The evolution of silicon wafer cleaning technology. J. Electrochem. Soc. **137**, 1887–1892 (1990)
51. Chaudhuri, S.K., Zavalla, K.J., Mandal, K.C.: Experimental determination of electron-hole pair creation energy in 4H-SiC epitaxial layer: an absolute calibration approach. Appl. Phys. Lett. **102**, 031109 (2013)
52. Bethe, H.A.: Theory of the boundary layer of crystal rectifiers. MIT Radiation Laboratory Report, pp. 43–12 (1942)
53. Rhoderick, E.: Metal-semiconductor contacts. IEEE Proceeding. **129**(1), 1–14 (1982)
54. Goldberg, Y., Levinshtein, M.E., Rumyantsev, S.L.: In: Levinshtein, M.E., Rumyantsev, S.L., Shur, M.S. (eds.) Properties of Advanced Semiconductor Materials GaN, AlN, SiC, BN, SiC, SiGe, p. 99. John Wiley & Sons, New York (2001)
55. Tung, R.T.: Electron transport at metal-semiconductor interfaces: general theory. Phys. Rev. B. **45**, 13509–13523 (1992)
56. Jang, M., Kim, Y., Shin, J., Lee, S.: Characterization of erbium-silicided Schottky diode junction. IEEE Electron Device Lett. **26**(6), 354–356 (2005)
57. Chen, Y.: Defects structures in silicon carbide bulk crystals, epilayers and devices. Dissertation, State University of New York, Stony Brook, New York (2008)
58. Mandal, K.C., Muzykov, P.G., Chaudhuri, S.K., Terry, J.R.: Assessment of 4H-SiC epitaxial layers and high resistivity bulk crystals for radiation detectors. Proc. SPIE. **8507**, 85070C (2012)
59. Authier, A.: Dynamical theory of X-ray diffraction. Int. Tables for Crystallogr. **B**, 534–551 (2006). Chapter 5.1.
60. MacGillavry, C.H., Rieck, G.D. (eds.): International Tables for X-Ray Crystallography. Kynoch Press, Birmingham (1968)
61. Scattering Factors. http://www.ruppweb.org/new_comp/scattering_factors.htm
62. Creagh, D.: Tables of X-ray absorption corrections and dispersion corrections: the new versus the old. Nucl. Instrum. Methods. Phys. Res. Sect. A. **295**, 417–434 (2013)
63. Lo Giudice, A., Fizzotti, F., Manfredotti, C., Vittone, E., Nava, F.: Average energy dissipated by mega-electron-volt hydrogen and helium ions per electron-hole pair generation in 4H-SiC. Appl. Phys. Lett. **87**, 222105 (2005)
64. Rogalla, M., Runge, K., Soldner-Rembold, A.: Particle detectors based on semi-insulating silicon carbide. Nucl. Phys. B Proc. Suppl. **78**(1–3), 516–520 (1999)
65. Labedev, A.A., Ivanov, A.M., Strokan, N.B.: Radiation hardness of SiC and hard radiation detectors based on the SiC films. Fiz.Tekh. Poluprovodn. **38**, 129–150 (2004)
66. Knoll, G.F.: Radiation Detection and Measurements, 3rd edn. Wiley, New York (2000)
67. Mandal, K.C., Chaudhuri, S.K., Nguyen, K.V.: An overview of application of 4H-SiC n-type epitaxial Schottky barrier detector for high resolution nuclear detection. 2013 IEEE Nucl. Sci. Symp. Medical Imaging Conf. (2013 NSS/MIC) (2013)
68. Breese, M.B.H.: A theory of ion beam induced charge collection. J. Appl. Phys. **74**, 3789–3799 (1993)
69. Ziegler, J.F., Biersack, J.P., Littmark, U.: The Stopping and Range of Ions in Solids. Pergamon, Oxford (1985)
70. Bertuccio, G., Pullia, A.: A method for the determination of the noise parameters in pre-amplifying systems for semiconductor radiation detectors. Rev. Sci. Instrum. **64**, 3294–3298 (1993)
71. Lang, D.V.: Deep-level transient spectroscopy: a new method to characterize traps in semiconductors. J. Appl. Phys. **45**, 3023–3032 (1974)
72. Dalibor, T., Pensl, G., Nordell, N., Schoener, A.: Electrical properties of the titanium acceptor in silicon carbide. Phys. Rev. B. **55**(20), 13618–13624 (1997)
73. Szmidt, Ł., Gelczuk, J., Dąbrowska-Szata, M., Sochacki, M.: Characterization of deep electron traps in 4H-SiC junction barrier Schottky rectifiers. Solid State Electron. **94**, 56–60 (2014)
74. Zhang, J., Storasta, L., Bergman, J.P., Son, N.T., Janzén, E.: Electrically active defects in n-type 4H–silicon carbide grown in a vertical hot-wall reactor. J. Appl. Phys. **93**(8), 4708–4714 (2003)

75. Castaldini, A., Cavallini, A., Polenta, L., Nava, F., Canali, C., Lanzieri, C.: Deep levels in silicon carbide Schottky diodes. Appl. Surf. Sci. **187**(3–4), 248–252 (2002)
76. Tawara, T., Tsuchida, H., Izumi, S., Kamata, I., Izumi, K.: Evaluation of free carrier lifetime and deep levels of the thick 4H-SiC epilayers. Mat. Sci. Forum. **457–460**, 565–568 (2004)
77. Klein, P.B., Shanabrook, B.V., Huh, S.W., Polyakov, A.Y., Skowronski, M., Sumakeris, J.J., O'Loughlin, M.J.: Lifetime-limiting defects in n−4H-SiC epilayers. Appl. Phys. Lett. **88**, 052110 (2006)
78. Danno, K., Kimoto, T., Matsunami, H.: Midgap levels in both n- and p-type 4H-SiC epilayers investigated by deep level transient spectroscopy. Appl. Phys. Lett. **86**, 122104 (2005)
79. Bartlett, R.J., Trela, W.J., Michaud, F.D., Southworth, S.H., Alkire, R.W., Roy, P., Rothe, R.P., Walsh, J., Shinn, N.: Characteristics and performance of the Los Alamos VUV beam line at the NSLS. Nucl. Instrum. Methods. Phys. Res. Sect. A. **266**, 199–204 (1988)
80. Day, R.H., Blake, R.L., Stradling, G.L., Trela, W.J., Bartlett, R.J.: Los Alamos X-ray characterization facilities for plasma diagnostics. Proc. SPIE. **689**, 208–217 (1986)
81. Wilson, M.D., Cernik, R., Chen, H., Hansson, C., Iniewski, K., Jones, L.L., Seller, P., Veale, M. C.: Small pixel CZT detector for hard X-ray spectroscopy. Nucl. Instrum. Methods Phys. Res. Sect. A. **652**, 158–161 (2011)
82. Chaudhuri, S.K., Krishna, R.M., Zavalla, K.J., Matei, L., Buliga, V., Groza, M., Burger, A., Mandal, K.C.: $Cd_{0.9}Zn_{0.1}Te$ crystal growth and fabrication of large volume single-polarity charge sensing gamma detectors. IEEE Trans. Nucl. Sci. **60**(4), 2853–2858 (2013)
83. Sajjad, M., Chaudhuri, S.K., Kleppinger, J.W., Mandal, K.C.: Growth of large-area $Cd_{0.9}Zn_{0.1}Te$ single crystals and fabrication of pixelated guard-ring detector for room-temperature γ-ray detection. IEEE Trans. Nucl. Sci. **67**(8), 1946–1951 (2020)
84. Krishna, R.M., Chaudhuri, S.K., Zavalla, K.J., Mandal, K.C.: Characterization of $Cd_{0.9}Zn_{0.1}Te$ based virtual Frisch grid detectors for high energy gamma ray detection. Nucl. Instrum. Methods Phys. Res. Sect. A. **701**, 208–213 (2013)

Room-Temperature Radiation Detectors Based on Large-Volume CdZnTe Single Crystals

Sandeep K. Chaudhuri and Krishna C. Mandal

Abstract Owing to their highly penetrating nature, detection of gamma rays has always remained a challenge. In spite of availability of high-resolution semiconductor detectors like silicon and germanium, their applicability in the field of medical diagnosis or homeland security is difficult because of the requirement of bulky cryogenic attachments. The requirement of energy resolution of 0.5% or less for 662 keV gamma rays at room temperature set by the US Department of Energy has urged the development of a variety of alternative materials, which are compact and operable at room temperature. Among all such materials, CdZnTe (CZT) has stood out as the most promising as it is a wide-bandgap semiconductor that can be operated at room temperature as high-resolution detector. Large-volume high-Z detectors are essential for efficient gamma-ray absorption and detection. Fabrication of large-volume radiation detectors operable at room temperature is a challenging task due to insufficient yield of detector-grade crystals from crystal growth operation. This chapter discusses the aspects of crystal growth to obtain large-volume $Cd_{0.9}Zn_{0.1}Te$ (CZT) single crystals and fabrication and characterization of large-volume detectors with small pixel, Frisch collar, and coplanar grid contact geometries.

1 Introduction

Radioactivity is one of those natural events known to human beings, which in spite of being present abundantly in nature is impossible to perceive with our natural senses. The presence of radioactivity was discovered accidentally by the famous works of Roentgen along with the first documented X-ray detectors in the form of fluorescence plates [1]. X-rays and their higher energy-relative γ-rays, identified as similar type of electromagnetic radiation with different mechanisms of emission, are probably the most invisible types of radiation owing to the fact that photons are electrically neutral, and their rest mass is zero. Fortunately, X-rays and γ-rays fall in

S. K. Chaudhuri · K. C. Mandal (✉)
Department of Electrical Engineering, University of South Carolina, Columbia, SC, USA
e-mail: chaudhsk@mailbox.sc.edu; mandalk@cec.sc.edu

© The Author(s), under exclusive license to Springer Nature Switzerland AG 2022
K. Iniewski (ed.), *Advanced Materials for Radiation Detection*,
https://doi.org/10.1007/978-3-030-76461-6_10

211

the class of radiation known as ionizing radiation which enabled discovery of techniques to fabricate appropriate detectors. As the name suggests, ionizing radiation produces ions or charged particles when they interact with matter. These charge carriers can be made to move efficiently in materials like single-crystalline semiconductors, to generate electronic signals.

Semiconductor detectors offer high resolution, operate at relatively lower voltages compared to scintillator detectors, and are compact enough to design handheld or portable devices. High-purity germanium (HPGe) and lithium-drifted silicon Si (Li) detectors offer unparallel energy resolution even for very-high-energy gamma rays but these detectors cannot be used at room temperature due to their narrow bandgaps. They are generally attached with bulky cryogenics, like liquid nitrogen dewars, for their successful operation. Detectors with high-atomic-number (high-Z) constituents and wide bandgap such as CdZnTe (CZT), HgI_2, TlBr, and PbI_2 are promising alternatives for room-temperature detection with a substantially low form factor.

This chapter addresses the applicability of large-volume CZT detectors with the stoichiometric formula $Cd_{0.9}Zn_{0.1}Te$. The chapter particularly focuses on the growth of large-volume detector-grade CZT crystals and fabrication of X/γ-ray detectors in single-polarity configuration which essentially bypasses the effect of poor hole transport. Digital correction techniques to further improve the quality of the spectral response have also been discussed.

2   Compact Room-Temperature *X/γ-Ray* Detector

2.1   *Requirement for Room-Temperature X/γ-Ray Detectors*

The requirement of energy resolution of 0.5% or less for 662 keV gamma rays as set by the US Department of Energy (DOE) and other practical requirements of field deployability have urged the development of a large number of alternative materials which are compact and operable at room temperature. The intrinsic energy resolution of a semiconductor detector primarily depends on three factors, viz. the Fano noise, electronic noise, and noise due to incomplete charge collection. The Fano noise is related to the number of electron-hole pairs (*ehp*) created due to the interaction of the incident radiation and the detector material. The more the *ehp*, the lesser the Fano noise and the higher the intrinsic energy resolution [2]. The *ehp* primarily depends on the bandgap of the crystal structure of the semiconductor. Semiconductors with narrow bandgap will result in the creation of more *ehp*. But that does not ensure a higher energy resolution as narrow bandgap leads to higher leakage currents as well. Wide bandgap ensures lower leakage currents at room temperature and hence enables fabrication of detectors, which can be operated at room temperatures and above. Hence, one prerequisite for well-resolved detection is a detector composed of semiconductor material with wide bandgap.

The detector material must have a high material density as it helps attenuate X-rays and gamma rays more efficiently. Apart from high material density, the elements comprising the semiconductor should also have high atomic numbers (high Z). X-rays and gamma rays interact with material in three primary ways depending on their energy: (1) photoelectric absorption (up to ~200 keV), (2) Compton scattering (up to few MeV), and (3) pair production (above ~6 MeV). Photoelectric absorption leads to full energy deposition of the photons but Compton interaction (inelastic scattering) does not. So, one would try to maximize the events of photoelectric interaction. The absorption cross section through photoelectric effect is proportional to NZ^5 and that of Compton scattering is only NZ where N is the number of atoms per unit volume [2]. Hence, from the above discussion it is clear that another prerequisite for well-resolved detection is a semiconductor material composed of high-Z elements.

The next thing of utmost concern is the efficient collection of charge carriers. Noise due to the incomplete charge collection in thick detectors is one of the highest contributing factors to detector noise [2]. The charge pairs created by the incident radiation need to be collected fast and efficiently. The created charge pairs drift in the opposite direction from the location of their creation across the detector, under the influence of the applied electric field or bias, and reach the electrodes with favorable polarity. In the course of drifting, charge is induced at the electrodes, which is a function of the instantaneous position of the charge carriers. Upon reaching the collecting electrodes, the charge carriers deposit their full charges. However, in real detectors, due to the presence of structural and intrinsic point defects, many charge carriers get trapped and may recombine or get detrapped. On recombination, which is usually the case for deep traps, the charge carriers are lost and deposit partial charges on the collecting electrode. Trapping followed by recombination results in deterioration of carrier lifetime. Detrapping, on the other hand, usually occurs for shallow-lying (close to valence or conduction band edge) defects and the detrapped charges continue their journey to the electrodes. Multiple trappings and detrappings slow down the drift motion of the charges which result in poor carrier mobility. The carrier lifetime (τ) and drift mobility (μ) define the charge transport properties of the material. A more useful parameter which defines the charge transport property is the mobility-lifetime ($\mu\tau$) product.

2.2   Cadmium Zinc Telluride as Detector Material

All the prerequisites for detector material selection that has been mentioned above are present in the ternary compound CdZnTe [3–5]. The constituent elements have high atomic numbers: $Z_{Cd} = 48$, $Z_{Zn} = 30$, and $Z_{Te} = 52$. The material density of CZT is 5.78 g/cm^3. Specific stoichiometry such as $Cd_{0.9}Zn_{0.1}Te$ exhibits a bandgap of ~1.6 eV which can result in bulk electrical conductivity as high as 10^{11} Ω-cm. CZT stands out in the class of compact, room-temperature, high-Z semiconductor X/γ-ray detector in almost all aspects compared to other detectors in the same class such as CdTe, TlBr, PbI_2, and HgI_2. In spite of all the success story, the performance of the CZT detectors is limited which is primarily due to three reasons: (1) poor hole transport properties due to the presence of hole-trapping defects, (2) poor crystal growth yield impeding fabrication of large-volume detectors, and (3) operation at elevated temperatures. Different macroscopic defects such as twin/grain boundaries and cracks, and microstructural defects such as mosaic structures, tilt boundaries, dislocations, point defects, impurities, and tellurium inclusions/precipitates, act as hole-trapping centers [3–9]. Tellurium (Te) inclusions are the major trapping centers for holes, and crystal faults like sub-grain boundaries and their networks are primary hosts to Te inclusions. Although the Te inclusions can be subdued using post-growth annealing under Cd vapor, the process may lead to formation of starlike defects which in turn can trap electrons [10–13]. The sub-grain boundary networks are generally distributed randomly in the CZT matrix causing severe spatial compositional inhomogeneity throughout the crystal volume adding to the cause of poor spectral quality. In addition, CZT crystals grown using methods like high/low-pressure Bridgman technique produce inhomogeneity in zinc (Zn) concentration along the ingot length due to Zn segregation coefficient of 1.35 [14]. Inclusion of 10% Zn in CdTe can increase the bandgap by 2%. The nonuniformity in the Zn concentration thus leads to very poor crystal growth yield [15, 16]. Only about 30% of the total length of a grown CZT ingot produces detector-grade crystalline quality which naturally increases the production cost of these detectors [17].

2.3 Single-Polarity Charge-Sensing Detection

Apart from the attempt of growing high-quality defect-free crystals [7, 18], detectors can be fabricated with single-polarity charge-sensing geometry, e.g., virtual Frisch grid configuration [19–21], coplanar grid structure [22, 23], and small-pixel geometry [24, 25]. These kind of detector geometries eliminate the effect of poor hole transport properties to a great extent by modifying the electric field gradient due to the applied bias in a fashion so as to resemble that of actual Frisch grids in gas detectors and hence are often referred to as virtual Frisch grid configuration [26–28]. Figure 1 shows the schematics of the variation of weighting potential as a function of detector depth in virtual Frisch grid configuration. Weighting potential indicates the amount of induced charge on the collecting electrode. It can be seen from the figure that the collecting electrode hardly sees any charge induced on them unless the charge carriers cross the *neutral region* and enter into the *active region*. Hence, the collecting electrode will "*see*" the electrons only, and that too when they are in the active region. In an ideal situation the detectors in these configurations will not sense the hole transits as long as they are created in the neutral region.

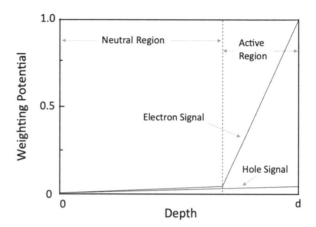

Fig. 1 Variation of the weighting potential as a function of depth of interaction in the detector

3 Growth of Large-Volume $Cd_{0.9}Zn_{0.1}Te$ Single Crystals

There are basically two crystal growth methods prevalent for the growth of detector-grade large-volume CdZnTe single crystals: travelling heater method (THM) [29–31] and Bridgman growth method (BGM) [8, 32–38]. The basic idea of any melt-growth process is to melt the constituent elements and control the solidification rate in such a manner so as to facilitate crystallization. Care should be taken so that the solid-state reaction (synthesis) takes place properly to achieve the targeted stoichiometry in a homogeneous manner and the growth is unidirectional to achieve a high degree of single crystallinity, and most importantly check the formation of defects which affect the charge transport properties adversely. A relatively new crystal growth method known as high-pressure electrodynamic gradient technique (EGT) [39], which is again a variant of the directional solidification technique, is reported to facilitate detector-grade crystal growth with reduced defects and crystal imperfections. EGT addresses the most important issue encountered during melt-growth process like BGM, which is the uncontrolled heat transport (radiative) in the solid-liquid interface. Travelling heater method (THM) is a combination of zone-melting and growth-from-solution techniques wherein a CZT solvent zone is migrated through the solid source material using a thermal gradient. THM enables growth of high-quality large-volume single crystals; however, the rate of crystal growth is very slow, typically 1–2 mm/day. Bridgman technique and its modified versions on the other hand rely on the growth of single crystals from a melt by progressively freezing the melt from one end to the other. The crystal growth rate is typically an order of magnitude higher (1–4 mm/h) than the THM. The BGM suffers from the typical issues of melt growth, the most common being low crystal growth yield, extended defects, and low electron mobility-lifetime ($\mu\tau$) product. The following section discusses the large-volume single crystals grown using a modified vertical Bridgman technique.

We have used a variation of vertical Bridgman technique. The method is a low-temperature growth using tellurium (Te) as a solvent while maintaining crystal homogeneity and advantages of other modern CZT growth techniques. The melting temperatures of Cd, Zn, and Te at a pressure of 1 atm are 321.1 °C, 419.6 °C, and 449.5 °C, respectively. In order to form the compounds CdTe and CdZnTe, the crystal growth temperature must be above the melting points of the compounds. CdTe has a melting point of 1096 °C for a 1:1 atomic ratio of Cd and Te. From the phase diagram consideration, it can be noticed that as the atomic ratio of Te is increased (excess Te), the growth temperature will decrease.

3.1   Details of the Crystal Growth Arrangement

The purity of the precursor materials used for CZT crystal growth has been shown to impact the resulting CZT ingot quality [40]. Zone refining is a process through which ultrapure precursor materials could be obtained for crystal growth. The principle of zone refining is basically melting a small zone of a material while the remaining portion is maintained in solid phase and moving the molten zone across the material [41]. If the segregation coefficient of the impurity, defined as the ratio of impurity concentration in solid phase (C_s) to the impurity concentration in liquid phase (C_l), $k = C_s/C_l$, is less than 1, impurities dissolve in the liquid phase and move to the end of the tube containing the material, leaving rest of the material highly pure. The cycle is repeated from one end of the tube to the other several times to obtain the desired purity. For the crystal growth mentioned in this chapter, commercially available 5 N (99.999%) purity Cd, Zn, and Te precursor materials were purified by an in-house two-ring zone refiner to achieve ~7 N purity. For the zone-refining process, the furnace temperature was increased slightly above the melting point of each precursor material. The furnace was horizontally moved along the ampoule length using a track actuator controlled by a programmed Arduino microcontroller. The ring heaters slowly shifted from one end to the other at a speed of ~25 mm/h. A total of 40 passes were carried out which left the remainder of the material highly pure (~7 N). After the zone-refining process, the material was removed from the ampoule under argon-controlled environment and the impure end of the ingot was separated. The removed "pure" precursor materials were stored in argon-filled polyethylene bottles for further steps. Glow discharge mass spectroscopy (GDMS) analysis was performed on the zone-refined (ZR) precursor materials (Cd, Zn, and Te). It was found that the precursor materials have common impurities on the order of parts per billion (ppb) or less [42].

After zone refining, the polycrystalline materials were loaded in a carbon-coated conically tipped quartz ampoule under a vacuum of 10^{-6} Torr. Conically tipped ampoules facilitate nucleation at a well-defined point and prevent multiple spurious nucleation point [43] and thus eliminate the need of seed crystal. The quartz ampoule was carbon coated by passing n-hexane (HPLC grade, 95+%) vapors at about 750 °C within a furnace under argon flow in order to protect the precursor Cd material from

Fig. 2 (**a**) A portion of $Cd_{0.9}Zn_{0.1}Te$ crystal cut from the grown ingot. (**b**) A 19.0 mm × 19.0 mm × 5.0 mm crystal cut out for large-area pixelated detector (Crystal A). (**c**) A 11.1 cm × 11.1 cm × 10.0 cm block cut out from the ingot for the fabrication of virtual Frisch grid detectors (Crystal B)

reacting with the quartz surface. The ampoule was inserted into a three-zone tube furnace. To reduce the off-stoichiometry of CZT polycrystalline charges, the materials were heated and synthesized slowly at three different temperatures. Heat treatments were performed at 750, 850, and 950 °C for 3 h at each stage to synthesize the charges from the precursors. During synthesis, the ampoule was constantly rotated at a speed of 15 rpm to get the compositional homogeneity. Axial temperature gradient of 3 °C/cm was achieved inside the furnace which is our optimized setting for minimizing stress at the solid-liquid interface resulting from thermal expansion coefficients. The grown crystal directionally solidified by moving the ampoule downward at a constant velocity of ~1 mm/h. All the axial and rotational movements of the ampoule were achieved using stepper motors automated using microprocessor (Arduino) controlled drivers coupled with PC-based user interface. Figure 2 shows a large-volume ingot cut into a cylindrical shape, from which detector blocks are cut out in desired shapes and sizes.

3.2 Fabrication of Small-Pixel, Multipixel, Frisch Collar, and Coplanar Grid Detector

From the grown ingot, several single-crystal blocks were cut out which were grounded, lapped (down to 1500 grit SiC paper), polished (down to 0.05μm alumina

Fig. 3 Detectors in various
geometries: (**a**) Single small
pixel with a guard ring; (**b**)
coplanar grid; (**c**) Frisch
collar; and (**d**) multipixel

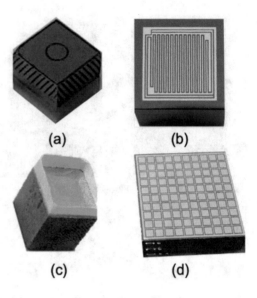

powder), and chemo-mechanically polished using 2% Br_2-MeOH (bromine-methanol) solution. The polished crystals were thoroughly cleaned following standard semiconductor cleaning procedures. Detectors in various geometries were fabricated using the polished CZT crystals. A large-area 19.0 mm × 19.0 mm × 5.0 mm sized crystal (Crystal A) was used to fabricate a 10 × 10 pixelated detectors, while cubic crystals with physical dimensions ~11.1 mm × 11.1 mm × 10.0 mm (Crystal B) were used to fabricate detectors in small pixel, Frisch collar, and coplanar geometry.

Figure 3a shows a detector in single small-pixel configuration accomplished by depositing circular gold contact and a guard-ring structure on the Te-rich face (parallel to (111) crystallographic planes), and a full planar contact on the opposite face (not shown in the picture). The coplanar grid structure was an interdigitated two-grid gold pattern as shown in Fig. 3b. The coplanar geometry was equipped with a guard ring as well. Frisch collar configuration was achieved with a copper sheath tightly wrapped around the crystal. The height of the copper sheath was ~9 mm and was flush with the cathode. The sheath was electrically insulated from the CZT material by lining the crystal side surfaces with insulating Teflon or Kapton tape. The insulation reduces the surface leakage current and prohibits leakage current from the sheath. The copper sheath contained a projected tab which was used to connect it to the cathode. Figure 3c shows the photograph of the Frisch collar detector fabricated and used in this work. A 10 × 10 pixelated structure (anode) was fabricated on the Te-rich face (parallel to (111) crystallographic planes) of the thoroughly cleaned 19.0 mm × 19.0 mm × 5.0 mm sized crystal using photolithography as shown in Fig. 3d. Gold was evaporated to form ~80 nm thick metallic electrical contacts. The dimension of each pixel is 1.3 mm × 1.3 mm pitched at 1.8 mm. A square back contact (cathode) using gold was made on the opposite surface. To minimize the inter-pixel and inter-electrode leakage current, a 0.05 mm wide grounding grid was also fabricated on the anode side.

4 Experimental Setup for Alpha Particle and Gamma-Ray Detection

4.1 Pulse Height Spectroscopy (PHS)

The most common way to evaluate the performance of radiation detectors is to obtain pulse height spectrum (PHS), a histogram of the distribution of the pulse heights, by exposing the detectors to radioactive sources. A standard benchtop radiation spectrometer which comprises a charge-sensitive preamplifier coupled to a shaping amplifier has been used to obtain the PHS. A charge-sensitive preamplifier connects to the detector, receives the charge produced by the detector, and produces an output voltage proportional to the amount of charge received at its input. The preamplifier also allows to apply bias to the detector necessary to collect the charges. Although the amplitude of the preamplifier pulse is proportional to the amount of charge produced by the detector, the precise pulse height determination requires filtering out the electronic noise. Shaping amplifiers do the filtering of the preamplifier signals through filtering circuits. The most common filtering circuit involves one stage of differentiation (high pass) using a CR circuit followed by four stages of integration (low pass) using RC circuits, and is commonly known as $CR\text{-}RC^4$ filtering. The filtering process changes the steplike preamplifier pulse shape to semi-Gaussian shape. The shaping amplifier produces a voltage at its output proportional to the amplitude of the shaped pulse. The actual amplitude of output pulse depends on the gain setting of the amplifier. The shaping amplifier also allows to change the shaping time constant of the filter. It often happens that the detector produces long-duration pulses, and to obtain the complete information, the pulse needs to be processed with higher shaping times. The shaping time of the amplifier also determines the electronic noise of the detection system. The shaping time needs to be optimized for each detector by considering both the average pulse durations and the electronic noise of the system. The output of the shaping amplifier is fed to a multichannel analyzer unit (MCA) which generally consists of an analog-to-digital converter (ADC also known as digitizer) and comparator circuits. The digitizer digitizes the shaped amplifier signals and the comparator circuit determines the amplitude voltage (pulse height) of the digitized pulse. The higher the ADC resolution is, the higher is the energy resolution of the spectrometer. At this stage, the measured pulse height is directly proportional to the incident energy. The pulse heights thus determined are binned to form a histogram known as the pulse height spectrum. The centroid of the histogram gives the mean energy of the incident radiation [44].

It often happens that the obtained PHS using an analog spectrometer is affected by the poor charge transport properties so much so that the basic spectral features such as the photopeak, Compton edge, and escape peaks are obscured. It might happen as well that the peaks are so broad that two closely separated peaks could not be identified. Also, sometimes the risetime of the detector pulses is so long that the limited choice of shaping times provided with the analog shaping amplifier is not sufficient. In such cases, digital spectrometer provides a powerful alternative wherein it is possible to acquire unusually long-duration pulses. In addition, there is practically no limitation on the choice of shaping times. Moreover, digital correction techniques could be applied to regenerate PHS where most of the spectral features could be recovered with remarkable improvement in the resolution, peak-to-valley ratio, peak-to-background ratio, etc. The digital spectrometer used in the following studies consists of the detector-preamplifier assembly connected to a digitizer card (National Instrument PCI-5122). The digitizer card is driven by a graphical user interface (GUI) which allows the user to control the data acquisition process using a PC. A separate program was coded to generate the PHS and apply the correction algorithm. Special rearrangements were made in the abovementioned configuration for detectors with special geometry and will be discussed in the relevant sections.

4.2   Calibration of the Spectrometer

In order to calculate the energy corresponding to the detected peaks in a PHS, the spectrometer needs to be calibrated appropriately. For the present study, a calibration technique using a precision electronic pulse generator has been adopted [45]. The calibration was accomplished by injecting pulses of various known amplitudes (V_{pulser}) from a precision pulser through a calibrated feed-through capacitor with capacitance C to the preamplifier input. The corresponding peak positions of the shaped pulses in the MCA were noted in channel number units (P_C) for each pulser amplitude. The material-equivalent pulse height of the pulse amplitudes, E_{pulser}, is given by the equation below:

$$E_{pulser} = \frac{V_{pulser} \times \varepsilon \times C}{q} \tag{1}$$

Here q is the electronic charge and ε the electron-hole pair creation energy of the detector material. Equation (1) means E_{pulser} would be the energy of the peak in a PHS had the detector material with electron-hole pair creation energy ε produced $V_{pulser} \times C$ amount of charge. The MCA peak positions P_C are then plotted as a function of E_{pulser}. Assuming a linear behavior of the spectrometer, a linear regression of the data points gives the calibration parameters.

4.3 Definition of Spectral Quality

The spectral quality of a pulse height spectrum is evaluated in terms of several factors. The primary factor is the full width at half maximum (FWHM) of the photopeak(s). The *energy resolution* of the spectrometer is a function of the incident radiation energy and is defined qualitatively as the ability of the spectrometer to distinguish between closely spaced peaks. Quantitatively, the energy resolution is quoted as either the FWHM value in energy units for a particular photopeak energy or the percentage energy resolution expressed as the ratio of the FWHM of the photopeak to the energy of the photopeak times 100. The photopeak, in the case of detectors with poor charge transport properties, often tails at the lower energy side (will be discussed in detail) leading to poor energy resolution. The *peak-to-valley ratio* (*P/V*) is calculated as the ratio of full-energy peak height to the average height at a distance of 3σ from the peak centroid (σ being the standard deviation of the Gaussian fit of the peak). Valley describes the region between the end of the Compton edge and the beginning of the photopeak. The *P/V* value depicts the extent to which the photopeak tailing has occurred. The higher the *P/V* ratio is, the higher is the extent of tailing and the lower will be the energy resolution. Further, when the photon scatters at the defect centers through inelastic scattering, it leads to the increase in the *peak-to-background (Compton)* ratio (*P/B*) which is calculated as the ratio of full-energy peak height to the Compton height measured approximately 100 keV below the Compton edge. A higher *P/B* value indicates that higher fraction of the incident photons is interacting via elastic photoelectric absorption, which results in lower Fano noise and higher energy resolution.

5 Characterization of Physical Properties of the Crystal

Prior to the detector fabrication and characterization, it is essential to evaluate the physical properties of the crystal itself. The routine investigations include resistivity measurements, Te inclusion/precipitation size, uniformity in the internal electric field due to the applied bias, and charge transport properties. All the above measurements help to ensure whether the grown crystal is of detector-grade quality or not. These properties are directly or indirectly governed by the presence of electrically active defects.

5.1 Resistivity Measurements

The electrical bulk resistivity of the crystals is determined in planar geometry. In this study a Keithley 237 source-measure unit has been used to measure the current flow (I) through the thickness of the detector as a function of DC bias voltage (V) ($I - V$

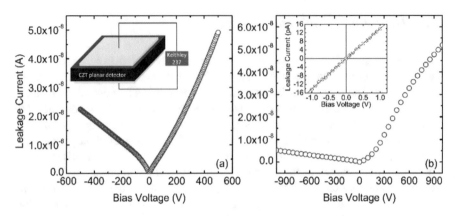

Fig. 4 Variation of leakage current as a function of bias voltage in planar configuration for Crystal A (**a**) and Crystal B (**b**); insets in (**a**) show the arrangement of *I-V* measurements and in (**b**) show the low-range *I-V* used for resistivity calculations

Table 1 Physical properties of the Crystal A and Crystal B used for detector fabrication

Detector	Sizes (mm³)	Resistivity (Ω-cm)	Drift mobility cm²/V.s	$\mu_e \tau_e$ cm²/V
Crystal A	$19.0 \times 19.0 \times 5.0$	5×10^{10}	918 ± 2	6.2×10^{-3}
Crystal B	$11.1 \times 11.1 \times 10.0$	8.5×10^{10}	838 ± 8	2.7×10^{-3}

characteristics). The measurements were carried out at room temperature under dark condition. The detector was housed in a lighttight metal box capable of shielding out any electromagnetic interference. Figure 4 shows the $I - V$ measurements in the Crystal A and Crystal B. It could be noted that the $I - V$ curves are not symmetric with respect to the bias polarity. This is mostly due to the different surface conditions. To determine the electrical bulk resistivity, linear $I - V$ characteristics are obtained with low range of bias sweep (inset of Fig. 4b). A straight-line fit of the linear $I - V$ curves gives the resistance R of the crystal.

Assuming the planar detector to be a uniform ohmic conductor, the resistivity ρ is given by $R = \rho d/A$, with d being the thickness of the detector and A the area of the planar contacts. Both the crystals show a bulk resistivity on the order of 10^{10} Ω-cm which is favorably high enough for detector fabrication. The resistivity values of the crystals are given in Table 1 along with other crystal properties.

5.2 Tellurium Inclusion

As has been mentioned before, Te inclusions act as major hole-trapping centers, and crystal faults like sub-grain boundaries and their networks are primary hosts to Te inclusions. While CZT is largely transparent to infrared (IR) rays, Te-rich volumes inside the CZT crystals are almost opaque to IR rays and hence can be visualized

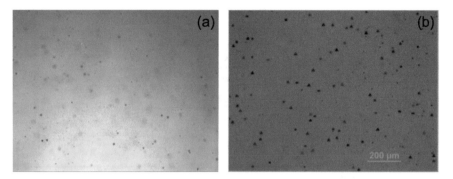

Fig. 5 IR transmission image showing the Te inclusions in Crystal A (**a**) and Crystal B (**b**)

using IR transmission microscopy [46]. Te inclusions with diameters greater than 10μm can act as potential charge-trapping centers and significantly degrade the performance of the detectors [47, 48]. Due to the narrow bandgap of Te (~0.3 eV), tellurium-rich zones exhibit higher electrical conductivity which may distort the internal electric field distribution and hence the carrier transport properties. Figure 5 shows the IR transmission images of the Crystals A and B. The images revealed an average tellurium inclusion/precipitate size of ~8μm or less.

5.3 Charge Transport Measurements

The charge transport properties for the electron transit, viz. the mobility-lifetime product and the drift mobilities, were calculated using radiation detection-based techniques. The $\mu\tau$ product was calculated using the analog spectrometer calibrated following the aforementioned technique for $Cd_{0.9}Zn_{0.1}Te$ by choosing the permittivity value (ε) equal to 10 [2]. A radioisotope source emitting alpha particles of energy 5486 keV was used to irradiate the detectors fabricated in planar configuration. Alpha particles have very little penetration depth inside most detector materials, which is negligible compared to the thickness of the detectors. Being a monoenergetic source, a single peak corresponding to the alpha particles is registered in the PHS. The ratio of the location of the peak (E_{alpha}) on the MCA and the actual energy emitted by the source gives the charge collection efficiency (CCE). The variation of CCE as a function of the applied bias could be modelled according to the formalism developed by Hecht [49]. Equation 2 below shows the Hecht equation modified for a single-polarity charge transit:

$$CCE = \frac{\mu\tau V}{d^2}\left[1 - \exp\left(\frac{-d^2}{\mu\tau V}\right)\right] \tag{2}$$

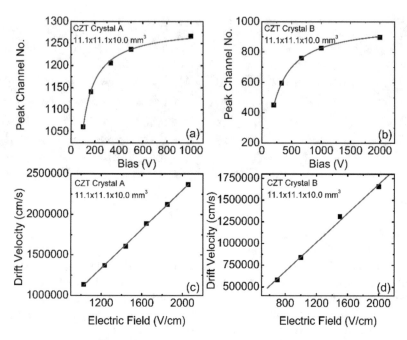

Fig. 6 Hecht plots for the Crystal A (**a**) and Crystal B (**b**) for electron transit. Drift velocity of electrons plotted as a function of applied electric for the Crystal A (**c**) and Crystal B (**d**). The solid dots show the experimental data and the solid lines are the regression plots

The regression of the variation of CCE vs. V plot with Eq. 2 returns the $\mu\tau$ product. Figure 6a, b shows the Hecht plot of the Crystals A and B, respectively. A single-polarity charge transit scenario could be easily arranged as alpha particles have very low penetration depth. Hence, the charge carriers can be considered to be generated just beneath the detector surface facing the source. An application of a bias with proper polarity makes either the electrons or the holes to transit across the detector thickness.

The drift mobility was calculated using the digital spectrometer and applying a time-of-flight technique [50]. The detector in planar configuration is exposed to a ^{241}Am alpha-emitting source. Preamplifier charge pulses were digitized and recorded in large numbers and the risetime of each pulse was calculated as the time taken to reach 90% from 10% of the maximum pulse height. The distribution of the risetime resembles a Gaussian shape whose centroid gives the mean risetime. The mean risetime is linearly extrapolated to 100% to obtain the transit time by assuming that the time evolution of the charge pulse depicts the transit of the charge carriers from the point of creation to the collecting electrode. The assumption generally holds good when the risetime response of the preamplifier and the digitizer card is faster than the transit time of the charge carriers. The drift velocity v_d is calculated by dividing the detector thickness by the transit time. In this case it is assumed that the charge carriers transverse the entire thickness in straight lines with constant velocity under the action of the applied electric field (E). It should be noted that the application of an electric field does not accelerate the charge particles as the charge carriers undergo successive scattering, trapping, and detrapping from the impurity and trap sites. The drift velocities are calculated with the detector biased at different applied voltages (electric field). The variation of the drift velocities is generally directly proportional to the applied electric field and follows a linear relation as shown in Eq. 3 below:

$$v_d = \mu E \tag{3}$$

where μ is the proportionality constant and is defined as the drift mobility. The drift mobility is thus calculated from the slope of a linear regression of the plot of drift velocity as a function of the electric field. The electric field is calculated as the ratio of applied bias to the detector thickness. Figure 6c, d shows the drift mobility plot of the Crystals A and B, respectively. The experimentally determined drift mobility values have been listed in Table 1.

6 Gamma-Ray Spectroscopy Using Large-Volume/Area CdZnTe Detectors

Unlike charged particle interaction with matter, where the charged particles interact with many absorber atoms continuously and lose their energies gradually, gamma rays abruptly lose their energies and disappear entirely or scatter through large angles. The point of interaction in a given thickness of the detector is also random for a gamma photon even with the same incident energy [44]. In detectors with large thickness, i.e., $\lambda_{e/h} \ll L$, where $\lambda_{e/h}$ is the drift length of the carrier and L is the detector thickness, the uncertainty in the pulse height measurements due to the incomplete charge collection is dominant [2]. While large thickness ensures high efficiency of gamma photon absorption and detection, the noise contribution due to incomplete charge collection worsens the energy resolution of the detector. Hence, large-volume detectors need special arrangements to minimize or compensate for the broadening due to the incomplete charge collection. As has been discussed in the Sect. 2.3, single-polarity charge-sensing electrode geometries help to minimize the effect of the poor hole transport properties. Large-volume detectors in small pixel, coplanar grid, and Frisch collar geometries were fabricated and evaluated. Except for the Frisch collar, all the detectors had guard rings. Detectors with guard ring require the guard ring to be connected to the ground. The guard ring prevents any electrical connection between the cathode side and the anode side through the surface of the detector [51–53]. Had there been any current flow due to the surface leakage current, the charges will flow to the ground through the guard ring. As the guard rings in the small pixel, coplanar, and pixelated detectors are close to the anode structure, the high-voltage bias was applied on the opposite face (planar side). In the case of the

Frisch collar detector the bias was applied to the top surface and the bottom contact was grounded along with the Frisch collar. For the coplanar detector two Cremat CR-110 preamplifiers were used to collect the signals from the collecting grid (CG) and the non-collecting grid (NCG). The charge pulses were digitally acquired using a high-speed NI PCI-5122 digitizer card and processed by a custom-made program built in LabVIEW platform to obtain pulse height spectrum (PHS) in real time. For the case of multipixel detector, the guard ring was connected to the ground using a 25μm (dia) gold wire bonded using conductive epoxy. A movable arm was used to connect the pixels to the preamplifier in turn for collecting spectra from each pixel. The cathode was placed on a metallic pad on a PCB which was connected to filter circuit for biasing. The radiation source was placed under the detector facing the cathode. The detection system is a hybrid system to record analog as well as digital data simultaneously. Both the systems share a charge-sensitive CR110 (Cremat) preamplifier. The analog system uses an Ortec 671 shaping amplifier to filter the preamplifier pulses and obtain pulse heights, which are then binned using a Canberra Multiport II multichannel analyzer driven by Genie 2000 user interface to obtain a pulse height spectrum. The digital spectrometer on the other hand uses the PCI 5122 digitizer card to digitize and store the raw preamplifier charge pulses.

For the coplanar grid detector, the CG and NCG signals were subtracted digitally to obtain the difference signal which is free from the effects of hole transit. The difference signal, which is effectively the electron signal, was filtered using a digital CR-RC^4 shaping algorithm [54, 55]. The pulse heights of the shaped difference pulses were digitally measured and plotted in a histogram to obtain the pulse height spectra. Figure 7a shows a ^{137}Cs PHS obtained using the small guarded pixel on Te-rich face and solid cathode on Cd-rich face biased at −3500 V. The guard ring was connected to ground. The 662 keV gamma peak was well resolved along with the rest of the spectral features like the backscattered peak and the Compton edge. The energy resolution for the 662 keV gamma rays was found to be ~3.7% with a *P/V* and *P/B* values of 17 and 1.3, respectively. Figure 7b shows the ^{137}Cs spectrum obtained in the coplanar grid geometry with the cathode and the collecting grid biased at −3000 V and +200 V, respectively, with the guard ring connected to the ground as usual. The detector was irradiated from the cathode side. The best energy resolution that could be obtained for the 662 keV gamma rays was ~8%. The detector in a virtual Frisch grid geometry provided a much better energy resolution of ~4.3% for the 662 keV gamma rays. Figure 7c shows the ^{137}Cs spectrum obtained using the detector in virtual Frisch grid configuration. Figure 7d shows a pulse height spectrum obtained using a ^{137}Cs (662 keV) gamma source from a random pixel from the 10 × 10 array. A bias of −1500 V (3000 V/cm) was applied to the planar contact and the guard ring was connected to ground during measurements. The photopeak corresponding to the 662 keV gamma ray was found to be well resolved from the Compton background. The Compton edge was well defined, and the backscattered peak was also identified in the PHS. The energy resolution was calculated from the FWHM of the photopeak and was found to be ~1.6% for 662 keV gamma rays which can be categorized as very high resolution for CZT-based detectors. The *P/B* and the *P/V* ratios were calculated to be ~2 and ~10, respectively. The photopeak, although

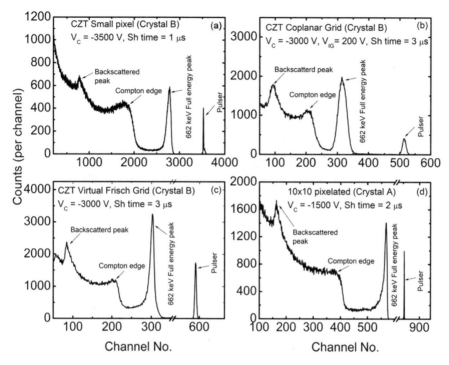

Fig. 7 Pulse height spectra obtained from small pixel (**a**); coplanar grid (**b**); Frisch collar (**c**); and multipixel (**d**); detector configuration

well resolved, appears to be slightly asymmetric as a tailing at the lower energy side was observed. Figure 7 also shows the pulser peaks recorded along with the ^{137}Cs PHS. The FWHM of the pulser peak, which shows the overall electronic noise of the spectrometer, was calculated to be 3.6 keV for the abovementioned pixel. The overall electronic noise was seen to vary from pixel to pixel with the values ranging from 3.34 to 4.34 keV.

Ideally, a gamma photon pulse height spectrum should present a peak-to-valley ratio equal to the number of counts at the photopeak position after proper background subtraction. An actual PHS obtained from CZT detectors however always shows a lower P/V ratio because of the tailing at the lower energy side of the photopeak. Such tailing of photopeaks is generally attributed to ballistic deficit due to trapping of holes in defect centers [3]. It should be noted that in the present situation, the detector being a single-polarity (electron) sensing device is not supposed to be affected by the hole movement. With a pixel dimension of $\varepsilon \times \varepsilon$, a transiting electron is not supposed to induce any charge on the collecting pixel until it is within a distance of $z \approx \varepsilon$, which in this case is 1.3 mm. The degree to which a small pixel effectively works depends on the pixel dimension-to-detector width ratio (ε/L), with L being the detector thickness. The smaller the ε/L ratio is, the higher is the effectiveness of single-pixel effect [24]. For the present detector the ε/L ratio is

0.26. According to the calculations given in Ref. [24], it can be estimated that with a ε/L ratio of 0.26, the charge induced by an electron created as close as 1 mm from the anode pixel is 90% of the total charge induced, with the rest 10% being due to the holes. Thus, a fraction of the signal induced at the anode, even though small, is due to hole transit also, which accounts for the tailing at the lower energy side of the 662 keV photopeak given in Fig. 7d.

Not all the pixels however produced equal quality of gamma-ray spectra for ^{137}Cs source. In fact, nonuniformity in CZT crystals is a common problem and is known to cause nonuniform charge collection efficiency [56], especially in large-area crystals [57]. The pixelated detector showed a spatial variation of energy resolution as observed from pixel-by-pixel gamma spectroscopic studies. The average value of the energy resolution was found to be 3.8% at 662 keV. The plot of the spatial variation of the detector parameters could be found elsewhere [58].

7 Digital Spectrum Recovery Methods

The general notion of interaction of gamma rays with matter is that gamma rays can penetrate deep inside the bulk crystal and have the probability to interact at any point and generate charge pairs; even the photons have the same energy. As a result, in the case of gamma rays, the radiation-induced charge pairs of each polarity traverse different distances depending on the location of interaction. For interactions near the cathode, the electrons traverse the detector thickness, and for interactions near the anode the holes do. Problems arise, owing to their poor transport properties, when the holes have to traverse a longer distance. It might happen that the holes get trapped in defects and recombine resulting in pulses with lower heights as the collecting electrodes do not register full charge leading to what is known as ballistic deficit. Moreover, the output pulse height varies depending on where the interaction has taken place. For interactions closer to the cathode the holes have a better chance to reach the cathode before recombining compared to the case of intearction closer to the cathode. This leads to variation of output pulse heights even with monoenergetic gamma rays interacting through photoelectric interactions.

Biparametric (BP) plots are a convenient way to correlate the pulse heights to the risetimes of each pulse obtained from detectors [55, 59]. In a BP plot the interaction events are mapped onto a 3D plot where the x-axis and the y-axis depict the pulse height (energy) and risetime (depth of interaction), respectively, and the z-matrix shows the intensity of correlation. Thus, each point in the plot shows how many interactions occur with a given pulse height and risetime. The plots normally show the events with similar incident energy bunched up in one region of the BP plot. In an ideal case, where none of the charges are lost, the events from the interactions of monoenergetic gamma photons appear as a vertical bunch of points in the BP plot with a spread coherent with the uncertainty in the measurements. In the case of recombination of generated charge pairs, the bunch of events shows an inclination (deviation from the vertical trend) implying that the detector registers different pulse heights for different depths of interaction even for monoenergetic gamma photons interacting through photoelectric effect. Digital corrections can be used to move the points showing deviation from ideal behavior to locations where they should have been, had there been no charge loss. This is referred to as the corrected BP plot. A pulse height spectrum can then be regenerated from the corrected BP plot where the effect of charge loss has been mostly eliminated or compensated for [60, 61].

The correction technique works fine for detectors in planar configuration. In the present situation, however, the BP plot is obtained in a detector with a single-polarity charge sensing configuration. But as has been argued in the previous section, the resultant signal still has some charge transition in the neutral region as shown in Fig. 1, a phenomenon commonly referred to as grid inefficiency in Frisch grid detectors [62]. A more relevant plot of weighting potential as a function of normalized interaction depth in pixelated detector showing the grid inefficiency could be found in Ref. [3]. The higher the grid inefficiency, the higher the effect of the transit of charge carriers in the neutral region. Hence, the following discussion will be under the premise that the movement of electrons (and holes) affects the signals from this particular pixel throughout the detector thickness because of finite grid inefficiency. Figure 8a shows the biparametric plot obtained from a pixel in the abovementioned 10×10 pixelated detector, which has exhibited a poor energy resolution. The events from the 662 keV gamma photon interactions could be easily identified as the curved bunch of events visible at the right side of the plot. These events are well distinguished from the rest of the points which are due to the Compton interaction within the detector volume. The 662 keV events show that a substantial portion of the collected pulse has lower risetimes as well as pulse heights indicating permanent hole trapping. These events, in the bent portion of the correlated events, are responsible for the lower energy tailing. In BP plots obtained from CZT detectors, it is generally seen that for the 662 keV events, the pulses with higher risetimes

Fig. 8 (**a**) Biparametric plot obtained from the multipixel detector, fabricated from Crystal A, exposed to a ^{137}Cs from a pixel with lower resolution. (**b**) Depth-dependent binning of ^{137}Cs pulse height spectra obtained from the same pixel

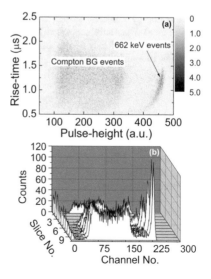

exhibit relatively lower pulse heights. The higher risetime is indicative of the low hole mobility and the corresponding lower pulse heights indicate ballistic deficit. In the present case, the 662 keV events however showed a rather different behavior. The pulses with lower risetimes had lower pulse heights for the 662 keV events. Such pulses could result from interactions in the active region and very close to the anode. In those cases, the charge pulses primarily consist of hole movements, that too in the active region only, resulting in pulses with very low pulse heights and naturally short risetimes. The interactions of 662 keV gamma rays in the active region are only natural as it could be estimated that a substantial fraction of these gamma photons could pass the entire 5 mm thickness of the detector without being completely stopped.

More insights into the trapping of charge carriers can be obtained from depth-wise binning of pulse height spectra obtained from the BP plot. Figure 8b shows such a plot obtained for the present pixel. The BP plot was sliced into ten slices each for a certain interval of risetimes. Slice #10 represents the interval with the highest risetimes and slice # 1 represents the interval with the lowest ones. It could be seen from Fig. 8b that the slices #1 and #4 showed very broad photopeaks with very low *P/B* ratios. The broadening of the photopeak in these slices could be attributed to the uncertainty in pulse height measurement due to lower signal-to-noise ratio for pulses with lower pulse heights. In addition, the average photopeak positions of these slices are also centered at lower channel number compared to slices #5–10, which can be attributed to the effect of ballistic deficit. The slices #2 and #3 on the other hand showed only noise and Compton background but hardly any trace of the photopeak. All these factors led to the lower energy tailing of the photopeak. A percentage resolution of 2% at 662 keV and *P/V* and *P/B* ratio of ~14 and ~2, respectively, were found from the pulse height plot of the slice #9, which is an indicative of the quality of ^{137}Cs spectra had there been no hole trapping.

Acknowledgments The authors acknowledge the financial support provided by the DOE Office of Nuclear Energy's Nuclear Energy University Programs (NEUP), Grant No. DE-NE0008662. The work was also partially supported by the Advanced Support Program for Innovative Research Excellence-I (ASPIRE-I) of the University of South Carolina (UofSC), Columbia, Grant Nos. 15530-E419 and 155312N1600.

References

1. Stanton, A.: Wilhelm Conrad Röntgen on a new kind of rays: translation of a paper read before the Würzburg physical and medical society, 1895. Nature. **53**(1369), 274–276 (1896)
2. Owens, A., Peacock, A.: Compound semiconductor radiation detectors. Nucl. Instrum. Methods Phys. Res. Sect. A. **531**, 18–37 (2004)
3. Sordo, S.D., Abbene, L., Caroli, E., Mancini, A.M., Zappettini, A., Ubertini, P.: Progress in the development of CdTe and CdZnTe semiconductor radiation detectors for astrophysical and medical applications. Sensors. **9**(5), 3491–3526 (2009)

4. Schlesinger, T.E., Toney, J.E., Yoon, H., Lee, E.Y., Brunett, B.A., Franks, L., James, R.B.: Cadmium zinc telluride and its use as a nuclear radiation detector material. Mater. Sci. Eng. **32**, 103–189 (2001)
5. Milbrath, B., Peurrung, A., Bliss, M., Weber, W.: Radiation detector materials: an overview. J. Mater. Res. **23**, 2561–2581 (2008)
6. Mandal, K.C., Krishna, R.M., Muzykov, P.G., Hayes, T.C.: Fabrication and characterization of $Cd_{0.9}Zn_{0.1}Te$ Schottky diodes for high resolution nuclear radiation detectors. IEEE Trans. Nucl. Sci. **59**(4), 1504–1509 (2012)
7. Bolotnikov, A.E., Ackley, K., Camarda, G.S., Cui, Y., Eger, J.F., De Geronimo, G., Finfork, C., Fried, J., Hossain, A., Lee, W., Prokesch, M., Petryk, M., Reiber, J.L., Roy, U., Vernon, E., Yang, G., James, R.B.: High-efficiency CdZnTe gamma-ray detectors. IEEE Trans. Nucl. Sci. **62**(6), 3193–3198 (2015)
8. Mandal, K.C., Kang, S.H., Choi, M., Bello, J., Zheng, L., Zhang, H., Groza, M., Roy, U.N., Burger, A., Jellison, G.E., Holcomb, D.E., Wright, G.W., Williams, J.A.: Simulation, modeling, and crystal growth of $Cd_{0.9}Zn_{0.1}Te$ for nuclear spectrometers. J. Electron. Mater. **35**, 1251–1256 (2006)
9. Mandal, K.C., Kang, S.H., Choi, M., Kargar, A., Harrison, M.J., McGregor, D.S., Bolotnikov, A.E., Carini, G.A., Camarda, G.S., James, R.B.: Characterization of low-defect $Cd_{0.9}Zn_{0.1}Te$ and CdTe crystals for high-performance Frisch collar detector. IEEE Trans. Nucl. Sci. **54**, 802–806 (2007)
10. Yang, G., Bolotnikov, A., Fochuk, P., Kopach, O., Franc, J., Belas, E., Kim, K., Camarda, G., Hossain, A., Cui, Y., Adams, A., Radja, A., Pinder, R., James, R.B.: Post-growth thermal annealing study of CdZnTe for developing room-temperature X-ray and gamma-ray detectors. J. Cryst. Growth. **379**, 16–20 (2013)
11. Mandal, K.C., Hang, S.H., Choi, M., Wei, L., Zheng, L., Zhang, H., Jellison, G.E., Groza, M., Burger, A.: Component overpressure growth and characterization of high-resistivity CdTe crystals for radiation detectors. J. Electron. Mater. **36**, 1013–1020 (2007)
12. Koley, G., Liu, J., Mandal, K.C.: Investigation of CdZnTe crystal defects using scanning probe microscopy. Appl. Phys. Lett. **90**, 102121 (2007)
13. Liu, J., Mandal, K.C., Koley, G.: Investigation of nanoscale properties of CdZnTe crystals by scanning spreading resistance microscopy. Semicond. Sci. Tech. **24**, 045012 (2009)
14. Roy, U.N., Weiler, S., Stein, J., Cui, Y., Groza, M., Buliga, V., Burger, A.: Zinc mapping in THM grown detector grade CZT. J. Cryst. Growth. **347**(1), 53–55 (2012)
15. Fougères, P., Chibani, L., Hageali, M., Koebel, J., Hennard, G., Zumbiehl, A., Siffert, P., Benkaddour, M.: Zinc segregation in HPB grown nuclear detector grade $Cd_{1-x}Zn_xTe$. J. Cryst. Growth. **197**(3), 641–645 (1999)
16. Rodríguez, M.E., Gutiérrez, A., Zelaya-Angel, O., Vázquez, C., Giraldo, J.: Influence of crystalline quality on the thermal, optical and structural properties of $Cd_{1-x}Zn_xTe$ for low zinc concentration. J. Cryst. Growth. **233**(1–2), 275–281 (2001)
17. Roy, U.N., Camarda, G.S., Cui, Y., Gul, R., Yang, G., Zazvorka, J., Dedic, V., Franc, J., James, R.: Evaluation of CdZnTeSe as a high-quality gamma-ray spectroscopic material with better compositional homogeneity and reduced defects. Sci. Rep. **9**, 7303 (2019)
18. Pak, R.O., Mandal, K.C.: Defect levels in nuclear detector grade $Cd_{0.9}Zn_{0.1}Te$ crystals. ECS J. Solid State Sci. Technol. **5**(4), P3037–P3040 (2016)
19. Krishna, R.M., Muzykov, P.G., Mandal, K.C.: Electron beam induced current imaging of dislocations in $Cd_{0.9}Zn_{0.1}Te$ crystal. J. Phys. Chem. Solids. **74**(1), 170–173 (2013)
20. Krishna, R.M., Hayes, T.C., Muzykov, P.G., Mandal, K.C.: Low temperature crystal growth and characterization of $Cd_{0.9}Zn_{0.1}Te$ for radiation detection applications. Proc. Mater. Res. Soc. Symp. **1314**, 39–44 (2011)
21. McGregor, D.S., He, Z., Seifert, H.A., Wehe, D.K., Rojeski, R.A.: Single charge carrier type sensing with a parallel strip pseudo-Frisch grid CdZnTe semiconductor radiation detector. Appl. Phys. Lett. **72**(7), 792–794 (1998)

22. Luke, P.N.: Single-polarity charge sensing in ionization detectors using coplanar electrodes. Appl. Phys. Lett. **65**(22), 2884–2886 (1994)
23. Chaudhuri, S.K., Krishna, R.M., Zavalla, K.J., Matei, L., Buliga, V., Groza, M., Burger, A., Mandal, K.C.: Performance of $Cd_{0.9}Zn_{0.1}Te$ based high-energy gamma detectors in various single polarity sensing device geometries In: IEEE NSS Conference Record (2012)
24. Berrett, H.H., Eskin, J.D., Barber, H.B.: Charge transport in arrays of semiconductor gamma-ray detectors. Phys. Rev. Lett. **75**(1), 156–159 (1995)
25. Bell, S.J., Baker, M.A., Duarte, D.D., Schneider, A., Seller, P., Sellin, P.J., Veale, M.C., Wilson, M.D.: Performance comparison of small-pixel CdZnTe radiation detectors with gold contacts formed by sputter and electroless deposition. J. Instrum. **12**, 06015 (2017)
26. McGregor, D.S., Rojeski, R.A.: High-resolution ionization detector and array of such detectors. United States of America Patent 6,175,120 (1998)
27. Montemont, G., Arques, M., Verger, L., Rustique, J.: A capacitive Frisch grid structure for CdZnTe detectors. IEEE Trans. Nucl. Sci. **48**(3), 278–281 (2001)
28. Kargar, A., Brooks, A.C., Harrison, M.J., Kohman, K.T., Lowell, R.B., Keyes, R.C.: The effect of the dielectric layer thickness on spectral performance of CdZnTe Frisch collar gamma ray spectrometers. IEEE Trans. Nucl. Sci. **56**(3), 824–831 (2009)
29. Roy, U.N., Burger, A., James, R.B.: Growth of CdZnTe crystals by the traveling heater method. J. Cryst. Growth. **379**, 57–62 (2013)
30. Chen, H., Awadalla, S.A., Iniewski, K., Lu, P.H., Harris, F., Mackenzie, J., Hasanen, T., Chen, W., Redden, R., Bindley, G., Kuvvetli, I., Budtz-Jørgensen, C., Luke, P., Amman, M., Lee, J.S., Bolotnikov, A.E., Camarda, G.S., Cui, Y., Hossain, A., James, R.B.: Characterization of large cadmium zinc telluride crystals grown by traveling heater method. J. Appl. Phys. **103**, 014903 (2008)
31. Amman, M., Lee, J.S., Luke, P.N., Chen, H., Awadalla, S.A., Redden, R., Bindley, G.: Evaluation of THM-grown CdZnTe material for large-volume gamma-ray detector applications. IEEE Trans. Nucl. Sci. **56**(3), 795–799 (2009)
32. Zha, M., Zappettini, A., Calestani, D., Marchini, L., Zanotti, L., Paorici, C.: Full encapsulated CdZnTe crystals by the vertical Bridgman method. J. Cryst. Growth. **310**(7–9), 2072–2075 (2008)
33. Szeles, C., Bale, D., Grosholz, J. Jr, Smith, G.L., Blostein, M., Eger, J.: Fabrication of high performance CdZnTe quasi-hemispherical gamma-ray CAPture (TM) plus detectors. In: Proceedings of SPIE - The International Society for Optical Engineering, Bellingham, WA (2006)
34. James, R.B., Schlesinger, T.E., Lund, J.C., Schieber, M.: In: Schlesinger, T.E., James, R.B. (eds.) $Cd_{1-x}Zn_xTe$ Spectrometers for Gamma and X-Ray Applications in Semiconductor and Semimetals, vol. 43, pp. 336–378. Academic Press, San Diego (1995)
35. Li, Q., Jie, W., Fu, L., Wang, T., Yang, G., Bai, X., Zha, G.: Optical and electrical properties of indium-doped $Cd_{0.9}Zn_{0.1}Te$ crystal. J. Cryst. Growth. **295**(2), 124–128 (2006)
36. Hermon, H., Schieber, M., Lee, E.Y., McChesney, J.L., Goorsky, M., Lam, T., Meerson, E., Yao, H., Erickson, J., James, R.B.: CZT detectors fabricated from horizontal and vertical Bridgman-grown crystals. Nucl. Instrum. Meth. Phys. Res. Sect. A. **458**(1–2), 503–510 (2001)
37. Schieber, M., James, R.B., Hermon, H., Vilensky, A., Baydjanov, I., Goorsky, M., Lam, T., Meerson, E., Yao, H.W., Erickson, J., Cross, E., Burger, A., Ndap, J.O., Wright, G., Fiederle, M.: Comparison of cadmium zinc telluride crystals grown by horizontal and vertical Bridgman and from the vapor phase. J. Cryst. Growth. **231**(1–2), 235–241 (2001)
38. Zappettini, A., Marchini, L., Zha, M., Benassi, G., Zambelli, N., Calestani, D., Zanotti, L., Gombia, E., Mosca, R., Zanichelli, M., Pavesi, M., Auricchio, N., Caroli, E.: Growth and characterization of CZT crystals by the vertical Bridgman method for X-ray detector applications. IEEE Trans. Nucl. Sci. **58**(5), 2352–2356 (2011)
39. Szeles, C., Cameron, S.E., Soldner, S.A., Ndap, J., Reed, M.D.: Development of the high-pressure electro-dynamic gradient crystal-growth technology for semi-insulating CdZnTe growth for radiation detector applications. J. Electron. Mater. **33**, 742–751 (2004)

40. Mandal, K.C., Noblitt, C., Choi, M., Rauh, R.D., Roy, U.N., Groza, M., Burger, A., Holcomb, D.E., Jellison, J.E. Jr.: Crystal growth, characterization, and testing of $Cd_{0.9}Zn_{0.1}Te$ single crystals for radiation detectors. In: Hard X-Ray and Gamma-Ray Detector Physics VI, Denver, Colorado, USA (2004)
41. Pfann, W.G.: Zone melting. Int. Mater Rev. **2**(1), 29–76 (1957)
42. Krishna, R.M.: Crystal Growth, Characterization and Fabrication of CdZnTe-Based Nuclear Detectors. Doctoral Thesis, University of South Carolina (2013)
43. Benz, K.W., Fiederle, M.: In: Triboulet, R., Siffert, P. (eds.) In CdTe and Related Compounds; Physics, Defects, Hetero- and Nano-structures, Crystal Growth, Surfaces and Applications Part II: Crystal Growth, Surfaces and Applications, vol. IC, p. 80. Elsevier, Amsterdam (2010)
44. Knoll, G.F.: Radiation Detection and Measurements, 3rd edn. Wiley, New York (2000)
45. Chaudhuri, S.K., Zavalla, K.J., Mandal, K.C.: Experimental determination of electron-hole pair creation energy in 4H-SiC epitaxial layer: an absolute calibration approach. Appl. Phys. Lett. **102**, 031109 (2013)
46. Rai, R.S., Mahajan, S., McDevitt, S., Johnson, C.J.: Characterization of CdTe, (Cd,Zn)Te, and Cd(Te,Se) single crystals by transmission electron microscopy. J. Vac. Sci. Technol. B. **9**, 1892–1896 (1991)
47. Carini, G.A., Bolotnikov, A.E., Camarda, G.S., Wright, G.W., James, R.B., Li, L.: Effect of Te precipitates on the performance of CdZnTe detectors. Appl. Phys. Lett. **88**, 143515–143517 (2006)
48. Bolotnikov, A.E., Camarda, G.S., Carini, G.A., Cui, Y., Li, L., James, R.B.: Cumulative effects of Te precipitates in CdZnTe radiation detectors. Nucl. Instrum. Meth. Phys. Res. Sect. A. **571** (3), 687–698 (2007)
49. Hecht, K.: Zum mechanismus des lichtelektrischen primärstromes in isolierenden kristallen. Z. Phys. **77**, 235–245 (1932)
50. Sellin, P.J., Davies, A.W., Lohstroh, A., Ozsan, M.E., Parkin, J.: Drift mobility and mobility-lifetime products in CdTe:cl grown by the travelling heater method. IEEE Trans. Nucl. Sci. **52** (6), 3074–3078 (2005)
51. Jaklevic, J.M., Goulding, F.S.: Semiconductor detector x-ray fluorescence spectrometry applied to environmental and biological analysis. IEEE Trans. Nucl. Sci. **19**(3), 384–391 (1972)
52. Belcarz, E., Chwaszczewska, J., Słapa, M., Szymczak, M., Tys, J.: Surface barrier lithium drifted silicon detector with evaporated guard ring. Nucl. Instrum. Methods. **77**(1), 21–28 (1970)
53. Nakazawa, K., Oonuki, K., Tanaka, T., Kobayashi, T., Tamura, Y., Mitani, K., Sato, T., Wtanabe, G., Takahashi, S., Ohno, R., Kitajima, A., Kuroda, A., Onishi, Y.: Improvement of the CdTe diode detectors using a guard-ring electrode. IEEE Trans. Nucl. Sci. **51**(4), 1881–1885 (2004)
54. Nakhostin, M.: Recursive algorithms for real-time CR-$(RC)^n$ digital pulse shaping. IEEE Trans. Nucl. Sci. **58**(5), 2378–2381 (2011)
55. Chaudhuri, S.K., Zavalla, K.M., Krishna, R.M., Mandal, K.C.: Biparametric analyses of charge trapping in $Cd_{0.9}Zn_{0.1}Te$ based virtual Frisch grid detectors. J. Appl. Phys. **113**, 074504 (2013)
56. Sellin, P.J., Davies, A.W., Gkoumas, S., Lohstroh, A., Özsan, M.E., Parkin, J., Perumal, V., Prekas, G., Veale, M.: Ion beam induced charge imaging of charge transport in CdTe and CdZnTe Nucl. Instrum. Methods Phys. Res. Sect. B. **266**, 1300–1306 (2008)
57. Yin, Y., Chen, X., Wu, H., Komarov, S., Garson, A., Li, Q., Guo, Q., Krawczynski, H., Meng, L.-J., Tai, Y.-C.: 3-D spatial resolution of 350 µm pitch pixelated CdZnTe detectors for imaging applications. IEEE Trans. Nucl. Sci. **60**(1), 9–15 (2013)
58. Chaudhuri, S.K., Nguyen, K., Pak, R.O., Matei, L., Buliga, V., Groza, M., Burger, A., Mandal, K.C.: Large area Cd0.9Zn0.1Te pixelated detector: fabrication and characterization. IEEE Trans. Nucl. Sci. **61**(4), 793–798 (2014)

59. Verger, L., Boitel, M., Gentet, M.C., Hamelin, R., Mestais, C., Mongellaz, F., Rustique, J., Sanchez, G.: Characterization of CdTe and CdZnTe detectors for gamma-ray imaging applications. Nucl. Instrum. Methods Phys. Res. Sect. A. **458**(1–2), 297–309 (2001)
60. Chaudhuri, S.K., Lohstroh, A., Nakhostin, M., Sellin, P.J.: Digital pulse height correction in HgI_2 γ-ray detectors. J. Instrum. **7**, T04002 (2012)
61. Sajjad, M., Chaudhuri, S.K., Kleppinger, J.W., Mandal, K.C.: Growth of large-area Cd0.9Zn0.1Te single crystals and fabrication of pixelated guard-ring detector for room-temperature γ-ray detection. IEEE Trans. Nucl. Sci. **67**(8), 1946–1951 (2020)
62. Göök, A., Hambsch, F.-J., Oberstedt, A., Oberstedt, S.: Application of the Shockley–Ramo theorem on the grid inefficiency of Frisch grid ionization chambers. Nucl. Instrum. Methods Phys. Res. Sect. A. **664**, 289–293 (2012)

Phase Diagram, Melt Growth, and Characterization of $Cd_{0.8}Zn_{0.2}Te$ Crystals for X-Ray Detector

Ching-Hua Su and Sandor L. Lehoczky

Abstract In this study, the solidus curve of the $Cd_{0.8}Zn_{0.2}Te$ homogeneity range was constructed from the partial pressure measurements by optical absorption measurements which provided the information of the melt-growth parameters to achieve crystals with the required electrical resistivity. The melt growth of $Cd_{0.8}Zn_{0.2}Te$ crystals was then processed by directional solidification under controlled Cd overpressure. During the growth experiments, several procedures have been developed to improve the crystalline quality: (1) reducing the structural defects from wetting by HF etching of fused silica ampoule, (2) minimizing the contamination of impurities during homogenization, (3) promoting single-crystal growth by mechanical pulsed disturbance, and (4) maximizing electrical resistivity and minimizing Te precipitates by controlling Cd overpressure during growth and post-growth cooling. Additionally, the thermal conductivity, electrical conductivity, and Seebeck coefficient of a vapor-grown CdTe and two melt-grown $Cd_{0.8}Zn_{0.2}Te$ crystals were measured between 190 °C and 780 °C to provide an in-depth understanding of the thermal and electrical conduction mechanisms of the crystals as well as the prospect of its thermoelectric applications.

1 Introduction

In the application of room-temperature high-energy radiation detector, there are two critical requirements for the detecting semiconductor such as CdZnTe [1–4]. The first one is high electrical resistivity, greater than 10^9 Ω-cm, to reduce the bulk leakage current. This requirement can only be met with materials of low carrier concentrations which are controlled by the concentrations of intrinsic defects, i.e., native point defects such as Cd vacancy, and extrinsic point defects, including intentional and unintentional dopants. The information on the solidus curve and

C.-H. Su (✉) · S. L. Lehoczky
Materials and Processing Laboratory, Engineering Directorate, EM31 NASA/Marshall Space Flight Center, Huntsville, AL, USA
e-mail: ching.h.su@nasa.gov

© The Author(s), under exclusive license to Springer Nature Switzerland AG 2022
K. Iniewski (ed.), *Advanced Materials for Radiation Detection*,
https://doi.org/10.1007/978-3-030-76461-6_11

235

the equilibrium vapor pressures over this narrow homogeneity range compound is therefore an important database in the control of the concentration of native point defects. The other critical requirement is the reduction of structural defects which act as trapping and recombination centers. Various defects that have been shown to adversely affect charge transport in CdZnTe [5, 6] are grain boundaries, twins, Te inclusions, dislocations, and subgrain boundaries [7]. Due to the strong trapping at grain boundaries, high-performance radiation detectors are almost exclusively fabricated from CdZnTe single crystals. Even so, defects within the single crystals such as Te precipitates/inclusions [8], subgrain boundaries, slip planes, and dislocation can also cause charge transport problems. In this case, the knowledge of homogeneity range is also important in that the solid solubility limit determines the boundary of the formation of Te-rich precipitates in the grown crystals.

The solidus curve delineates the solubility of group II or VI elements in the II-VI compound and the existence of homogeneity range is the material's nature to reduce the Gibbs energy of formation by increasing its entropy term [9]. Therefore, the solubility, i.e., the maximum amount of solute that can be incorporated into a solid compound, is zero both at the temperature of absolute zero and at the maximum melting temperature (due to the phase transition). Along the temperatures in between, the solubility goes through a maximum as a function of temperature, known as retrograde solubility. The maximum non-stoichiometry was estimated for the $Cd_{1-x}Zn_xTe$ solid solutions, for $x \leq 0.15$, based on the projection of the pressure-temperature diagram [10–12] and it was found that the $Cd_{1-x}Zn_xTe$ solidus gradually shifts toward Te-rich side with increasing ZnTe content, x. With most of the studies focusing on the processing of CdTe and $Cd_{0.9}Zn_{0.1}Te$ crystals by melt-growth techniques, such as vertical gradient freeze (VGF) [13], high-pressure Bridgman (HPB) [6, 7], and traveling heater method (THM) [14] for the application of X-ray detectors [15], it is worthwhile to investigate other composition of the pseudo-binary, such as $Cd_{0.8}Zn_{0.2}Te$ which has been grown by traditional vertical Bridgman but needed improvements in several areas as compared to $Cd_{0.85}Zn_{0.15}Te$ [16]. Additionally, the higher energy bandgap of $Cd_{0.8}Zn_{0.2}Te$ increases the maximum achievable electrical resistivity than that of $Cd_{0.9}Zn_{0.1}Te$. The extra Zn also promotes solution hardening which helps to reduce the dislocation density. In this study, the solidus curve of $Cd_{0.8}Zn_{0.2}Te$ was determined from the partial pressure measurements by optical absorption technique which provided the information for the melt-growth parameters to process crystals with the required electrical resistivity.

During the practice of melt-growth experiments, several procedures have been developed to improve the crystalline quality by minimizing the contamination of impurities as well as improving the structural defects within the grown crystals to enhance charge transport properties. These procedures are listed briefly below and will be given in detail later.

- A procedure was developed to homogenize pure elements through localized eutectic reaction using traveling zone process.
- The Cu and other elemental contaminations were minimized by homogenizing the CdZnTe ampoule under external vacuum condition.

- The internal surface of the empty growth ampoule was etched by HF before the loading of starting material to reduce the interaction between sample and fused silica ampoule during growth.
- By controlling the Cd overpressure during growth, high-resistivity CdZnTe crystals have been consistently grown with the In dopant concentration of 4–6 ppm, atomic.
- A mechanical pulsed perturbation was applied to growth ampoule during the nucleation stage of crystal to promote the yield of single crystal.
- An optimal temperature for Cd reservoir was established to minimize the density of Te inclusions.
- A cooling schedule of Cd reservoir temperature during and after the growth process was developed to improve the radial uniformity of electrical properties in the grown crystal.

2 Phase Diagram of $Cd_{0.80}Zn_{0.20}Te$ Solid Solution

The data on the composition–temperature–partial pressures $(x_{Te}-T-P_{Te2})$, corresponding to the Te-saturated CdTe solid, have been established by the partial pressure measurements using the optical absorption method [17]. Recently, the same method has been conducted for the pseudo-binary of $Cd_{0.80}Zn_{0.20}Te$ [18]. Four samples with known masses of Cd, Zn, and Te were reacted in fused silica optical cells of known volume profile. The partial pressures of Te_2 and Cd, in equilibrium with the samples between 485 and 1160 °C, were determined by measuring the optical density of the vapor phase from the ultraviolet to the visible range. The composition of the condensed phase or phases was then calculated from the original masses and the amount of material in the vapor phase to establish the corresponding $x_{Te}-T-P_{Te2}$ data, including five Te-rich solidus points.

The detailed description of the experimental setup and procedures has been presented in Ref. [18]. A brief summary is given here. The T-shaped optical cells were made of fused silica. The top of the T-shaped cell consisted of an 18 mm OD and 15 mm ID cylindrical tube with flat, parallel quartz windows at both ends. The optical path length is either 10 or 5 cm. The bottom of the T-cell was a sidearm made by attaching a 6 cm long, 12 mm OD, and 8 mm ID tube to the midpoint of the cell proper, referred to as stem, which was joined coaxially by a 10 cm long 18 mm OD and 15 mm ID tube referred to as reservoir. The volumes of the empty cells were measured as a function along the length of the cell by adding distilled water. The cells were cleaned and then baked at 1180 °C for 18 h under vacuum.

The starting elemental materials were 99.9999% purity Cd rod, Te bar, and Zn teardrops. The elements were weighed by a Mettler AT-201 balance, with a resolution of 10μg, except for CZT-1, which was weighed by a Chyo Jupiter M1-20 microbalance with a resolution of 1μg. Four CdZnTe cells, CZT-1 to -4 and a calibration cell each for the Cd and Te element, as listed in Table 1, were prepared. The weighed elements were loaded directly into the baked-out optical cells, which

Table 1 The sample name, optical path length, total mass, and atomic fractions of Zn, x_{Zn}, and Te, x_{Te}, of each optical cell

Sample	Optical path (cm)	Total mass (g)	x_{Zn}	x_{Te}
CZT-1	9.83	20.539284	0.099954	0.500206
CZT-2	9.90	20.84812	0.099917	0.500494
CZT-3	9.90	20.50930	0.099995	0.500015
CZT-4	5.00	20.47089	0.100016	0.499990
Cd	9.90	2.1345	0	0
Te	9.90	3.316	0	1.0

were sealed off at a vacuum level less than 2×10^{-8} atm. The sealed cells were placed inside a five-zone T-shaped furnace. The furnace with the optical cell inside was then placed in a double-beam reversed-optics spectrophotometer (OLIS Inc., model 14H) with the optical cell proper in the path of the sample beam, and the reference beam passed under the furnace. The optical density, defined as $D = \log_{10}(I_{reference}/I_{sample})$, where I is the intensity (of reference or sample beam), was measured between the wavelengths of 200 and 700 nm. The light source was a deuterium lamp for wavelengths below 275 nm and a xenon lamp for wavelengths above 275 nm. The typical instrument band pass was 0.2, 1.2, and 2.7 nm, respectively, at the wavelengths of 600, 290, and 200 nm. The temperature of the optical cell proper, $T_{O.C.}$, was kept at 1050, 1100, or 1150 $\pm$ 2 °C while the temperatures of the stem and the reservoir sections decreased monotonically to T_R, i.e., the temperature of the reservoir for the sample. A baseline spectrum was measured first for each $T_{O.C.}$ with the reservoir temperature below 400 °C.

The typical spectra of optical density vs. wavelength on a CZT sample for a series of runs with increasing sample temperatures are presented in Fig. 1, which shows the absorption peak of Cd atom at 228.7 nm and a series of vibronic absorption peaks from diatomic Te molecule, Te_2, between 350 and 550 nm. The partial pressures of Te_2, P_2, and Cd, P_{Cd}, were derived as the following. First, the partial pressure of Te_2 was determined from the measured optical density using the calibration constants from a pure Te cell between 400 and 600 nm. Usually, the pressure was taken as the average of the values measured at 4–6 wavelengths where the optical density was between 0.1 and 2.5 with the upper limit being 3.0 for the detector. To calculate the partial pressure of Cd, the measured optical density in the UV region was assumed to be the result of a linear superposition of Cd and Te_2 absorption. The contribution of Te_2 to the optical density in the UV region was calculated from the measured Te_2 partial pressure and the calibration constants of pure Te. This contribution was subtracted from the measured optical density and the partial pressure of Cd, P_{Cd}, was calculated from the remaining optical density using the calibration results from an optical cell of pure Cd.

The measured values of P_2 for all of the four CZT runs, together with the pressure over pure Te, P_2°, are plotted against 1000/T (T is T_R in K) in Fig. 2a. The measured P_2 shows that, for CZT-1, -2, and -3, the samples started out as Te saturated at low temperature. As the temperature increased, sample CZT-1 and -3 moved into the

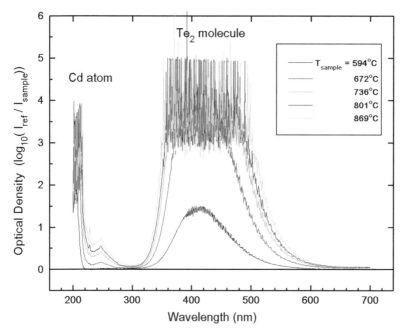

Fig. 1 Five spectra of optical density vs. wavelength for a CZT sample from a series of runs with different sample temperatures show the absorption of Cd atom at 228.7 nm and a series of vibronic absorption peaks from diatomic Te molecule, Te_2, between 350 and 550 nm

homogeneity range as the measured P_2 started to flatten out and, eventually, joined the Te-saturated loop at a higher temperature. For CZT-2, the sample composition remained Te saturated as the measured P_2 followed the so-called three-phase loop throughout most of the measuring temperatures. For CZT-4, only high-temperature measurements were performed and the data show that the sample composition was inside the homogeneity range at the low end of the measured temperatures and moved onto the Te-saturated loop at a higher temperature. The compositions of CZT-1, -2, and -3 were initially inside the two-phase field of Te-rich $Cd_{0.80}Zn_{0.20}Te$ (s) + Te-rich melt, and this condition remained for CZT-2 throughout the runs. As the temperature and the equilibrium P_2 increased, samples CZT-1 and -3 were losing Te to the vapor phase faster than the other two components. Therefore, their compositions became progressively less Te rich and moved inside the homogeneity range of $Cd_{0.80}Zn_{0.20}Te$ solid. As the temperature kept increasing, the measure P_2 flattened inside the homogeneity range as the partial pressure of the group II components kept on increasing. The sample composition eventually started to move back toward the Te-rich direction and finally crossed the Te-rich homogeneity range and became a mixture of Te-rich $Cd_{0.80}Zn_{0.20}Te$ (s) + Te-rich melt again. Sample CZT-4 started inside the homogeneity range and crossed the solidus temperature into the Te-rich mixture as the temperature increased. As the temperature further increased, the measured P_2 for all samples decreased rapidly following the

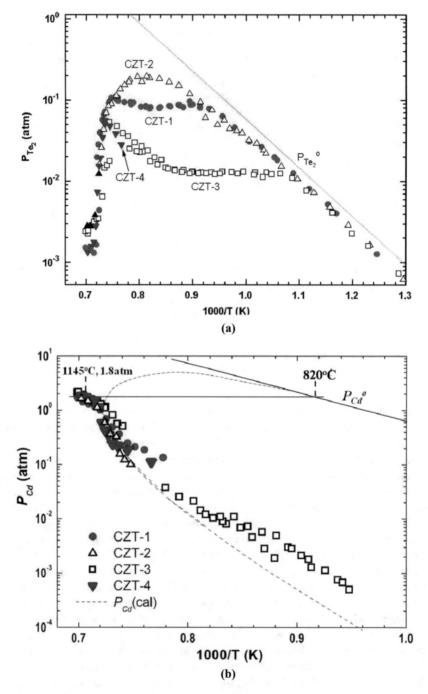

Fig. 2 Measured partial pressures of Te$_2$ (**a**) and Cd (**b**) for four samples of $(Cd_{1-x}Zn_x)_{1-y}Te_y$. The upper dashed curve in (**b**) is the approximated P_{Cd} for the Cd-saturated condition (taken from Fig. 3 in Ref. [18])

three-phase loop. At even higher temperature, all four samples broke away from the three-phase loop, and the measured P_2 flattened as the temperature increased. Supposedly, the break point temperature should be close to the liquidus temperature of the nonstoichiometric $Cd_{0.80}Zn_{0.20}Te$ samples with the Te content higher than 0.50.

The measured P_{Cd} for all four samples is shown in Fig. 2b together with P_{Cd} over pure Cd element, P^o_{Cd}. The data are consistent with the P_2 shown in Fig. 2a in that all samples went through the homogeneity range except CZT-2, which was Te saturated throughout the runs. As shown in Fig. 2b, the small variation in the sample compositions did not affect the measured P_{Cd} in the melt as much as the large differences in the measured P_{Cd} over the homogeneity range of the solid. For the growth condition of directional solidification, to be described in detail later, the melt was maintained at a temperature of 1145 ± 5 °C and the measured P_{Cd} in equilibrium with the melt at this temperature, 1.8 atm, can be provided by a pure Cd reservoir at 820 °C with the Cd pressure over pure Cd, P^o_{Cd}, expressing as a function of temperature by [19]

$$\log_{10} P^o_{Cd}(\text{atm}) = -5317/T(K) + 5.119 \qquad (1)$$

Assuming that the vapor phase is ideal, the Gibbs energy of formation of CdTe in $Cd_{0.80}Zn_{0.20}Te(s)$ from $Cd(g)$ and $Te_2(g)$ at 1 atm was calculated from the measured P_{Cd} and P_2 for the data points inside the homogeneity range as well as on the three-phase loop and they can be fitted well using the equation [18]

$$\Delta G_{f,CdTe}(x = 0.20) = RT\ln(P_{Cd}P_2^{1/2}) = -68642 + 44.4956T(\text{cal/mole}) \qquad (2)$$

where R is the gas constant and T is in K. Using Eq. (2), the Cd partial pressure for the Te-saturated section of $Cd_{0.80}Zn_{0.20}Te(s)$ was calculated from the corresponding measured P_2 and is plotted in Fig. 2b as the lower dotted line which is consistent with the measured P_{Cd}.

The partial pressure of Zn was then derived from the Gibbs energy of formation of ZnTe in $Cd_{0.80}Zn_{0.20}Te(s)$ which was approximated from the best-fit quasi-regular interaction parameters determined in Ref. [19] for the CdTe-ZnTe solid solution. As the temperature increased the composition of the condensed phase, or phases, in equilibrium with the partial pressures changed because of the incongruent sublimation of the sample. Assuming that the gas phase is ideal, the masses of Cd, Zn, and Te in the vapor phase can be calculated from their partial pressures and the measured volume and temperature profiles for each data point [18]. By subtracting the mass of each element in the vapor phase from the original masses, the composition of the solid sample at each data point can be calculated and is shown in Fig. 3. The solidus curve on the Te-rich side of the homogeneity range gives the estimated maximum limit to be $x_{Te} = 0.50012$ at 977 °C.

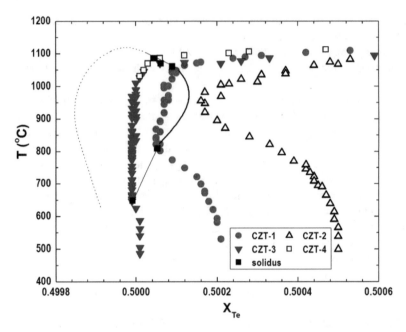

Fig. 3 The sample composition, x_{Te}, calculated from each partial pressure data point and the five solidus points, shown as solid black squares, determined from these data. The maximum limit is estimated to be $x_{Te} = 0.50012$ at 977 °C. The dotted curve on the Cd-rich region is arbitrarily drawn (taken from Fig. 7 in Ref. [18])

3 Procedures to Improve Quality of Melt-Grown CdZnTe Crystals

3.1 Melt Growth Under Controlled Cd Overpressure

The conditions for the growth of inclusion-free CdZnTe single crystals have been previously investigated. In the vertical gradient freeze growth of $Cd_{1-x}Zn_xTe$ ($x = 0.04$) as the lattice-matched substrate for the epitaxial growth of HgCdTe IR detectors, the Cd pressure at the congruent melting point was estimated and the corresponding amounts of Cd were added to the ampoules to reach 1–1.3 atm of Cd [20]. Asahi et al. [21] and Koyama et al. [22] have grown $Cd_{1-x}Zn_xTe$ ($x = 0.03$–0.04) by VGF under a Cd overpressure to minimize the size of precipitates in the grown crystals. Szeles et al. [23, 24] have grown semi-insulating $Cd_{0.9}Zn_{0.1}Te$ crystals for the room-temperature X- and gamma-ray radiation detectors using an electrodynamic gradient technique with a controlled Cd overpressure.

Under the crystal growth environments, to be presented later, the partial pressures of Te_2 and Cd over the melt at 1145 °C differ by three orders of magnitude. During the crystal growth of CdZnTe by directional solidification, a starting material of stoichiometric composition, i.e., $x_{Te} = 0.5000$, will lose more Cd to the vapor phase than Te, and the composition of the condensed phase will shift toward the Te-richer direction. The amount of this shift depends on the total sample mass, temperature, and volume profiles. There are two consequences the shift creates: (1) Cd vacancy and (2) Te-rich second phase. Based on a defect chemistry analysis to interpret the x_{Te}–T–P_{Te2} data for $Cd_{0.8}Zn_{0.2}Te$, the best-fit parameters [25] showed that the native defects associated with a Te-rich crystal are believed to be mainly doubly ionized Cd-vacancy acceptor with essentially no ionized Te anti-site donor [26]. Secondly, if the composition of Te becomes high than 0.50012, i.e., the maximum of homogeneity range, the solidified crystal will consist of a mixture of CdZnTe solid and a Te-rich second phase, which was usually stated as Te inclusion/precipitate. Even if the shift in Te composition is not larger than 0.50012, with the usually slow cooling rate after the growth, the solidified crystal might still have the Te-rich precipitates embedded inside because of the retrograded nature of the solidus curve. Therefore, it is recommended to provide the equilibrium partial pressures over the melt during crystal growth to maintain the condensed phase as close as possible toward stoichiometric composition. Strictly speaking, one needs to investigate the thermodynamics of the Cd-Zn-Te system to determine the composition and temperature of a ternary Cd, Zn, and Te reservoir that can provide exactly the respective partial pressures of Cd, Zn, and Te_2 [27]. However, since the partial pressure of Cd is two orders of magnitude higher than the others, it is usually sufficient to provide a constant Cd overpressure using a pure Cd reservoir. The precise Cd reservoir temperature will be determined by the temperature of the $Cd_{0.8}Zn_{0.2}Te$ melt during the crystal growth and the post-growth cooling. The measured Cd overpressure, over the melts at different temperatures as shown in Fig. 2b, can be fit with the equation

$$\log_{10} P_{Cd}(atm) = -10560/T(K) + 7.702 \qquad (3)$$

For the melt growths employing different thermal profiles, e.g., for the melt surface at temperatures of 1132, 1150, and 1170 °C, the Cd overpressure can be derived from Eq. (3) to be, respectively, 1.53, 1.91, and 2.42 atm, which correspond to a pure Cd reservoir of 805, 826, and 850 °C, respectively, from Eq. (1).

3.2 Reduction of Interaction Between Samples and Fused Silica Ampoules by HF Etching

During the processing of electronic/optic compound semiconductors at elevated temperature, the sample has the tendency to interact with the fused silica container and form chemical bonding, the so-called wetting, which makes the sample attaching to the ampoule wall. During the cooldown of the sample after the processing, as the bulk of the sample goes through larger thermal contraction than fused silica, the wetting area remains attached to the fused silica wall and, consequently, causes the sample to separate apart which causes cracks and other structural defects. These defects are detrimental to the electronic/optic performance of materials, especially on the applications of high-quality compound semiconductors.

The severity of the chemical interaction between the sample and fused silica inner wall depends on several factors: (1) the chemical activity of the sample and the fused silica, (2) the processing temperature, (3) the duration at the temperature, and (4) the contact area of the interaction. For example, the heating schedule in the homogenization of HgCdTe was designed to reduce the chemical activity of Cd by annealing the sample first at lower temperature to eliminate the wetting [28]. A dewetted growth of CdTe was practiced on earth, provided by a gas pressure in the crucible which equalized the hydrostatic pressure in the melt [29]. But the contactless growth was stable only for the first 25 mm of the growth because of the changing of the solid-liquid interface shape. In general, the first three of these factors are case dependent; that is, they are different for various material systems as well as different processing conditions. To find a general solution for each different case, it is desired to focus on the fourth factor, i.e., to reduce the contact area between the sample and the container. The total area of the grown crystal contacting the container depends on the surface features of the inner wall. A very flat and smooth surface provides a large contact area whereas a rough surface, with peaks and valleys, limits the contact area to the peaks due to the surface tension of the melt/crystal. With this concept, a simple method to modify the surface morphology of the fused silica inner wall was developed by etching it with HF acid before any crystal growth activity. Typically, the HF rinse of 30 s to 2 min has been widely adopted in the cleaning of the surface of fused silica inner tubing before any processing activities. However, a long-time HF etching of 30 min can modify the surface features of fused silica and affect the interaction between the sample and its container. The effects of the HF etching have been demonstrated [30] in the crystal growth by physical vapor transport of ZnSe and Fe-doped ZnSe as well as melt growth of PbTe compound semiconductors. The method is applicable to all processes involving interaction between materials and fused silica container at elevated temperatures.

3.3   Homogenization of Starting Materials Under External Vacuum Environment

The starting materials were prepared from 99.99999% purity Cd and Zn and 99.9999% purity Te. The weighed elements were loaded inside fused silica ampoules, which have been previously cleaned, HF etched, and baked out under vacuum. The loaded ampoules were evacuated and sealed under vacuum condition. Because of the large mass of the sample, more than 300 g, the ampoules are sometimes explored inside the rocking furnace during heating up. It was concluded that the exothermic Te eutectic reaction between metals and Te released a large amount of energy, which rapidly raised the sample temperature and caused the instantaneous jump of vapor pressure inside the ampoule. To mitigate this problem, the loaded homogenization ampoule was first gone through a localized eutectic reaction by a traveling zone process. The procedure was similar to a zone-melting process except that the zone temperature was set to be just high enough, about 510 °C, to induce localized Te eutectic reaction (about 450 °C). The zone traveled at 15–20 cm/h and a spongelike sample was formed after the zone process.

For the first few runs, the homogenization ampoules were loaded inside a resistance heated tubular furnace for the melting and mixing of the charge. The spongy-looking material was heated up to 1070 °C, soaked for 36 h, then raised to 1160 °C, and rocked for 3–5 h before casting by turning off the furnace power vertically. However, the electrical properties and the impurity analysis of the grown crystals, as shown in the later section of characterization, implied the contamination of Cu, C, and O impurity. During the later runs, the contamination was successfully minimized by loading the homogenization ampoules inside another closed tubing of fused silica with a larger diameter, which was pumped to provide an exterior vacuum of 10^{-3} Torr during the homogenization process. Even with the modified homogenization process and the controlled Cd overpressure, the electrical property of undoped grown crystals was not consistent, with the electrical resistivity, ρ, varying from 10^3 to 10^8 Ω-cm, presumably caused by the residual impurities. By adding the intentional indium dopant, 4–6 ppm (atomic), to the pure elements during homogenization, the electrical resistivity of the grown samples was consistently above 10^8 Ω-cm when the Cd reservoir was maintained between 785 and 825 °C, with details given in later sections.

3.4 Mechanical Pulsed Disturbance to Promote Single-Crystal Growth

For certain semiconductors with important applications, the existing unseeded bulk growth of directional solidification from the melt usually results in poor-quality multi-crystalline section in the first-grown section which causes the low yield of the commercial growth process. The multi-grained crystal growth was partially caused by the large supercool of the melt, which not only results in a large section of ingot solidifying uncontrollably under spontaneous nucleation [31] but also prohibits the ideal growth condition for single-crystal nuclei forming at the very tip of the ampoule and growing into large single grains. The DTA measurements on a CdTe sample [32] started with heating the sample up rapidly to above its melting point of 1092 °C and, after soaking for 9 h, a furnace cooling rate of about 10 °C/min is commenced. During the cooling of the furnace, the sample temperature stayed at 1092 °C and eventually started to decrease when the furnace reached 1040 °C, i.e., a supercooling of 50 °C. The degree of supercooling was also reported [33] to be correlated to the Cd partial pressure over the CdTe melt where the supercooling was studied by measuring the electrical conductivity during solid-liquid transition. It was found that the degree of supercooling decreases from 23 °C to 13 °C and to 8 °C when the Cd overpressure was raised from 1.3 atm to 1.4 atm and to 1.6 atm, respectively.

To prevent the undesired formation of a large multi-grained spontaneous nucleation, i.e., to promote nucleation under the condition of small supercooling, a short-time mechanical perturbation was applied to the growth ampoule at a critical time during growth when the melt at the ampoule tip just reached below the liquidus temperature. The technique was implemented to the directional solidification process of $Cd_{0.80}Zn_{0.20}Te$ crystals [32] by adding a solenoid AC vibrator, which was bound to the extension of the growth ampoule, with a frequency of 60 Hz and an adjustable magnitude. The high-frequency shaking of the ampoule for 10–30 s causes local inhomogeneity in the supercooled melt, which promotes the nucleation in the melt and, consequently, a small section of solid, usually single or double grain, was formed at the growth tip which grew continuously throughout most of the ingot length. The effects of acoustic wave from the vibrations have also been studied analytically through the classical nucleation theory [34]. It was found that the proximate effect of acoustic pressure is to reduce both the size of the critical nucleus and the work required to form it from monomers. As the work serves to be the activation energy, the ultimate effect of acoustic pressure is to increase the rate of nucleation. However, if the atomic structure of the nucleus is the same as that of an ordinary solid, the compressibility is too small for acoustic vibration effects to be noticeable. If on the other hand, the structure is similar to that of a loosely bound colloidal particle, then the effects of acoustic vibration become potentially observable.

3.5   Crystal Growth and Post-growth Cooling

The growth ampoule was made of fused silica with a diameter ID × OD from 20 mm × 25 mm to 40 mm × 45 mm and a tapered length of 2.5 cm at the growth tip. The ampoules were cleaned, HF etched, and baked at 1180 °C under vacuum condition for 16 h. The homogenization ampoules were opened and the starting materials were ground into particles with dimension less than 5 mm. After the starting material has been loaded inside the growth ampoule, a basket holding about 2 g of pure Cd was inserted on the top of the ampoule as the Cd reservoir, which was fixed beneath the seal-off cup. The ampoules were then sealed under a vacuum lower than 10^{-5} Torr.

The Bridgman growth furnace was set up vertically with four independently controlled electrical resistance heating zones. A typical thermal profile and the initial ampoule position are shown schematically in Fig. 4. The thermal profile was provided by the heating zones of, from top to bottom, the Cd reservoir zone (which was equipped with a heat pipe—the isothermal furnace liner), the hot zone, the booster zone, and the cold zone. The liquidus temperature of $Cd_{0.8}Zn_{0.2}Te$ was

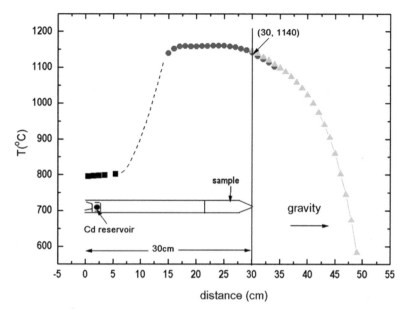

Fig. 4 The thermal profile and the initial ampoule position for a typical crystal growth

determined to be between 1115 and 1132 °C [19, 35, 36]. Figure 4 shows that the starting melt was about 7 cm long and a total length of 32 cm is required for the growth ampoule so as to maintain the starting melt above 1130 °C with a positive thermal gradient (to prevent bubbling in the melt) and an independently controlled Cd reservoir on the top of the ampoule. The melt was soaking for at least 48 h at 1145 °C in order to completely dissolve any local clusters of Te-rich chains as reported in Ref. [37–39] that clusters, in the forms of branched chains of Cd_nTe_{3n+1} or Te atoms, were observed in $Cd_{1-x}Zn_xTe$ ($0 \leq x \leq 0.1$) during the dynamic viscosity measurements. The Cd reservoir temperature from 750 to 935 °C was employed with the range of 785–820 °C used for most of the runs. The furnace translation rates ranged from 0.75 to 2 mm/h (as mostly used).

After the growths were completed, with a furnace translation time of 100–125 h, the solidified sample and the Cd reservoir were cooled down to room temperature with constant rates over a period of 96–144 h. However, after slicing the ingots perpendicularly, the measured electrical resistivity of the sliced disc was not uniform across the radial section [40]. For instance, on the wafer cut at 2.5 cm from the tip of the ingot CZT-26, with a Cd reservoir of 785 °C, the center of the ingot showed electrical resistivity of 2×10^9 Ω-cm, whereas the resistivity at the edge was 40 Ω-cm. An adjustment of cooling schedule was adopted for the CZT-29 run. The environments of the CZT-26 and -29 ingots at the end of crystal growth and during the 2 days' cooling are shown in Fig. 5 as red and black boules, respectively, relative to the three-phase diagram of Cd given in Fig. 2b. As shown in the top-left corner of the figure, at the end of the growth, when the top of the ingots was at the

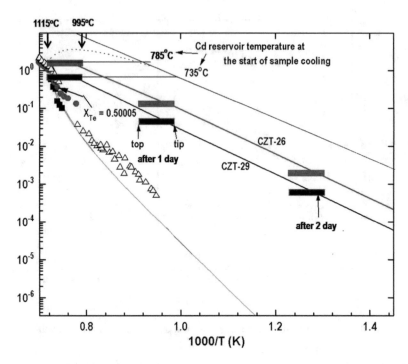

Fig. 5 The Cd three-phase loop for $Cd_{0.80}Zn_{0.20}Te$ and the stoichiometric positions for CZT-26 and CZT-29 during post-growth cooling (taken from Fig. 3 in Ref. [40])

solidus temperature, 1115 °C, the tip of the 8 cm long crystals was at 995 °C, as deduced from the thermal profile of the growth furnace. During the late stage of growth and long duration of cooling, the Cd overpressure of CZT-26 was too high, especially for the earlier grown section, which caused the Cd vapor to diffuse into the ingot and shifted the stoichiometry of the surface region. To optimize the environment of Cd overpressure, the Cd reservoir temperature was programmed to cool down for 50–80 °C in the last 55–75 h of furnace translation. Figure 5 shows that, for the growth of CZT-29, the Cd reservoir temperature cooled down from 785 to 735 °C in the last 70 h of growth. So at the end of the furnace translation, while the whole CZT-26 sample was in equilibrium with a Cd pressure of 1.7 atm the Cd overpressure over CZT-29 was 0.69 atm. The electrical property measurements confirmed the radial uniformity of the CZT-29 crystal. The black solid line in Fig. 5 for CZT-29 corresponds to the optimal temperatures for the crystals and the Cd reservoir during the post-growth cooling. In an effort to grow inclusion-free single crystal of CdTe, Franc et al. [20] have reported the stoichiometric line of Cd overpressure, defined by $x_{Cd} = x_{Te} = 0.5$, during the growth of CdTe as P^S (atm) $= 8 \times 10^5$ exp.$(-1.76 \times 10^4/T)$. A comparison between their P^S and our optimal P_{Cd} for CZT-29 shows that the two lines are in the same pressure range with the slope of their line a little steeper than ours. For instance, at the temperatures of

995, 743, and 510 °C, respectively, the values for P^S are 0.75, 0.024, and 1.4×10^{-4} atm and our optimal pressures are 0.46, 0.032, and 5.2×10^{-4} atm. However, the differences between these two lines should be pointed out. Firstly, the P_{Cd} line for CZT-29 is the optimal line to maintain a constant stoichiometry of the grown crystal during the cooling process, whereas the P^S from Ref. [20] was the stoichiometric line for $x_{Cd} = x_{Te} = 0.5$. Secondly, the compositions of the crystals are different, i.e., $Cd_{0.8}Zn_{0.2}Te$ in our case and CdTe in their study.

4   Characterizations

The crystalline quality and morphology of the grown crystal were first examined by visual observation on the as-grown ampoule. Then the cylindrical crystal was sliced perpendicularly by a wire saw at specific locations from the first freeze tip and a 2 mm thick wafer was obtained. A 2 mm × 2 mm × 20 mm prism was sliced from the wafer for the chemical analysis of glow discharge mass spectroscopy (GDMS) provided by Shiva Technologies. Other samples were sliced, lapped, and mechanical polished for the subsequent Hall measurements. Slices from crystals grown under different Cd reservoir temperatures were prepared by lapping and polishing for the examinations under IR microscope to study the structure defects, especially for the size and density of Te inclusions.

4.1   Crystalline Quality and Morphology by Visual Observations

4.1.1   Effects of the HF Etching Pretreatment

The effects of HF etching on the reduction of interaction between samples and fused silica ampoules were visually examined on the as-grown ampoules and the results are shown in the pictures below. Figure 6a shows the grown crystal without the pretreatment of HF etching. Although the grown ingot was a single crystal, it broke into three long axial pieces as shown in the inset at lower left corner. Presumably, at least three wetting areas caused the sample surface to attach to the inner wall of the ampoule and separate the ingot axially into three pieces during cooling. Usually, the HF-treated samples detached from the wall intact and slid freely inside the ampoules after applying gentle horizontal movements as shown in Fig. 6b.

4.1.2   Effects of Mechanical Disturbance

The growth environments and procedures for two 20 mm diameter ingots of CZT-37 and CZT-38 were implemented as similar as possible except that the mechanical

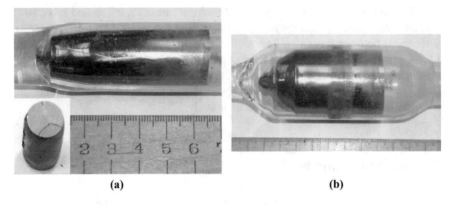

Fig. 6 The pictures of two as-grown CdZnTe ampoules. (**a**) CZT-39 ampoule: without HF treatment resulted in the sample separating axially into three pieces as shown by the inset; (**b**) CZT-36 ampoule: with HF treatment, as-grown sample detached and slid inside the ampoule

tapping was introduced to the latter but not the former. They were both grown with furnace translation of 1.25 mm/h, growth time of 125 h, and cooling time of 96 h except the values for total mass and Cd reservoir temperature (95.2 g, 820 °C and 91.9 g, 805 °C) for CZT-37 and -38, respectively. Figure 7a shows that the cross section of CZT-37 at 1.5 cm from the tip has at least five grains and some twinnings and the cross sections at 1.8 cm and 2.5 cm, given in Fig. 7b, also exhibit multiple grains. Figure 7c shows the slices cut at 1.5 cm and 1.8 cm from the ingot CZT-38 which exhibit a monocrystalline structure. The cross section at 1.5 cm has some twinnings which disappeared at the cross section of 1.8 cm. The longitudinal cut of the remaining CZT-38 ingot shown in Fig. 7d extends the monocrystalline structure through the ingot till the last 1.0 cm section which resulted in the coverage of single grain, more than 70% of the ingot. The single crystallinity has been confirmed from the X-ray diffraction spectra.

Several 40 mm diameter ingots have also been processed with and without mechanical disturbance. Without the mechanical perturbation, as shown in Fig. 8a, the as-grown ingot of CZT-16 shows multiple crystalline grains with twins. Another grown ingot without mechanical tapping, CZT-20, as shown in Fig. 8b, also exhibits multiple grains. On the other hand, the ingot of CZT-36, grown with the mechanical perturbation, which slid inside the growth ampoule as shown in Fig. 6b, was sliced axially from the tip to 2.5 cm. The first-grown section shows two grains in the tapered shoulder area as given in Fig. 8c. Twinnings occurred in one of the grains but they were limited to the first 1.5 cm section. The ingot developed into one major and one minor crystalline grain, as shown in the cross-section area at 2.5 cm given in Fig. 8d, with the major grain covering more than 70% of the sample.

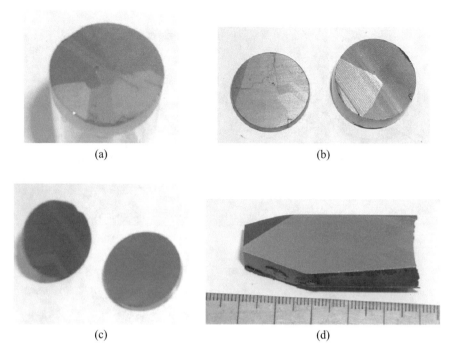

(a)   (b)

(c)   (d)

Fig. 7 CZT-37, grown without the mechanical perturbation: (**a**) shows the cross section at 1.5 cm from the tip and (**b**) shows cross sections at 1.8 cm and 2.5 cm. CZT-38, grown under the same conditions as CZT-37 except that the mechanical perturbation was applied: (**c**) shows the slices cut at 1.5 cm and 1.8 cm from the ingot and (**d**) shows the longitudinal cut of the remaining ingot

4.2 Chemical Analysis

The effect of homogenization under external vacuum condition was evaluated by chemical analysis. The impurity concentrations of samples cut from the crystals were analyzed by GDMS and the results from two undoped ingots are described here. The growth conditions for the two runs were similar: the furnace translation rate was 1 mm/h for both runs and the Cd reservoir temperatures were 770 °C and 805 °C for CZT-2 and CZT-9, respectively. The major difference was that CZT-2 was homogenized in an atmospheric environment, whereas CZT-9 was homogenized in an external pressure of less than 10^{-3} Torr vacuum condition. Table 2 presents the analyses of two samples cut at 2.5 cm from their first-to-freeze tips. The results show that homogenization in the vacuum environment significantly reduces the impurity concentrations in the grown crystals, especially in the concentrations of the major impurities Cu, C, O, and Fe. The electrical conductivity measurements were performed on the remains of these two wafers. The results were consistent with the GDMS analysis in terms of the measured p-type electrical resistivity of 8×10^4 Ω-cm and 3×10^9 Ω-cm, respectively, for the CZT-2 and CZT-9 2.5 cm samples.

Fig. 8 Two 40 mm diameter ingots grown without the mechanical perturbation: (**a**) multiple crystalline grains with twins in the grown ingot of CZT-16 and (**b**) multiple grains in the grown ingot of CZT-20. The 40 mm diameter ingot, CZT-36, grown with the mechanical perturbation: (**c**) the axially sliced section of the first 2.5 cm sample showing two grains with twins in the tapered shoulder area and (**d**) the cross-section area at 2.5 cm from the grown tip with the major grain covering more than 70% of the sample

4.3 Electrical Properties' Measurements

Lake Shore, Model 7500 Hall Effect Measurement Systems with a magnetic field strength up to 10 kg were used for the measurements of the electrical properties. The cut wafers were chemical-mechanical polished to about 1 mm in thickness and electrical contacts were made by depositing Au film, using RF sputtering. The Hall data measurements confirmed that the contacts were ohmic. The measured

Table 2 The impurity concentrations (in ppb, atomic) measured by GDMS on samples cut from the grown crystals of CZT-2 and -9

Sample Element	CZT-2; 2.5 cm Concentration	CZT-9; 2.5 cm Concentration
C	980	20
N	45	15
O	980	35
Na	12	54
Mg	170	27
Al	26	7
Si	11	8
S	230	280
Cl	56	49
Cr	28	67
Mn	<10	60
Fe	480	93
Ni	23	46
Cu	3300	110
Se	160	210

Table 3 The measured Hall coefficients and resistivity, ρ, at room temperature, and the derived electron and hole concentrations, n and p, from two grown ingots, CZT-26 and CZT-29

Sample	Hall coeff. [cm^3/C]	ρ [Ω cm]	n [cm^{-3}]	p [cm^{-3}]
CZT-26; 2.5 cm	8.9×10^9	2.1×10^9	3×10^5	2.8×10^6
CZT-29a; 2.5 cm	-1.84×10^{13}	8.4×10^9	$\sim 1.7 \times 10^6$	$<10^6$
CZT-29b; 4.0 cm	-1.2×10^{11}	1.2×10^9	1×10^6	4×10^7

Taken from Table 1 of Ref. [40]

resistivities of the In-doped samples grown with Cd reservoir between 785 and 825 °C were consistently higher than 10^8 Ω-cm and up to 2×10^{11} Ω-cm. Two quantities, electrical resistivity and Hall coefficient, were measured which are determined by four material parameters, namely, electron concentration, n; hole concentration, p; electron mobility, μ_n; and hole mobility, μ_p. The governing equation for resistivity, ρ, is given by

$$\rho = 1/(ne\mu_n + pe\mu_p), \tag{4}$$

and the equation for Hall coefficient, R_H, under the condition $\mu_n \gg \mu_p$ (as is the case in CdZnTe) can be expressed as

$$n = \frac{1}{\rho e \mu_n}\left(\frac{\mu_p}{\mu_n} - \frac{R_H}{\rho \mu_n}\right). \tag{5}$$

To obtain the electron and hole concentrations, the values of 1000 and 100 cm^2/Vs for the electron and hole mobilities, respectively, measured on $Cd_{0.8}Zn_{0.2}Te$ samples at room temperature [41], were used. Table 3 lists the measured Hall coefficients and resistivities at room temperature, and the calculated electron and hole concentrations for two ingots, CZT-26 and CZT-29; both were doped with In (4.2 ppm, atomic) and grown with a Cd reservoir at 785 °C. Since the CZT-26 wafer showed radial nonuniformity in electrical properties, the values given here are for the center section of a wafer cut at 2.5 cm from its tip. The CZT-29a and -29b samples were sliced at 2.5 cm and 4.0 cm, respectively, from the tip of CZT-29.

4.4   Infrared (IR) Transmission Microscopy

The IR transmission images were taken on sliced and polished wafers, about 2 mm thick, by an IR camera and IR microscope to study the structural defects, such as cracks, grain boundary, twins, as well as Te precipitates. The IR transmission image of a typical 20 mm diameter single-crystal wafer, presented in Fig. 9a, shows no evidence of crack, grain boundary, and twin. Some foggy line segments might have been the images of polishing lines on the surfaces. Figure 9b shows the image from IR microscope taken on a wafer sliced from a crystal grown under a Cd reservoir of 717 °C. With the scale of 1 mm, the size of the small dots, presumably Te precipitates, ranges from 10 to 20μm. The identity of the three larger particles, with the size of the lower right corner at 65μm, is not clear. They might have been the Te inclusions from the engulfment of Te-rich liquid droplet due to the fluctuation of the growth interface. Figure 9c–f show the IR micrographs from crystals grown under Cd overpressure with increasing Cd reservoir temperature (given at the bottom of each image). From the trend of the Te precipitate density, it was concluded that a Cd reservoir temperature of 820 ± 10 °C resulted in the lowest precipitate density.

From the above results on the optimal Cd reservoir temperature for the directional solidification of a $Cd_{0.8}Zn_{0.2}Te$ melt maintaining at 1145 °C, it is summarized that the temperature range of 785–825 °C will fulfill the requirement of a high level of resistivity (section "Electrical Properties' Measurements") and 820 ± 10 °C will minimize the structural defects (section "Infrared (IR) Transmission Microscope"). Therefore, it is claimed that the employment of a Cd reservoir temperature of 820 ± 5 °C during the growth process will provide the optimal Cd pressure over the melt to maximize the electrical resistivity as well as minimize the structural defects, including Te precipitates/inclusions of the grown $Cd_{0.8}Zn_{0.2}Te$ crystals. Since the solids of different compositions, x in the $Cd_{1-x}Zn_xTe$ system, have different liquidus/solidus temperatures as well as different homogeneity ranges, the procedure presented here for the $Cd_{0.8}Zn_{0.2}Te$ solid may not be applicable to other compositions. As a comparison, inclusion-free CdTe crystal was reported to have been grown by vertical Bridgman technique using a Cd reservoir of 850 °C with a melt temperature of 1118 °C [42].

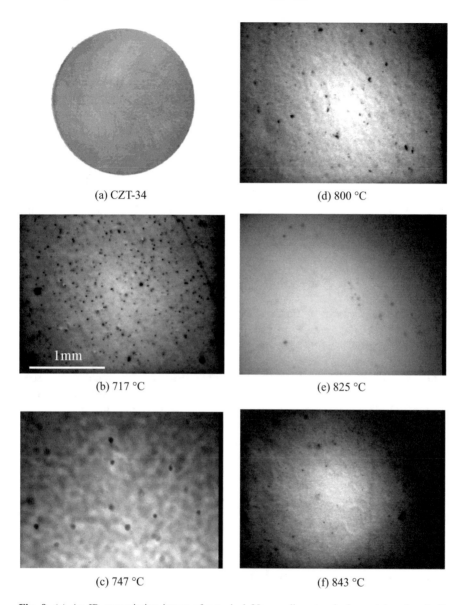

(a) CZT-34

(d) 800 °C

(b) 717 °C

(e) 825 °C

(c) 747 °C

(f) 843 °C

Fig. 9 (**a**) An IR transmission image of a typical 20 mm diameter single-crystal wafer; (**b–f**) transmission images of IR microscope taken from wafers sliced from crystals processed under different Cd reservoir temperatures

4.5   Thermoelectric Properties of CdZnTe

The thermal conductivity is one of the thermophysical properties needed for any meaningful simulation of the growth process. The extreme high level of electrical resistivity at room and cryogenic temperatures makes it difficult for the measurements and interpretation of the results. The measurements of a lower electrical resistivity in the elevated temperature together with Seebeck coefficients can provide an in-depth understanding of the electrical conduction mechanisms in the crystals. Therefore, the thermal conductivity, electrical conductivity, and Seebeck coefficient of a vapor-grown CdTe and two melt-grown $Cd_{0.8}Zn_{0.2}Te$ crystals, including In-doped CZT-30 and un-doped CZT-34 samples, were measured between 190 °C and 780 °C to study the thermoelectric properties of CdZnTe. Additionally, from the three sets of measurements it would be interesting to investigate the thermoelectric properties of CdZnTe since there have been very few evaluations [43] on the potential thermoelectric applications of this wide-bandgap semiconductor system.

4.5.1   Crystal Growth by Physical Vapor Transport (PVT)

The CdTe crystals have been grown by a horizontal seeded PVT process [44, 45]. The present measurements were performed on the crystal of CdTe-15 which was grown with the following procedures. After a CdTe single-crystal seed with growing face of (112) was positioned in a fused silica growth ampoule of 16 mm ID, 22.7 g of homogenized CdTe ($x_{Te} = 0.50002$) starting material was loaded. The loaded ampoule was baked under vacuum at 880 °C for 5 min to adjust the stoichiometry of the starting material before sealing off under vacuum. During the growth, the CdTe source was maintained at 880 °C with a temperature bump and gradient of 35 °C/cm at the supersaturation location where the seed surface was positioned. After 281.5 h of growth with furnace translation rate of 0.158 mm/h, the furnace was cooled to room temperature in 32 h. All of the starting material has been transported.

4.5.2   Thermoelectric Property Measurements

The single-crystal samples, in the shape of 20 mm diameter and 3.0 ± 0.5 mm thick disc, were cut perpendicularly to the growth axis from the grown crystals by a wire saw. The slices were 2.4 cm, 2.6 cm, and 2.4 cm from the first grown location for CdTe-15, CZT-30, and CZT-34, respectively, and were characterized by the following.

Thermal Conductivity

The thermal conductivity, κ, was determined from the following equation:

$$\kappa = \rho C_p \alpha \qquad (4)$$

where ρ, C_p, and α are density, heat capacity, and thermal diffusivity, respectively. The thermal diffusivity was measured by the Flashline 3050 System from Anter Corp. (now TA Instruments) between room temperature and 760 °C. In the flash technique, when the sample was heated and in equilibrium with a preset temperature, the radiant energy of a high-intensity light pulse was absorbed on the front surface of the disc and the resultant temperature rise on the rear face was recorded. The diffusivity was calculated from the thermogram using the Clark and Taylor analysis [46]. For each temperature, the diffusivity was taken from the average of three successful measurements.

The density of CdZnTe was calculated from the lattice constants, a, of CdTe and ZnTe from the Vegard's law:

$$a(Cd_{1-x}Zn_xTe) = (1 - x)a(CdTe) + x\ a(ZnTe) \qquad (5)$$

The room-temperature lattice constants have been reviewed by Williams [47] to be 0.6481 and 0.6103 nm, respectively, for CdTe and ZnTe. The density at elevated temperature was determined by the lattice constant values at room temperature and their linear thermal expansion coefficient values reviewed in Ref. [47] and in Ref. [48] for CdTe and ZnTe, respectively.

As described earlier, the partial pressure of Zn can be derived from the results of the best-fit quasi-regular interaction parameters determined in Ref. [19] for the CdTe-ZnTe solid solution. In a quasi-regular solution of $Cd_{1-x}Zn_xTe$, the heat capacity, C_p, is given by

$$C_p(Cd_{1-x}Zn_xTe) = (1 - x)\ C_p(CdTe) + x\ C_p(ZnTe) \qquad (6)$$

where the heat capacity for CdTe was adopted from K.C. Mills [49] to be $0.1668 + 1.377 \times 10^{-4}\ T(K)$ (J/g-K) and $C_p(ZnTe)$ is given by $0.2407 + 5.638 \times 10^{-5}\ T(K)$ (J/g-K) [50].

The Electrical Conductivity and Seebeck Coefficient

The electrical conductivity and Seebeck coefficient were measured simultaneously by the ULVAC ZEM-3 instrument from room temperature to 800 °C. The samples were sliced into the shape of roughly 2 mm × 2 mm × 15 mm rectangular prism. The electrical conductivity was determined from ten measured I-V points and the

Seebeck coefficient was derived from the values of Seebeck voltages measured over 6 mm distance along the sample under three different applied thermal gradients.

4.5.3 Results and Analysis

Thermal Conductivity

The measured thermal conductivities from room temperature to 750 °C for the three samples are given in Fig. 10. The thermal conductivity of CdTe was slightly above 0.08 W/cm-K at room temperature, decreased to and stayed around 0.035 W/cm-K when temperature reached above 200 °C, and then slowly decreased when temperature reached above 700 °C. The thermal diffusivity of CdTe has been measured by a similar laser-flash method between 930 and 1085 °C [51]. Their thermal diffusivity data were multiplied by the values of heat capacity and density at the temperatures to obtain the values for thermal conductivity which are also shown in Fig. 10. Those data are in line with the extension of present data from the lower temperature range.

The thermal conductivity for the In-doped CdZnTe, CZT-30, stayed below 0.03 W/cm-K in the low-temperature range with a shallow minimum at about 250 °C. The un-doped CdZnTe sample, CZT-34, has the thermal conductivity similar to that of CZT-30 in the low-temperature range and increased to a slight

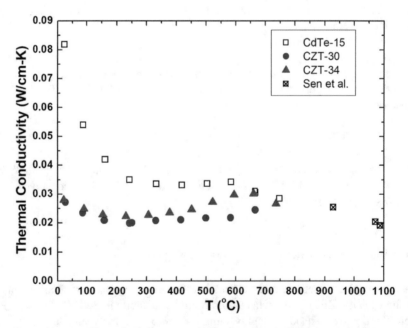

Fig. 10 The measured thermal conductivity for CdTe, In-doped (CZT-30), and un-doped (CZT-34) Cd$_{0.80}$Zn$_{0.20}$Te samples plotted against temperature. The three data for CdTe (Sen et al.) at temperature above 900 °C are from Ref. [51] (taken from Fig. 1 of Ref. [43])

maximum at about 650 °C. All three sets of the thermal conductivity data merged together at about 750 °C—an implication that the samples became intrinsic as the phonon contribution to thermal conductivity started to dominate at elevated temperatures.

Electrical Conductivity

The measured electrical conductivities for the three samples are plotted in log scale vs. 1000/T(K) as shown in Fig. 11. The CdTe crystal, grown from a starting material being heat treated to approach congruent sublimation condition [45], contains very small amount of native defects (donor or acceptor) and hence is close to be intrinsic as evidenced from the almost linear dependence of its electrical conductivity (in log scale) vs. 1000/T, especially for temperature above 440 °C, or 1000/T <1.4. High-temperature electrical conductivity of CdTe single crystals has also been measured previously in the temperature range from 400 to 1200 °C [52], shown as a line segment in the figure, which is about 50% higher than the present data.

The data for the In-doped CdZnTe, CZT-30, showed low values of electrical conductivity in the low-temperature range and also showed the intrinsic character-istics by forming a linear line roughly parallel but lower than those data for CdTe due

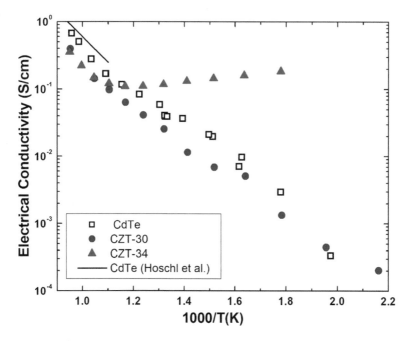

Fig. 11 The measured electrical conductivities plotted in a scale of $\log_{10}$ vs. 1000/T(K). The line segment at the upper left corner shows data from Ref. [52] (taken from Fig. 3 of Ref. [43])

to the larger bandgap of CdZnTe. It has been shown that the intrinsic carrier concentration of CdTe can be fitted well [53] by the following form for a nondegenerate semiconductor with parabolic valence and conduction bands:

$$n_i = NT^{3/2} \exp\left(-E_g/2kT\right) \tag{7}$$

where N is a constant (related to band structure parameters), T is in K, E_g is the effective bandgap energy, and k is the Boltzmann constant. From the measurements of optical transmission at 300 K on CdZnTe [54], the measured cut-on energy for $x = 0$ and 0.20 was 1.464 and 1.572 eV, respectively. Assuming that the bandgap difference, 0.108 eV, remains the same and similar band structure parameters at elevated temperatures, the ratio of intrinsic carrier concentration of $Cd_{0.8}Zn_{0.2}Te$ to that of CdTe would be

$$n_i(Cd_{0.8}Zn_{0.2}Te)/n_i(CdTe) = \exp(-0.108 \ eV/2kT) \tag{8}$$

Assuming that the mobilities for the carriers are the same for $Cd_{0.8}Zn_{0.2}Te$ and CdTe at a fixed temperature, this ratio in Eq. (8) will also be the ratio for their individual electrical conductivity. At the temperature of 1000 K, this ratio of n_i would be 0.534 from Eq. (8) which is comparable to the ratio of the measured electrical conductivity at 1000 K, as interpolating from Fig. 11 to be 0.54. The electrical conductivity of the un-doped CZT-34 sample started with a value of 0.3 S/cm at around 300 °C and decreased to a minimum of about 0.1 S/cm and then merged with the In-doped sample at a temperature above 650 °C.

Seebeck Coefficient

The types of electrical conduction of CdTe-15 were revealed from the Seebeck coefficient measurements as shown in Fig. 12. In the low-temperature range, the Seebeck coefficients were positive—indicating a p-type dominant conduction, presumably from native acceptors of Cd vacancy. The measured Seebeck coefficients showed a rather high value of 1.1 mV/K at the lowest measured temperature of 346 °C. As temperature increased, the measured Seebeck coefficients decreased slowly as the sample approached to be intrinsic and the conduction type converted to n-type around 680 °C due to the mobility difference—about 300 cm^2/Vs for electrons and 15 cm^2/Vs for holes at 680 °C as reported by Smith [55].

The Seebeck coefficients of CZT-30 indicated n-type conduction at low temperature, converted to p-type at about 250 °C, reached a maximum of 0.9 mV/K between 400 and 500 °C, and converted back to n-type at 770 °C. The small amount of n-type In doping compensated the residual native acceptor Cd/Zn vacancy and showed n-type characteristics in the low-temperature range. As the temperature increased, the ionization of native acceptors converted the sample into the p-type conduction until at even higher temperatures when the sample approached to be

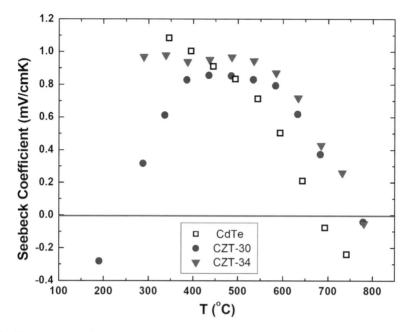

Fig. 12 The measured Seebeck coefficient of CdTe, In-doped (CZT-30), and un-doped (CZT-34) $Cd_{0.80}Zn_{0.20}Te$ samples plotted against temperature. Positive and negative Seebeck coefficients represent, respectively, p-type and n-type dominant electrical conduction (taken from Fig. 4 of Ref. [43])

intrinsic and the conduction type converted back to n-type due to the high mobility of electrons.

The Seebeck coefficients measured on un-doped CZT-34 sample in Fig. 12 showed p-type conduction at low temperatures with a plateau of 0.97 mV/K between 250 and 550 °C, then merged with In-doped CZT-30, and converted to n-type at 770 °C. The trend confirms that, for the un-doped sample, the native acceptor was formed from the Te-rich range of stoichiometry during the solidification. The figure of merit for thermoelectric application, zT, is defined as

$$zT = \alpha^2 \sigma T / \kappa \qquad (9)$$

where α is the Seebeck coefficient, σ the electrical conductivity, T the temperature in K, and κ the thermal conductivity. The calculated values of zT, as shown in Fig. 13, for the three samples between 190 and 780 °C were orders of magnitude lower than the state-of-the-art p-type thermoelectric materials mainly due to the low values of electrical conductivity. As a simple estimate, the value of zT for CdTe can be improved to 1.0 at 500 °C if the carrier concentration at 500 °C reaches p-type 1×10^{18} cm^{-3}, either from native acceptors or intentional dopants. For an estimate on the n-type thermoelectric properties for CdTe at elevated temperature, the large uncertainty in extrapolating the Seebeck coefficient gives a value of zT to be

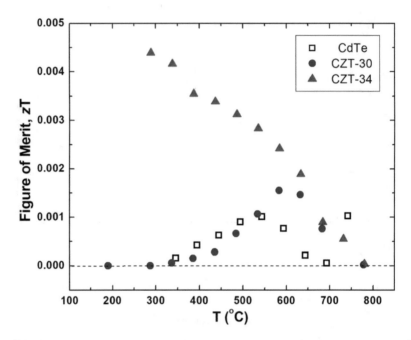

Fig. 13 The calculated figure of merit for thermoelectric application, zT. All of the data are p-type conduction except the data at the two highest temperatures for CdTe as well as the data at the highest temperature for both CZT samples and the lowest temperature for CZT-30 are n-type (taken from Fig. 5 of Ref. [43])

1.2 ± 0.4 for n-type CdTe at 1050 °C, compared to the state-of-the-art value of around 1.0 for the n-type SiGe alloy.

5 Summary

The data on the composition–temperature–partial pressures (x_{Te}–T–P_{Te2}), corresponding to the Te-saturated $Cd_{0.80}Zn_{0.20}Te$ solid, have been established by measuring the partial pressure using the optical absorption technique. The solidus curve on the Te-rich side of the homogeneity range gives the maximum limit to be $x_{Te} = 0.50012$ at 977 °C. During crystal growth, the measured P_{Cd} in equilibrium with a $Cd_{0.80}Zn_{0.20}Te$ melt, at a temperature of 1145 ± 5 °C, is 1.8 atm, which can be provided by a pure Cd reserved of 820 °C.

Several procedures have been developed to improve the crystalline quality: (1) a homogenization procedure to minimize the contamination of Cu, C, O, and Fe; (2) a HF etching treatment of the empty fused silica ampoule to reduce the interaction between sample and fused silica ampoule during growth; (3) a Cd reservoir at 820 ± 10 °C over the melt to consistently grow high-resistivity CdZnTe crystals

(In dopant concentration of 4–6 ppm, atomic) with minimal amount of Te inclusions/ precipitates; (4) a mechanical pulsed perturbation to the growth ampoule during the nucleation stage of crystal to promote the yield of a single crystal; and (5) a cooling schedule of Cd reservoir temperature after the growth process to improve the radial uniformity of electrical properties in the grown crystal. The measured thermal conductivity, electrical conductivity, and Seebeck coefficient of a vapor-grown CdTe and two melt-grown $Cd_{0.8}Zn_{0.2}Te$ crystals between 190 °C and 780 °C have provided an in-depth understanding of the thermal and electrical conduction mechanisms and the prospect of its thermoelectric applications.

Acknowledgements The authors would like to acknowledge the supports of Biological and Physical Sciences Division, Science Mission Directorate, NASA Headquarters, and NASA ROSS (Research Opportunities in Space Science) project.

References

1. Rudolph, P.: Fundamental studies on Bridgman growth of CdTe. Prog. Cryst. Growth Charact. **29**, 275 (1994)
2. James, R.B., Schlesinger, T.E., Lund, J., Schieber, M.: Semiconductors and Semimetals, vol. 43, p. 335. Academic Press, New York (1997)
3. Schlesinger, T.E., Toney, J.E., Yoon, H., Lee, E.Y., Brunett, B.A., Franks, L., James, R.B.: Mater. Sci. Eng. **32**, 103 (2002)
4. Del Sordo, S., Abbene, L., Caroli, E., Mancini, A.M., Zappettini, A., Ubertini, P.: Sensors. **9**, 3491 (2009)
5. Szeles, C., Driver, M.C.: SPIE Proc. Ser. **3446**, 1 (1998)
6. Amman, M., Lee, J.S., Luke, P.N.: J. Appl. Phys. **92**, 3198 (2002)
7. Szeles, C.: IEEE Trans. Nucl. Sci. **51**, 1242 (2004)
8. Carini, G.A., Bolotnikov, A.E., Camarda, G.S., Wright, G.W., James, R.B.: Appl. Phys. Lett. **88**, 143515 (2006)
9. R. F. Brebrick, Progress in Solid State Chemistry, vol.13, Ed. H. Reiss, (Pergamon Press, Oxford 1967) p. 213 . Ch. 5
10. Guskov, V.N., Greenberg, J.H., Fiederle, M., Benz, K.-W.: J. Alloys Compd. **371**, 118 (2004)
11. Greenberg, J.H.: Prog. Cryst. Growth Charact. Mater. **47**, 196 (2003)
12. Greenberg, J.H., Guskov, V.N.: J. Crystal Growth. **289**, 552 (2006)
13. Franc, J., Moravec, P., Hlidek, P., Belas, E., Hoschl, P., Grill, R., Sourek, Z.: J. Electron. Mater. **32**, 761 (2003)
14. Chen, H., Awadalla, S.A., Iniewski, K., Lu, P.H., Harris, F., Mackenzie, J., Hasanen, T., Chen, W., Redden, R., Bindley, G., Kuvvetli, I., Budtz-Jørgensen, C., Luke, P., Amman, M., Lee, J.S., Bolotnikov, A.E., Camarda, G.S., Cui, Y., Hossain, A., James, R.B.: J. Appl. Phys. **103**, 014903 (2008)
15. Fougeres, P., Siffert, P., Hageali, M., Koebel, J.M., Regal, R.: Nucl. Instrum. Methods Phys. Res. Sect A. **A428**, 38 (1999)
16. Li, G., Jie, W., Hua, H., Zhi, G.: Prog. Cryst. Growth Charact. Mater. **46**, 85 (2003)
17. Fang, R., Brebrick, R.F.: J. Phys. Chem. Solids. **57**, 443 (1996)
18. Su, C.-H.: J. Cryst. Growth. **281**, 577 (2005)
19. Yu, T.-C., Brebrick, R.F.: J. Phase Equilib. **13**, 476 (1992)
20. Franc, J., Grill, R., Hlidek, P., Belas, E., Turjanska, L., Hoschl, P., Turkevych, I., Toth, A.L., Moravec, P., Sitter, H.: Semicond. Sci. Technol. **16**, 514 (2001)

21. Asahi, T., Oda, O., Taniguchi, Y., Koyama, A.: J. Cryst. Growth. **161**, 20 (1996)
22. Koyama, A., Hichiwa, A., Hirano, R.: J. Electron. Mater. **28**, 587 (1999)
23. Szeles, C., Eissler, E., Reese, D.J., Cameron, S.E.: SPIE Conf. Proc. **3768**, 98 (1999)
24. Szeles, C., Cameron, S.E., Ndap, J.-O., Chalmers, W.C.: IEEE Trans. Nucl. Sci. **49**, 2535 (2002)
25. Brebrick, R.F., Fang, R.: J. Phys. Chem. Solids. **57**, 451 (1996)
26. Brebrick, R.F.: Private Communication
27. Sang, W., Qian, Y., Shi, W., Wang, L., Yang, J., Liu, D.: J. Crystal Growth. **214/215**, 30 (2000)
28. Su, C.-H., Lehoczky, S.L., Szofran, F.R.: J. Appl. Phys. **60**, 3777 (1986)
29. Fiederlea, M., Duffar, T., Garandet, J.P., Babentsov, V., Fauler, A., Benz, K.W., Dusserre, P., Corregidor, V., Dieguez, E., Delaye, P., Roosen, G., Chevrier, V., Launay, J.C.: J. Cryst. Growth. **267**, 429 (2004)
30. Su, C.-H.: Cryst. Res. Technol. 1900208 (2020)
31. Su, C.-H., Lehoczky, S.L., Szofran, F.R.: J. Cryst. Growth. **86**, 87 (1988)
32. Su, C.-H.: J. Cryst. Growth. **410**, 35 (2015)
33. Belas, E., Grill, R., Franc, J., Turjanska, L., Turkevych, I., Moravec, P., Hoschl, P.: J. Electron. Mater. **32**, 752 (2003)
34. Baird, J.K., Su, C.-H.: J. Cryst. Growth. **487**, 65 (2018)
35. Steininger, J., Strauss, A.J., Brebrick, R.F.: J. Electrochem. Soc. **117**, 1305 (1970)
36. Haloui, A., Feutelais, Y., Legendre, B.: J. Alloys Compd. **260**, 179 (1997)
37. Shcherbak, L., Feychuk, P., Plevachuk, Y., Dong, C., Kopach, O., Panchuk, O., Siffert, P.: J. Alloys Compd. **371**, 186 (2004)
38. Shcherbak, L., Feichouk, P., Panchouk, O.: J. Cryst. Growth. **161**, 16 (1996)
39. Shcherbak, L.: J. Cryst. Growth. **197**, 397 (1999)
40. Su, C.-H., Lehoczky, S.L.: J. Crystal Growth. **319**, 4 (2011)
41. Burshtein, Z., Jayatirtha, H.N., Burger, A., Butler, J.F., Apotovsky, B., Doty, F.P.: Appl. Phys. Lett. **63**, 102 (1993)
42. Rudolph, P., Engel, A., Schentke, I., Grochocki, A.: J. Crystal Growth. **147**, 279 (1995)
43. Su, C.-H.: AIP Adv. **5**, 057118 (2015)
44. Chattopadhyay, K., Feth, S., Chen, H., Burger, A., Su, C.-H.: J. Cryst. Growth. **191**, 377 (1998)
45. Su, C.-H., Sha, Y.-G., Lehoczky, S.L., Liu, H.-C., Fang, R., Brebrick, R.F.: J. Cryst. Growth. **183**, 519 (1998)
46. Clark, L.M., Taylor, R.E.: J. Appl. Phys. **46**, 714 (1975)
47. Williams, D.J.: Properties of Narrow Gap Cadmium-based Compounds. In: Capper, P. (ed.) EMIS Datareviews series, (INSPECT publication, 1994), p. 399 (1994)
48. Bhargava, R.N.: Properties of Wide Bandgap II-VI Semiconductors. In Bhargava, R. (ed.) EMIS Datareviews series, (INSPECT publication, 1997), p. 27 (1997)
49. Mills, K.C.: Thermodynamic Data for Inorganic Sulphides, Selenides and Tellurides. Butterworths, London (1974)
50. Barin, I., Knacke, O., Kubaschewski, O.: Thermochemical Properties of Inorganic Substances. Supplement. Springer-Verlag, Berlin (1977)
51. Sen, S., Konkel, W.H., Tighe, S.J., Bland, L.G., Sharma, S.R., Taylor, R.E.: J. Cryst. Growth. **86**, 111 (1988)
52. Hoschl, P., Belas, E., Tujanska, L., Grill, R., Franc, J., Fesh, R., Moravec, P.: J. Cryst. Growth. **220**, 444 (2000)
53. Su, C.-H.: J. Appl. Phys. **103**, 084903 (2008)
54. Johnson, S.M., Sen, S., Konkel, W.H., Kalisher, M.H.: J. Vac. Sci. Technol. **B9**, 1987 (1991)
55. Smith, F.T.J.: Met. Trans. **1**, 617 (1970)

Melt Growth of High-Resolution CdZnTe Detectors

Saketh Kakkireni, Santosh K. Swain, Kelvin G. Lynn, and John S. McCloy

Abstract CdTe and its alloy CdZnTe possess optimal physical properties for room-temperature high-resolution X-ray and gamma-ray detector applications in areas of homeland security and medical imaging. Melt growth and Te-rich flux growth are primary methods employed to produce bulk single crystals of CdTe/CZT. Reduction of performance-limiting defects including impurities, native defects and their complexes, and secondary phases, while ensuring high resistivity and single crystal yield, is desirable in this technology. Although CZT is the most successful semiconductor detector technology, further advances can be realized by exploring pathways for cost-effective growth of high-quality material. Melt growth methods can achieve faster growth rates; however, the transport properties and detector quality are typically inferior compared to crystals grown from Te-rich flux. The latter growth method can achieve very high purity due to solvent purification and low growth temperature. Recent studies indicate that similar benefits can be achieved in melt growth configuration by making certain modifications to allow faster growth from a Te-rich melt. The modifications included adjustments to melt composition, growth temperature, and forced melt convection during growth by accelerated crucible rotation technique (ACRT). Resulting crystals can attain electron lifetime of nearly 80 microseconds, without subjecting the crystals to post-growth annealing. The growth method, material characteristics in terms of secondary phases, purity, homogeneity, and performance of the detectors are discussed in this chapter. Potential approaches towards integration of the benefits of various known bulk growth methods in order to further advance CZT technology are discussed.

S. Kakkireni · S. K. Swain (✉) · K. G. Lynn · J. S. McCloy
Institute of Materials Research, Washington State University, Pullman, WA, USA
e-mail: swain@capesym.com; kgl@wsu.edu; john.mccloy@wsu.edu

© The Author(s), under exclusive license to Springer Nature Switzerland AG 2022
K. Iniewski (ed.), *Advanced Materials for Radiation Detection*,
https://doi.org/10.1007/978-3-030-76461-6_12

1   Introduction

Among the state-of-the-art technologies for room-temperature X-ray and gamma-ray detection, direct conversion devices based on CdTe and its alloy CdZnTe (typically up to 10% Zn) are superior compared to scintillators in terms of both energy and spatial resolution [1]. Favorable physical properties of these materials include high stopping power, high average atomic number (Z), high resistivity $>10^{10}$ Ω cm, and electron drift lengths reaching several centimeters [2]. These properties allow for compact, stable, and high-resolution devices for nuclear safeguards, medical imaging [3, 4], as well as astrophysics research [5]. CdTe/CdZnTe detectors also have advantages over other direct conversion technologies, such as silicon and high-purity germanium (HPGe), which require cooling and large volumes of material to achieve high spectroscopic efficiency and resolution. Due to these properties, CdZnTe is considered the most promising room-temperature detector available to date. Figure 1 shows the evolution of electron mobility-lifetime product $(\mu\tau)_e$ for various high-Z compounds considered for room-temperature detector applications over the past several decades [1].

Despite these excellent properties, compared to common scintillators (e.g., NaI and CsI) and elemental semiconductors (Si and Ge), it is relatively challenging to

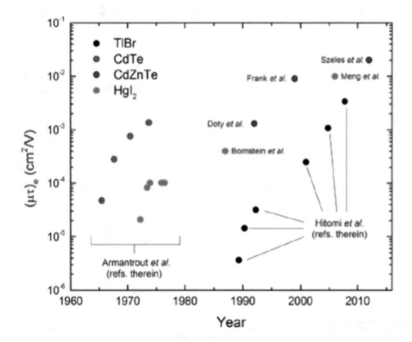

Fig. 1 Comparison of $(\mu\tau)_e$ and their time evolution for various high-Z wide-gap semiconductors in use today. Reprinted from P. Johns et al., J. Appl. Phys. 126, 040902 (2019), with the permission of AIP Publishing [4] and references therein

grow large single crystals of high-quality CdTe/CdZnTe at fast growth rates, which has been a bottleneck for large-scale commercial applications. Unfavorable thermophysical properties including poor thermal conductivity of solid that is lower than that of the melt, low critically resolved shear stress (CRSS), high symmetry of zinc blende crystal structure, and presence of Te inclusions make the crystal growth of high-quality material a challenging task [6]. While these challenges are well documented in literature and various techniques, each with its own set of advantages and disadvantages has been pursued to grow detector-grade CdTe/CdZnTe [7], producing high-quality detector-grade materials at considerable volumes, and lower cost is still a necessity. *Innovation in crystal growth is necessary to make this possible.* Major performance-limiting factors include deep-level point defects associated with impurities, native defects, as well as extended defects such as Te inclusions, precipitates, dislocations, twin, and subgrain boundaries [8]. Nevertheless, continued interest in the material due to the broad range of current and potential new applications has led to advancement in crystal growth [9–11], post-growth processing [12], and device designs [13–15]. These developments have made it possible to realize more than tenfold increase in electron lifetime over the past decade, detector volumes as high as ≈ 25 cm^3 [16], and energy resolution <1% @ 662 keV which is well suited for nuclear safeguard applications [17]. Similarly, CdTe/CdZnTe technology is currently increasingly adopted for medical imaging applications such as single-photon emission computed tomography (SPECT) and spectral CT due to recent improvements in the performance under high irradiation flux [3, 18]. Here better energy resolution and photon processing capability enable better quality image while allowing low dose, reduced exposure, patient comfort, and better safety of patients and healthcare workers [19].

At present, the travelling heater method (THM) is the most successful technique for commercial production of CdTe and CdZnTe, with Redlen Technologies (Canada) [20], Kromek (UK) [3], and Acrorad (Japan) [21] as the leading commercial vendors. The crystals grown by this method have a major advantage in achieving high-purity materials due to low growth temperatures and solvent purification effect of tellurium [22, 23], in addition to better compositional uniformity as the Te-rich flux zone composition and temperature remain constant during the entirety of the growth [24]. This technique also results in large-volume single-crystalline yield compared to other techniques that have been used for commercial production in the past [24, 25]. However, the technique has some well-known limitations that include slow growth rates in the order of 5 mm/day or less [24, 26] and presence of tellurium inclusions that require post-growth annealing for their elimination/reduction in order to realize optimum detector performance [24]. These factors influence the overall yield of high-quality crystals leading to increased cost, limited supply, and performance reliability issues.

On the other hand, techniques based on melt growth, such as vertical or horizontal Bridgman techniques, either under high pressure or in evacuated sealed volume, are generally inferior in terms of detector performance, yield, and homogeneity. The growth temperatures in melt growths are typically 300–400 °C higher than typical THM growths. This is a source of performance-limiting point defects and extended defects [6, 27]. In addition, nonuniformity due to segregation of Zn and dopants contributes to low yield of usable detector volume. However, melt growth techniques have advantages, such as fast growth rates in mm/h range and the ability to scale up to large crystal sizes, which are some of the reasons behind early commercial interest in these methods. For example, the high-pressure Bridgman (HPB) method had been used in the past by eV Products, Inc. (now under Kromek, Inc.), for commercial CdZnTe detector production, but the nonuniformity of the grown crystals resulted in very low detector yield, resulting in a shift from HPB to THM [3]. Orbotech Medical Solutions Ltd. (Israel), later acquired by General Electric (GE), uses the horizontal Bridgman technique for producing CdZnTe for medical and space applications [28–30]. In view of the demand for low-cost, large-volume, high-resolution devices, strategies combining the low growth temperature benefits of THM with the fast growth rate advantages of melt growth are shown to be promising [9, 11]. As discussed in this chapter, this is achieved by a forced melt convection technique, namely accelerated crucible rotation technique (ACRT). This technique was initially proposed in the 1970s [31] for overcoming the growth rate limitations in flux growth and later adapted to many crystal growth systems including CdTe/CdZnTe as discussed in the next sections.

At Washington State University (WSU), we have focused on developing strategies to produce high-resolution CdZnTe crystals at fast growth rates without requiring any post-growth treatment [9, 11, 32]. Modifications to the existing Bridgman growth system at WSU have been made to successfully reach this goal. These changes included (a) addition of high excess Te up to 12% to reduce growth temperature and (b) application of forced melt convection to allow for faster growth rate from off-stoichiometric melt and reduce the size of Te inclusions. Resulting crystals demonstrated electron lifetime values up to 80μs [11] and energy resolutions of ~1.1% @ 662 keV (Frisch collar), which is comparable to state-of-the-art commercial detectors. It is important to note that these results are achieved without subjecting the crystals to post-growth annealing and without applying any electronic corrections. The growth method, material characteristics in terms of electrical properties, secondary phases, and detector performance will be discussed. Potential approaches towards increasing the single-crystal volume while maintaining the detector performance are also discussed.

2   Accelerated Crucible Rotation Technique

Segregation of excess component at the growth interface (Te for CdTe/CdZnTe) during crystal growth imposes limitations on growth rates and causes particle trapping during solidification [33]. This phenomenon can be overcome by homogenizing the melt and redistributing the accumulated solute at the growth interface into the bulk of the melt. However, melt homogenization by diffusion and natural convection during crystal growth is a slow process. Additional stirring with the aid of forced convection is essential to accelerate the solidification process by enhancing the mass transfer to the solid-liquid interface and by suppressing deleterious phenomenon such as constitutional undercooling. Forced convection can induce flow velocities which are several orders of magnitude higher than those achieved by natural convection under typical thermal gradients imposed during growth [34]. Accelerated crucible rotation technique (ACRT) is one of the melt-stirring techniques involving periodic rotation of the crucible containing the melt. ACRT-induced flows can suppress thermal and compositional inhomogeneities that occur in static growth conditions [34]. ACRT was initially proposed by Scheel in the 1970s for flux growth of high-temperature solutions which resulted in much larger single crystals of $GdAlO_3$ that was otherwise difficult to achieve [35]. Since then, this technique has been utilized in various crystal growths encompassing vertical Bridgman (VB) [36], electrodynamic gradient (EDG) [11], travelling heater method (THM) [37], and Czochralski growth methods [38]. ACRT has been successfully applied to several materials including SiC [39], CdHgTe [40], CdZnTe [11], Rb_2MnCl_4 [41], YIG [42], InSb [43], ZnTe [44], and Nd:YAG [38], improving dopant uniformity, maximum possible stable growth rates, and single-crystalline yield. Typical parameters of an ACRT cycle are shown in Fig. 2; these include acceleration and deceleration rates, maximum rotation rate, and hold time at maximum rotation rate and at minimum rotation rate.

The optimal ACRT profile is a function of several system parameters such as melt viscosity, density, crystal diameter, and length of the melt column. The mathematical treatment for the hydrodynamics of the fluid flow under ACRT is detailed in the literature, and a brief summary is presented here [34, 45]. Three different flows are known to occur during different stages of ACRT rotation: spiral shearing, Ekman flow, and transient Couette flow. A spiral shearing force that is operational in the bulk of the melt causes a two-dimensional flow occurring during spin-up and spin-down cycles and is beneficial in bulk melt mixing and homogenization of solute. The acceleration of the crucible from rest causes shearing in the melt about the rotation axis, and as a result spiral arms are created in the melt. The concentration deviations

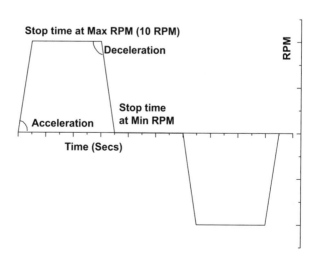

Fig. 2 Illustration of an ACRT rotation profile showing different stages during a full cycle

between the spiral arms are neutralized via diffusion if they are closely spaced, and it is estimated that the concentration differences disappear within a fraction of a second under typical flux growth conditions [34]. Ekman layer flow is the most dominant flow during the spin-up cycle, where the liquid near the solid-melt boundary is pushed away due to centrifugal forces, and this flow is essential for radial mixing of the melt. It also strongly affects the kinetics at the growth interface. This shearing effect at the growth interface is responsible for reducing the concentration boundary layer thickness at the interface and allowing higher stable growth rates and controls the formation of Te inclusions in CdTe/CdZnTe under optimized conditions. The influence of this flow, characterized by Ekman layer thickness, is limited to a region near the solid-melt interface [34]. The bulk of the fluid mixing is achieved during spin-down by the formation of Couette flow patterns. During the spin-down, the layer of fluid adjacent to the wall of the crucible decelerates faster than the fluid at the center (no-slip condition). Thus, the fluid in the central part of the crucible is pushed outward due to inertia, resulting in Couette flow patterns which are responsible for vertical mixing in the bulk of the fluid and eliminating the possible propagation of secondary wall nucleation into the bulk of the crystal [34]. Optimization of an ACRT profile to achieve enhanced mixing is possible by establishing stable flow patterns during different segments of the ACRT cycle. Although theoretical and computational studies are useful in terms of recommendations regarding the choice of ACRT parameters, experimental optimization is usually necessary. Extensive experimental work on the effect of different rotation parameters by Coates et al. suggests that rapid acceleration and declaration are critical for improved mixing and enhanced macrocrystallinity [46]. In practice, stop times and hold times and rotation-reversal are less important than predicted by theory [45, 46]. Horowitz et al. found that longer cycle time (smaller acceleration and deceleration rates) with higher maximum rotation rate yields a better crystal structure at similar imposed growth rates in Rb_2MnCl_4 [41]. Zhou et al. experimentally reported the change of mass transfer rate and interface velocity during an ACRT cycle in Te-doped InSb [43]. They found that enhanced mass transfer resulting from the Ekman flow persists much longer than the theoretical Ekman time. Also, periodic variation of growth rate is observed with a minimum occurring during the spin-up and maximum during spin-down of the ACRT cycle. Numerical simulation of ACRT applied to CdZnTe system shows counterintuitive results; the rotation profile designed based on theoretical criteria for stable flow results in poorer mixing compared to rotation profile designed to maximize the fluid velocities [47]. These discrepancies arise from the complexity in the assessment of the ACRT mechanisms in combination with crystal growth phenomena. Underlying thermally driven flows, three-dimensional nature of the flow structure in real systems, and time-dependent nature of the ACRT-induced flows make it difficult to accurately predict the outcome with present theories.

Hence, most of the work on rotation profile optimization to improve the second-phase distributions in CdZnTe performed at authors' laboratory at Washington State University (WSU) has been through experimentation over the past decade [32, 48, 49]. The following sections discuss growth, properties, and detector performance, reviewing some of our earlier published results and including recent results. The growths are performed in a multi-zone EDG furnace, capable of growing up to 85 mm diameter and ≈152 mm (≈6-in.) tall ingots. The boules reported here are typically ≈61 mm diameter and ≈75–85 mm tall. Growths are performed by loading the raw materials either directly into a graphite-coated quartz ampoule or with an additional pyrolytic boron nitride (pBN) crucible. Use of a pBN crucible generally resulted in relatively less sticking of the ingot to the crucible wall compared to growths directly in quartz ampoules. The raw materials CdTe, ZnTe, and elemental tellurium with 6 N5 (99.9995%) purity are purchased from 5 N Plus, Inc. (Canada). High-resistivity and detector-grade crystal is achieved using a co-doping scheme involving indium (In) [50]. Excess Te, in the range of ~1.5 to 12%, has been studied and crystals are grown under an imposed growth rate of 1–2 mm/h [9, 32]. The melting point of these compositions was estimated from the three-phase diagram and is in the range of 1000–1075 °C.

3   Electrical Properties and Detector Performance

High bulk resistivity $>10^{10}$ Ω cm is preferable for detector applications. The addition of Zn to CdTe widens the bandgap and can thus reduce the thermal noise due to increased resistivity compared to CdTe, in addition to other reported benefits of increasing the defect formation energy [51, 52]. Here the growths are performed with 10% Zn. As-grown undoped CdTe/CdZnTe grown under Te-rich conditions are known to result in p-type low-resistivity crystals due to cadmium vacancy (V_{Cd}) acceptors. Growths under Cd pressure control or post-growth annealing under Cd overpressure conditions can increase the resistivity to $\approx 10^8$–10^9 Ω cm [53], which is suboptimal for detector application.

Self-compensation by either indium (In) or chlorine (Cl) is a well-known strategy employed to achieve semi-insulating crystals. Since indium replaces V_{Cd} to form a donor, its solubility is higher under Te-rich conditions. Here indium concentrations of ~3000–5000 ppb atomic are used to reproducibly achieve bulk resistivity exceeding 10^{10} Ω cm under excess Te concentration up to ~57% [11]. Detectors with thicknesses from 1 mm up to 20 mm are fabricated from the grown boules. Thin detectors are tested for $\mu\tau_e$ and resolutions @ 122 keV ([57]Co) in planar electrode configuration. Most medical applications use energies few 10s of keV up to ~150 keV. Thick detectors are tested for $(\mu\tau)_e$ products and resolutions @ 662 keV (^{137}Cs) in Frisch-collar configuration [14] to evaluate the applicability in nuclear safeguard applications. $(\mu\tau)_e$ products are determined by fitting the electric field dependence of photopeak (32 keV of ^{137}Cs or 59.5 keV of [241]Am) position to Hecht equation. Since $(\mu\tau)_e$ value can vary depending on measurement conditions, it serves as a measure of relative crystal quality, whereas electron lifetimes and detector resolutions are ultimately the best indicators of performance [54, 55]. Impurities and Te-rich secondary phases are some of the key performance-limiting factors, especially for applications demanding high resolution in thick detectors. Here, by growing under a suitable Te-rich condition, purity is improved due to a

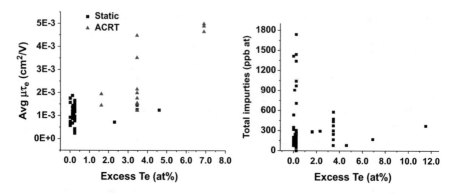

Fig. 3 Average $(\mu\tau)_e$ versus excess Te (left) and total impurities versus excess Te (right) for over 150 crystal growths performed at WSU

lower growth temperature as well as gettering effect of Te. Application of ACRT allowed us to reduce Te inclusions and simultaneously grow purer crystals at fast rates. Figure 3 shows the $(\mu\tau)_e$ products of numerous 1–3 mm detectors from over ~150 crystal growths at WSU performed with varying excess Te in the melt. Each data point corresponds to $(\mu\tau)_e$ of a particular growth averaged over the number of samples tested. Clearly seen is a trend of increasing $(\mu\tau)_e$ with increasing excess Te in the melt. While other differences in growth condition may also influence the transport properties, solvent purification effect appears to play a significant role as the superior transport properties were most reproducible in crystal growth experiments with high excess Te and lowered growth temperatures [11]. Uncertainties and interferences in GDMS measurements at low impurity concentrations make it difficult to establish a clear correlation between total impurities and $(\mu\tau)_e$ values as can be seen from Fig. 3, especially at such low concentrations of total impurities. Note that the GDMS data does not include impurities of carbon (C), nitrogen (N), and oxygen (O).

Although improvement in the $(\mu\tau)_e$ for thin detectors is achieved by increasing excess Te in the melt, it does not always translate to better performance in thick detectors due to variation in inclusion distribution. This is clearly elucidated in Figs. 4 and 5. Figure 4 compares thickness-dependent $(\mu\tau)_e$ and corresponding infrared (IR) micrographs for two different crystal growths. The main differences between the two growths are the melt composition (excess Te) and the implemented ACRT profile. The IR results indicate very different second-phase distributions in these crystals. For better comparison, the samples are selected from similar regions of the ingot. The effect of different excess Te in the melt is seen in the $(\mu\tau)_e$ values of thin detectors. For growth 1, which had higher fraction of large diameter inclusions, clear deterioration in $(\mu\tau)_e$ values for thick detectors is seen. On the other hand, growth 2 performed with optimized rotation profile exhibits smaller inclusion size, leading to a high $(\mu\tau)_e$ value of thick detectors, and shows less variation with increasing thickness. The detrimental influence of inclusion distributions is also

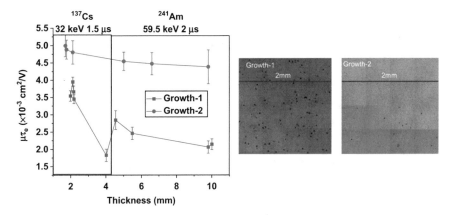

Fig. 4 Thickness dependence of $(\mu\tau)_e$ for two different ACRT crystal growth runs (left), corresponding IR images (right)

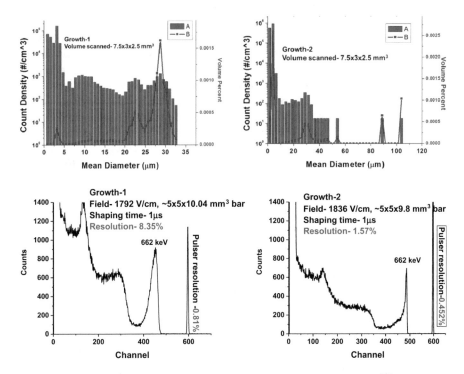

Fig. 5 Second-phase distribution in growth 1 (top left) and growth 2 (top right). ^{137}Cs spectrum showing resolution @ 662 keV for growth 1 (bottom left) and growth 2 (bottom right)

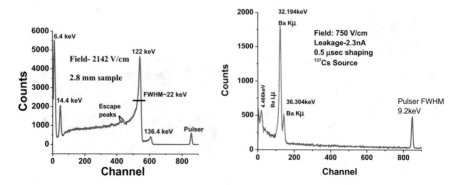

Fig. 6 ^{57}Co spectra (left) and ^{137}Cs spectra (right) for thin (~2–3 mm) planar detectors grown with optimized ACRT and excess Te composition

evident from the difference in resolution @ 662 keV measured for samples from the two growths, 8.35% for growth 1 versus 1.57% for growth 2.

The spectrometric performance of thin samples of ACRT-grown crystals is shown in Fig. 6, evaluated using a ^{57}Co and ^{137}Cs source. The irradiation is performed on the cathode side. All the major isotope lines are clearly resolved. Interestingly, the photo-peak corresponding to low-energy X-rays is well resolved, which is typically challenging, but could be achieved with good-quality surface and low-noise electronics given good carrier transport properties. The measured energy resolution of ~4% @ 122 keV, in planar electrode configuration, is better than that of scintillators, which makes the material quality readily suitable for low-count-rate medical imaging applications, although the performance under high irradiation flux necessary for computed tomography (CT) applications is yet to be tested in these materials. Furthermore, in imaging applications, where the anode is pixelated, further improvement in energy resolution is expected due to reduction in dark current and single-polarity charge sensing.

Several Frisch-collar detectors, with typical sample area of 5 mm^2 × 5 mm^2 and 10–15 mm thickness, are fabricated from ACRT-grown crystals for in-house evaluation. Although nonuniformity exists, energy resolution @ 662 keV for most detectors is measured to be ≈2–3% in as-grown crystals with the best resolution of ≈1.1% achieved without applying any electronic correction, as shown in Fig. 7. Such high-performing devices from as-grown crystals have never been reported by melt-growth methods, and this performance is comparable to state-of-the-art commercial detectors where post-growth crystal processing is typically employed.

In summary, in this section, the performance of ACRT-grown CdZnTe crystals and the effect of various aspects of crystal quality such as purity and extended defect distribution on detector performance are discussed. The most important takeaway is that the high-quality spectroscopic grade detector performance can be achieved in as-grown crystals without the need for post-processing while still maintaining fast growth rates.

Fig. 7 Energy resolution of 1.1% @ 662 keV for CdZnTe detector in Frisch-collar configuration

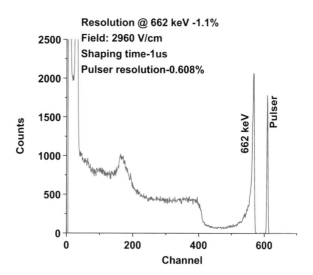

4 Secondary Phases in ACRT-Grown Crystals

Secondary-phase defects, including inclusions and precipitates, are common in CdTe/CdZnTe crystals. Nonstoichiometric growth conditions and combination of kinetic and thermodynamic factors lead to their formation, growth, and random distribution [56]. Inclusions originate from the trapping of melt at the solid-melt interface due to morphological instabilities which can be caused by factors such as constitutional undercooling. An example of this phenomenon is presented in Fig. 8 [57].

The crystal shown in Fig. 7 is grown from a melt composition of 53.5% Te in a CdZnTe with 10% ZnTe. Imposed growth rate and interface gradient are 0.5 mm/h and 20 °C/cm, respectively. The ingot was quenched after growing about 50% of the material. Capture of large Te droplets is clearly observed in the fast-frozen region, due to the imposed growth rate being very fast, which caused constitutional undercooling. The size of such inclusions can be up to 50μm or higher, and can deteriorate the resolutions even at very low densities [58]. The detrimental effect of poor inclusion distributions on detector performance was shown in the previous section. In general, the density of inclusions in CdTe/CdZnTe can vary from 10^4 cm^{-3} to 10^6 cm^{-3} which can be controlled by adjusting temperature gradient at the melt-crystal interface, growth rate, melt stoichiometry, post-growth cooling, and application of forced convection (e.g., ACRT) [9, 56, 59, 60]. WSU has previously studied the effects of post-growth cooling on inclusion distributions [61]. Slower cooling typically results in an increase in mean particle diameter and reduced inclusion density, with the opposite being true for faster cooling. Regardless of the growth method, the size and density of inclusions that are observed in Te-rich as-grown crystals by IR microscopy are observed in the range where it can degrade the energy resolution [58], since presence of these defects causes fluctuations in the

Fig. 8 IR image of
inclusion trapping (dark
regions) evident at the solid-
melt interface of a fast-
frozen boule [58]

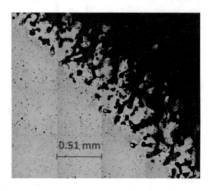

induced charge and degrades energy resolution especially in thicker devices. Various studies have correlated device performance and presence of these defects with observed electric field distribution by the Pockels effect and X-ray topography and correlated response maps [62, 63]. Inclusion formation due to constitutional undercooling increases with increasing off-stoichiometry and hence induces stricter limitations on permissible stable growth rates [33]. Precipitates, on the other hand, form due to retrograde solubility of Te in CdTe solid. Excess Te of ~530 ppm (atomic) can be accommodated into a single CdTe phase at ≈900 °C [64], although recent results indicate that the actual solubility limit is actually much lower [9, 65]. Upon cooling to room temperature in equilibrium, the excess Te precipitates into the matrix and is usually much smaller in size than inclusions [56] and does not significantly affect the charge transport. Unlike inclusion formation which occurs during growth, precipitation is a consequence of the retrograde solid solubility of Te in CdTe [56].

A common strategy to minimize the secondary phases is by annealing under Cd pressure. Establishing optimal parameter space for annealing and maintaining electrical compensation is important. Researchers have also applied two-step (Cd followed by Te) annealing to restore resistivity [12]. Temperature gradient annealing is also applied to thermally migrate Te inclusions [66]. It is reported that the impurities gettered at the inclusions release to the matrix upon annealing, which can deteriorate electrical properties [24]. It is therefore ideal if these particles can be controlled in size and density during growth. Various forced melt convection techniques including vibrational stirring, rotating magnetic field, and accelerated crucible rotation have been applied with promising results in terms of improvements in interface shape, second-phase particle size, and electrical property uniformity [67, 68]. Our lab has been pursuing ACRT for the past 10 years. Datta et al. [69] reported a decrease in mean diameter to ~6μm, and the distribution was reported to change from a bimodal, in the case of static growth, to a single peak distribution, in the case of ACRT growths, with the same composition of the melt. Recent optimization of ACRT parameters has led to significantly smaller mean diameter [9, 32, 69] as shown in Fig. 9. Static grown crystal under similar excess Te as ACRT exhibits significantly larger inclusion size.

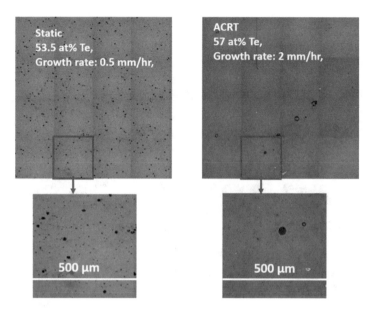

Fig. 9 Comparison of IR micrograph in static (left) and ACRT (right) crystals. The ACRT crystal was grown at a higher excess Te (57%) and faster growth rate of 2 mm/h.

The IR images shown here correspond to some of the best regions in the boules. Places near grain boundaries are seen to be decorated by a large density of inclusions [9]. Grain boundaries, twins, and dislocations make it energetically favorable to attract these defects. Even in single-crystal regions, when observed over a large region, larger diameter particles are observed but with low densities compared to static growth conditions. The density of large inclusions is fairly low and does not seem to significantly affect the detector resolutions. Some large-area scans of samples from ACRT-grown crystals are given in Fig. 10, along with corresponding resolution @ 662 keV. These inclusion distributions and resolutions are unprecedented for as-grown CdZnTe.

In addition, occasionally regions with subgrain boundaries are also encountered in ACRT-grown crystals that are decorated with higher density of inclusions and deteriorate the resolution (Fig. 11). One such example is given below; interestingly the photo-peak channel remains similar in all three samples at similar applied electric field, indicating similar values of $(\mu\tau)_e$.

Despite these nonuniformities, detectors with resolution <2% could be reproducibly obtained over different boules, as long as the excess Te and rotation profiles are kept the same. Interested reader can find more detailed analysis of the influence of rotation profiles and growth conditions on inclusion distributions of ACRT crystals in the references [32, 48]. These results indicate that a combined effect of purity and secondary-phase defects is responsible for detector performance. Results shown here demonstrate that both the purity and the secondary-phase particle size could be

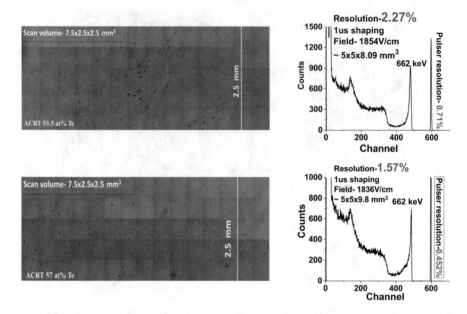

Fig. 10 Large-volume IR scans (left) of samples from ACRT-grown crystals (optimized rotation) and corresponding 662 keV resolution in Frisch-collar configuration (right)

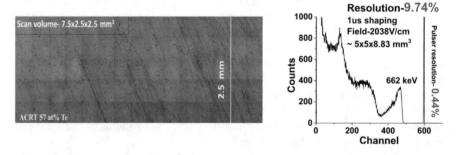

Fig. 11 Large-volume IR scan of a sample from ACRT-grown crystal (optimized rotation) showing decoration of inclusions with subgrain boundaries and the corresponding 662 keV resolution with Frisch-collar configuration (right)

controlled reproducibly by growing from an optimal excess Te melt in combination with forced melt convection by ACRT.

Despite these excellent results, further research is needed in order to determine ideal rotation parameters, excess Te, and growth rate to improve grain structure and nonuniformity for better yield of high-quality detectors [70]. Experiments in tandem with thermal modelling can be helpful. Modelling has been helpful in the past to provide guidance in terms of rotation parameter range and their effect on interface stability, although observations contrary to model predictions are also reported. For example, Divecha et al. [71] applied modelling to the WSU system and compared the impact of rotation parameters on flow patterns, interface stability, and undercooling. Their finding suggests that rotation patterns that maximize the flow and achieve homogeneous solute distribution in the melt do not reduce supercooling at the interface. Lower rotation rates determined as per flow stability criteria reduce undercooling in their study. Contrary to these numerical results, our experimental studies indicate that higher rotation rates are more effective in reducing second-phase particle size. Clearly, future studies will need to clearly understand these issues.

5 Discussion and Future Directions

To summarize the studies presented in the previous sections:

1. Thick detectors with performance suitable for high-resolution spectroscopy and imaging can be grown at fast rates of 1–2 mm/h.
2. The high-performing detector quality can be realized in as-grown crystals.
3. Extended defects related to Te inclusions could be controlled by applying ACRT.
4. ACRT in combination with excess Te melt appears to be an effective strategy in achieving fast purification of unknown deep-level impurities improving electron lifetime.

Increasing the overall volume of high-quality detector per boule and understanding and eliminating the observed inhomogeneity in second-phase distribution should be the next research goals. With increasing excess Te, the performance of detectors improved, provided that a rotation profile is used that is effective in reducing second-phase particle size. However, the grain structure is found to deteriorate [32, 48]. Despite these challenges, multiple high-resolution detectors with volume necessary for high-energy gamma spectroscopy could be obtained from these boules; however, the mining process necessary for obtaining single-crystalline areas from the boule is time intensive. As an immediate solution to the problem, increasing of the overall size of the crystal growth was performed. Results indicate that proportionately larger detectors could be fabricated from those boules while maintaining low second-phase content and maintaining detector performance.

This presents a potential opportunity to consider implementing ACRT to growth methods traditionally known to produce large crystals, such as the high-pressure Bridgman (HPB) technique [72]. HPB process typically uses a graphite crucible in a semi-sealed configuration under high inert gas pressure. Use of a sturdy graphite crucible allows for significantly larger boule sizes than sealed quartz ampoule-based growths. The HPB technique has in the past demonstrated the capability to grow a large-diameter ingot >5 in. and overall ingot weight exceeding 10 kg, which will be challenging in typical quartz ampoules [73]. Although the HPB process has produced high-quality detectors, the technique suffered from the common challenges of other high-temperature melt growths. The findings reported in this chapter suggest a pathway for large-volume detectors if ACRT in combination with excess Te could be implemented in the HPB technique. Modelling for such a system is reported [74].

Our recent studies on CdTe growth in HPB system show that the technique is suitable for integrated synthesis and growth of binary and ternary II-VI alloys [75]. This presents advantages, since separate pre-synthesized binaries will not be necessary as starting material; this is beneficial from both purity and cost perspectives, as elements are available at high purity and low cost compared to pre-compounded CdTe or ZnTe of the same purity. Purity comparison indicates that HPB-synthesized CdTe can be grown at a purity level comparable to high-quality gamma-ray detectors [75]. In addition, specific impurities such as oxygen and Si, whose role in CdZnTe is not clear, can be avoided, due to the use of graphite components.

As another approach to increase the yield of single crystals, seeding has been attempted [32]. From a growth configuration point of view, seeding in melt growth is relatively more challenging compared to seeding in the travelling heater method. CdTe/CdZnTe has fundamental physical properties that make it prone to grain and twin formation, and therefore experimental factors, including control over seed melt-back, interface shape at the seed-melt interface, and existence of sites for sidewall nucleation, play important roles in successful seeding. It is also important to maintain a convex shape of the interface to support outward grain growth. Thermal modelling has been employed to achieve these conditions in our seeding experiments. An ampoule support system was designed which consisted of high-thermal-conductive metal rods at the center and a low-thermal-conductivity outer cylinder, an idea initially proposed by Derby [76]. Our initial results indicate that successful seeding could be reproduced. Although all major grains and twins on the seed could be replicated, propagation of the seeded grain through the entirety of the crystal is still a challenge, especially in the presence of ACRT. Another major problem encountered in these experiments was sidewall nucleation due to use of an imperfect seed crystal, where multiple small grains were observed to form on the periphery of the seed and propagate into the bulk of the ingot. Future efforts should aim to address these issues by further understanding the thermal environment and understanding the effects of crystallographic orientations using a fully single-crystalline seed.

6 Conclusion

Application of forced melt convection by accelerated rotation of the crucible allows fast growth of CdZnTe from highly off-stoichiometric melt compositions. The use of Te-rich melt and corresponding reduction of growth temperature play a key role in increasing electron $(\mu\tau)_e$, likely due to reduction in the concentration of performance-limiting point defects. In addition, crystals that were grown with suitable ACRT conditions resulted in detectors with resolution of ~1.1% (@ 662 keV), without subjecting the crystals to post-growth annealing. The energy resolution is observed to correlate with the distribution of secondary phases as seen by IR microscopy. The growth method described is a potential pathway towards achieving cost reduction of high-resolution CdZnTe devices by overcoming the critical crystal growth challenges. Despite the excellent detector performance in as-grown crystals grown at fast rates, challenges including inhomogeneity in the boule and single-crystal yield remain that requires further research.

Acknowledgements This chapter is dedicated to late Professor Kelvin G. Lynn (1948–2020), who initiated and supervised CdZnTe research at WSU. The authors acknowledge the contributions made by former PhD students Dr. Jedidiah McCoy and Dr. Amlan Datta. We thank the members of IMR, WSU, Jasdeep Singh, and Becky Griswold for their assistance. We acknowledge the input provided by Dr. Mia Divecha, and Prof. Jeff Derby of the University of Minnesota towards implementing ACRT at WSU. Authors would also like to acknowledge the feedback provided by Dr. Aleksey Bolotnikov of Brookhaven National Laboratory. This work was supported by the National Nuclear Security Administration (NNSA) under Grant DE-NA0002565, U.S. Department of Energy.

References

1. Johns, P.M., Nino, J.C.: Room temperature semiconductor detectors for nuclear security. J. Appl. Phys. **126**(4), 40902 (2019)
2. Bolotnikov, A.E. et al.: Use of the drift-time method to measure the electron lifetime in long-drift-length CdZnTe detectors. J. Appl. Phys. **120**(10) (2016)
3. Prokesch, M., Soldner, S.A., Sundaram, A.G.: CdZnTe detectors for gamma spectroscopy and x-ray photon counting at 250×106 photons/(mm^2 s). J. Appl. Phys. **124**(4) (2018)
4. CZT detector technology for medical imaging Related content
5. Krawczynski, H.S., Jung, I., Perkins, J.S., Burger, A., Groza, M.: Thick CZT detectors for spaceborne x-ray astronomy. In: Hard X-Ray and Gamma-Ray Detector Physics VI, vol. 5540, no. 21, p. 1 (2004)
6. Rudolph, P., Mühlberg, M.: Basic problems of vertical Bridgman growth of CdTe. Mater. Sci. Eng. B. **16**(1–3), 8–16 (1993)
7. Schlesinger, T.E., et al.: Cadmium zinc telluride and its use as a nuclear radiation detector material. Mater. Sci. Eng. R Rep. **32**(4–5), 103–189 (2001)
8. Boiotnikov, A.E., Camarda, G.C., Wright, G.W., James, R.B.: Factors limiting the performance of CdZnTe detectors. IEEE Trans. Nucl. Sci. **52**(3 I), 589–598 (2005)
9. McCoy, J.J., Kakkireni, S., Gélinas, G., Garaffa, J.F., Swain, S.K., Lynn, K.G.: Effects of excess Te on flux inclusion formation in the growth of cadmium zinc telluride when forced melt convection is applied. J. Cryst. Growth. **535**, 125542 (2020)
10. Ivanov, Y.M.: Preparation of CdTe and CdZnTe single-crystalline ingots 100 mm in diameter by a modified Obreimov-Shubnikov method. Russ. J. Inorg. Chem. **60**(14), 1816–1823 (2015)
11. McCoy, J.J., Kakkireni, S., Gilvey, Z.H., Swain, S.K., Bolotnikov, A.E., Lynn, K.G.: Overcoming mobility lifetime product limitations in vertical Bridgman production of cadmium zinc telluride detectors. J. Electron. Mater. **48**(7), 4226–4234 (2019)
12. Kim, K., et al.: Two-step annealing to remove Te secondary-phase defects in CdZnTe while preserving the high electrical resistivity. IEEE Trans. Nucl. Sci. **65**(8), 2333–2337 (2018)
13. Luke, P.N.: Single-polarity charge sensing in ionization detectors using coplanar electrodes. Appl. Phys. Lett. **65**(22), 2884–2886 (1994)

14. Kargar, A., Jones, A.M., Mcneil, W.J., Harrison, M.J., Mcgregor, D.S.: CdZnTe Frisch collar detectors for g-ray spectroscopy. Nucl. Instrum. Methods Phys. Res. Sect. A. **558**, 497–503 (2006)
15. Bolotnikov, A.E., et al.: Performance of $8\times8\times32$ and $10\times10\times32$ mm3 CdZnTe position-sensitive virtual Frisch-grid detectors for high-energy gamma ray cameras. Nucl. Instrum. Methods Phys. Res. Sect. A. **969**, 164005 (2020)
16. Chen, H. et al.: Development of large-volume high-performance monolithic CZT radiation detector. In: Hard X-Ray, Gamma-Ray, and Neutron Detector Physics XX, vol. 10762, p. 107620N (2018)
17. Bolotnikov, A.E., et al.: CdZnTe position-sensitive drift detectors with thicknesses up to 5 cm. Appl. Phys. Lett. (9), 093504 (108, 2016)
18. Thomas, B., Veale, M.C., Wilson, M.D., Seller, P.: Characterisation of Redlen high-flux CdZnTe (2017)
19. Iniewski, K.: CZT sensors for computed tomography: from crystal growth to image quality. J. Instrum. **11**(12), C12034 (2016)
20. MacKenzie, J., Kumar, F.J., Chen, H.: Advancements in THM-grown CdZnTe for use as substrates for HgCdTe. J. Electron. Mater. **42**(11), 3129–3132 (2013)
21. Shiraki, H., Funaki, M., Ando, Y., Kominami, S., Amemiya, K., Ohno, R.: Improvement of the productivity in the THM growth of CdTe single crystal as nuclear radiation detector. IEEE Trans. Nucl. Sci. **57**(1 PART 2), 395–399 (2010)
22. Yang, G., Bolotnikov, A.E., Cui, Y., Camarda, G.S., Hossain, A., James, R.B.: Impurity gettering effect of Te inclusions in CdZnTe single crystals. J. Cryst. Growth. **311**(1), 99–102 (2008)
23. Audet, N., Cossette, M.: Synthesis of ultra-high-purity CdTe ingots by the traveling heater method. J. Electron. Mater. **34**(6), 683–686 (2005)
24. Roy, U.N., Burger, A., James, R.B.: Growth of CdZnTe crystals by the traveling heater method. J. Cryst. Growth. **379**, 57–62 (2013)
25. Prokesch, M., Szeles, C.: Accurate measurement of electrical bulk resistivity and surface leakage of CdZnTe radiation detector crystals. J. Appl. Phys. **100**(1) (2006)
26. Peterson, J.H., Yeckel, A., Derby, J.J.: A fundamental limitation on growth rates in the traveling heater method. J. Cryst. Growth. **452**, 12–248 (2016)
27. El, M.A., Triboulet, R., Lusson, A., Tromson-Carli, A., Didier, G.: Growth of large, high purity, low cost, uniform CdZnTe crystals by the 'cold traveling heater method. J. Cryst. Growth. **138** (1–4), 168–174 (1994)
28. Jung, I., Krawczynski, H., Burger, A., Guo, M., Groza, M.: Detailed studies of pixelated CZT detectors grown with the modified horizontal Bridgman method. Astropart. Phys. **28**(4–5), 397–408 (2007)
29. GE Healthcare Acquires CZT Detector Company I Imaging Technology News. [Online]. https:// www.itnonline.com/content/ge-healthcare-acquires-czt-detector-company.   Accessed 27 December 2020
30. To Infinity and Beyond: These Crystal Sensors Can See Blasts From Black Holes and Also Cancers I GE News." [Online]. https://www.ge.com/news/reports/infinity-beyond-crystal-sensors-see-black-holes-cancers. Accessed 21 July 2020
31. Scheel, H.J., Schulz-Dubois, E.O.: Flux Growth of Large Crystals by Accelerated Crucible-Rotation Technique (1971)
32. Kakkireni, S.: Strategies and Challenges to Improve Single Crystal Yield and Detector Properties of Cadmium Zinc Telluride Grown Via Accelerated Crucible Rotation. Washington State University, Washington, DC (2020)
33. Tiller, W.A., Jackson, K.A., Rutter, J.W., Chalmers, B.: The redistribution of solute atoms during the solidification of metals. Acta Metall. **1**(4), 428–437 (1953)
34. Schulz-Dubois, E.O.: Accelerated crucible rotation: hydrodynamics and stirring effect. J. Cryst. Growth. **12**(2), 81–87 (1972)

35. Scheel, H.J.: Accelerated crucible rotation: a novel stirring technique in high-temperature solution growth. Journal of Crystal Growth. **13–14**, 560–565 (1972)
36. Capper, P., et al.: Bridgman growth and assessment of CdTe and CdZnTe using the accelerated crucible rotation technique. Mater. Sci. Eng. B. **16**(1–3), 29–39 (1993)
37. Wald, F.V., Bell, R.O.: Natural and forced convection during solution growth of CdTe by the traveling heater method (THM). J. Cryst. Growth. **30**(1), 29–36 (1975)
38. Saleh, M., Kakkireni, S., McCloy, J., Lynn, K.G.: Improved Nd distribution in Czochralski grown YAG crystals by implementation of the accelerated crucible rotation technique. Opt. Mater. Express. **10**(2), 632 (2020)
39. Kusunoki, K., et al.: Solution growth of SiC crystal with high growth rate using accelerated crucible rotation technique. Mater. Sci. Forum. **527–529**, 119–122 (2006)
40. Capper, P., Gosney, J.J.G., Jones, C.L.: Application of the accelerated crucible rotation technique to the Bridgman growth of CdxHg1-xTe: simulations and crystal growth. J. Cryst. Growth. **70**(1–2), 356–364 (1984)
41. Horowitz, A., Gazit, D., Makovsky, J., Ben-Dor, L.: Bridgman growth of Rb2MnCl4 via accelerated crucible rotation technique. J. Cryst. Growth. **61**(2), 323–328 (1983)
42. Wende, G., Görnert, P.: Study of ACRT influence on crystal growth in high-temperature solutions by the 'high-resolution induced striation method. Phys. Status Solidi. **41**(1), 263–270 (1977)
43. Zhou, J., Larrousse, M., Wilcox, W.R., Regel, L.L.: Directional solidification with ACRT. J. Cryst. Growth. **128**(1–4), 173–177 (1993)
44. Yin, L., Jie, W., Wang, T., Zhou, B., Yang, F., Nan, R.: The effects of ACRT on the growth of ZnTe crystal by the temperature gradient solution growth technique. Crystals, **7**(3) (2017)
45. Brice, J.C., Capper, P., Jones, C.L., Gosney, J.J.G.: ACRT: a review of models. Prog. Cryst. Growth Charact. **13**(3), 197–229 (1986)
46. Coates, W.G., et al.: Effect of ACRT rotation parameters on Bridgman grown CdxHg1-xTe crystals. J. Cryst. Growth. **94**(4), 959–966 (1989)
47. Divecha, M.S., Derby, J.J.: Towards optimization of ACRT schedules applied to the gradient freeze growth of cadmium zinc telluride. J. Cryst. Growth. **480**, 126–131 (2017)
48. J. J. McCoy, "Implementation of Accelerated Crucible Rotation in Electrodynamic Gradient Freeze Method for Highly Non-stoichiometric Melt Growth of Cadmium Zinc Telluride Detectors," Washington State University: Washington, DC 2018
49. Datta, A.: Strategic Approaches towards Solving Critical Challenges in Crystal Growth of Detector Grade Cadmium Zinc Telluride (CZT) Including Melt Mixing (ACRT) Techniques by Amlan Datta a Dissertation Submitted in Partial Fulfillment of the Requirements for Th (2015)
50. Lynn, K., Jones, K., Ciampi, G.: Compositions of Doped, Co-Doped and Tri-Doped Semiconductor Materials. Google Patents (2011)
51. Bell, S.L., Sen, S.: Crystal growth of cd 1 − x Zn x Te and its use as a superior substrate for LPE growth of hg 0.8 cd 0.2 Te. J. Vac. Sci. Technol. A. **3**(1), 112–115 (1985)
52. Åberg, D., Erhart, P., Lordi, V.: Contributions of point defects, chemical disorder, and thermal vibrations to electronic properties of Cd1-xZnxTe alloys. Phys. Rev. B. **88**(4), 045201 (2013)
53. Krsmanovic, N., et al.: Electrical compensation in CdTe and Cd 0.9 Zn 0.1 Te by intrinsic defects. Phys. Rev. B. **62**, R16279 (2000)
54. Bolotnikov, A.E., et al.: Use of the drift-time method to measure the electron lifetime in long-drift-length CdZnTe detectors. J. Appl. Phys. **120**(10), 104507 (2016)
55. Jones, K.A., Datta, A., Lynn, K.G., Franks, L.A.: Variations in μτ measurements in cadmium zinc telluride. J. Appl. Phys. **107**(12), 123714 (2010)
56. Rudolph, P., Neubert, M., M:uhlberg, M.: Defects in CdTe Bridgman monocrystals caused by nonstoichiometric growth conditions. J. Cryst. Growth. **128**(1–4 PART 2), 582–587 (1993)
57. Datta, A., Swain, S., Bhaladhare, S., Lynn, K.G.: Experimental studies on control of growth interface in MVB grown CdZnTe and its consequences. In: IEEE Nuclear Science Symposium Conference Record, pp. 4720–4726 (2011)

58. Bolotnikov, A.E., Camarda, G.S., Carini, G.A., Cui, Y., Li, L., James, R.B.: Cumulative effects of Te precipitates in CdZnTe radiation detectors. Nucl. Instruments Methods Phys. Res. Sect. A. **571**(3), 687–698 (2007)
59. Dinger, R.J., Fowler, I.L., Fowled, I.L.: Te inclusions in CdTe grown from a slowly cooled Te solution and by the travelling solvent method. Rev. Phys. Appl. **12**(2), 135–139 (1977)
60. Roy, U.N., et al.: Size and distribution of Te inclusions in THM as-grown CZT wafers: the effect of the rate of crystal cooling. J. Cryst. Growth. **332**(1), 34–38 (2011)
61. Swain, S.K., Jones, K.A., Datta, A., Lynn, K.G.: Study of different cool down schemes during the crystal growth of detector grade CdZnTe. IEEE Trans. Nucl. Sci. **58**(5 PART 2), 2341–2345 (2011)
62. Wardak, A., et al.: Electric field distribution around cadmium and tellurium inclusions within CdTe-based compounds. J. Cryst. Growth. **533**, 125486 (2020)
63. Bolotnikov, A.E., et al.: Correlations between crystal defects and performance of CdZnTe detectors. IEEE Trans. Nucl. Sci. **58**(4 Part 2), 1972–1980 (2011)
64. Greenberg, J.H.: P-T-X phase equilibrium and vapor pressure scanning of non-stoichiometry in CdTe. J. Cryst. Growth. **161**(1–4), 1–11 (1996)
65. Li, B., Zhu, J., Zhang, X., Chu, J.: Effect of annealing on near-stoichiometric and non-stoichiometric CdZnTe wafers. J. Cryst. Growth. **181**, 204–209 (1997)
66. Kim, K.H., et al.: Temperature-gradient annealing of CdZnTe under Te overpressure. J. Cryst. Growth. **354**(1), 62–66 (2012)
67. Salk, M., Fiederle, M., Benz, K.W., Senchenkov, A.S., Egorov, A.V., Matioukhin, D.G.: CdTe and CdTe0.9Se0.1 crystals grown by the travelling heater method using a rotating magnetic field. J. Cryst. Growth. **138**(1–4), 161–167 (1994)
68. Capper, P., et al.: Interfaces and flow regimes in ACRT grown CdxHg1-xTe crystals. J. Cryst. Growth. **89**(2–3), 171–176 (1988)
69. Datta, A., Swain, S., Cui, Y., Burger, A., Lynn, K.: Correlations of Bridgman-grown Cd 0.9 Zn 0.1 Te properties with different ampoule rotation schemes. J. Electron. Mater. **42**, 3041–3053 (2013)
70. Bolotnikov, A.E., et al.: Performance-limiting defects in CdZnTe detectors. IEEE Trans, Nucl. Sci. **54**(4), 821–827 (2007)
71. Peterson, J.H., Fiederle, M., Derby, J.J.: Analysis of the traveling heater method for the growth of cadmium telluride. J. Cryst. Growth. **454**, 45–58 (2016)
72. Doty, F.P.: Properties of CdZnTe crystals grown by a high pressure Bridgman method. J. Vac. Sci. Technol. B. **10**(4), 1418 (1992)
73. Szeles, C.: Advances in the crystal growth and device fabrication technology of CdZnTe room temperature radiation detectors. IEEE Trans. Nucl. Sci. **51**(3 III), 1242–1249 (2004)
74. Yeckel, A., Derby, J.J.: Effect of accelerated crucible rotation on melt composition in high-pressure vertical Bridgman growth of cadmium zinc telluride. J. Cryst. Growth. **209**(4), 734–750 (2000)
75. Al-Hamdi, T.K., et al.: CdTe synthesis and crystal growth using the high-pressure Bridgman technique. J. Cryst. Growth. **534**, 125466 (2020)
76. Kuppurao, S., Derby, J.J.: Designing thermal environments to promote convex interface shapes during the vertical Bridgman growth of cadmium zinc telluride. J. Cryst. Growth. **172**(3–4), 350–360 (1997)

Solution Growth of CdZnTe Crystals for X-Ray Detector

Song Zhang, Bangzhao Hong, Lili Zheng, Hui Zhang, Cheng Wang, and Bo Zhao

Abstract This chapter presents an overview of our efforts on solution growth of large-size CdZnTe crystals for X-ray detectors. Firstly, a direct-mixing solution growth (DMSG) method is proposed in order to prepare high-purity CdZnTe crystals from 7N raw material during one single step, and the accelerated crucible rotation technique (ACRT) is used to get large-size and high-quality CdZnTe crystals. Secondly, unseeded THM process is studied to optimize the THM process. In this part, the THM furnace is firstly designed based on numerical investigation of the relationships among the ampoule diameter, the height of Te-rich solution, and the height of the heater. The thermal field stability during crystal growth is also improved by introducing a dummy crystal which has the same thermal conductivity with CdZnTe crystals. The influences of ampoule geometry and wall temperature gradient on nucleation rate as well as controlling growth interface by ampoule ration during unseeded THM are studied in order to get large crystals through unseeded THM growth. Finally, all the crystals are characterized by infrared transmission spectra, Te inclusion, composition, and I–V curve of the as-made CdZnTe detectors. A comparison between the DMSG and THM method is conducted based on numerical simulation and characterization of as-grown CdZnTe crystals, and some future research work to improve the crystal quality is discussed at the end of this chapter.

1 Introduction

The development of sensitive and portable X-ray and gamma radiation detectors needs progress in the production of large, high-quality crystalline cadmium zinc telluride (CdZnTe). The modified vertical Bridgman method has been widely used as

S. Zhang (✉) · B. Hong · L. Zheng · H. Zhang
Tsinghua University, Beijing, China
e-mail: zsthu@tsinghua.edu.cn; hbz17@mails.tsinghua.edu.cn; zhenglili@tsinghua.edu.cn; zhhui@tsinghua.edu.cn

C. Wang · B. Zhao
Ruiyan Technology Co. Ltd., Hangzhou, China

K. Iniewski (ed.), *Advanced Materials for Radiation Detection*,
https://doi.org/10.1007/978-3-030-76461-6_13

a melt growth method in preparation of CdZnTe crystals [1, 2]. However, this melt growth method suffers from high growth temperature and segregation of solute, which cause common problems in crystals, such as extended structural defects and inhomogeneity. Solution growth method is proposed as a low-temperature growth method, which is attracting more and more attentions during the studies of detector-grade CdZnTe crystals [3–5]. Commonly, the CdZnTe crystals are prepared from a Te-rich solution in solution growth method which has the potential to improve the problems during melt growth method [6]. There are two main solution growth methods widely used, i.e., the temperature gradient solution growth method and the traveling heater method.

The temperature gradient solution growth (TGSG) method as shown in Fig. 1a is proposed by combining solution growth method and vertical Bridgman method during which the structure of furnace is like a VB furnace while the feeding material is a Te-rich solution so that the temperature of growth interface can be reduced. The interface temperature is determined by constitution of the remaining solution based on the phase diagram. During TGSG process, all the feeding material is melted and mixed before pulling of the ampoule. The interface temperature is reduced with reduction of solute in the Te-rich solution, for which the temperature of heaters is also controlled to reduce so as to maintain the position of growth interface. The traveling heater method (THM) is quite similar to zone melting (see Fig. 1b), where the pre-synthesized polycrystalline feeding dissolves into Te-rich solution through diffusion and convective transport and finally deposits at the growth interface when the heater moves up (or more commonly, the ampoule moves downward). Compared with melt growth, the traveling heater method is a suitable solution growth method for growing crystals with high melting point or high vapor pressure and has its intrinsic advantages, such as low growth temperature, charge purification, and longitudinal homogeneity [6].

To date, the mass production of CdZnTe single crystals larger than 20 mm × 20 mm × 15 mm with low defects and uniform composition is still a big challenge and suffers from a low yield rate. To be noted, Te inclusion is still an important defect that degrades material properties during solution growth method, which is caused by constitutional supercooling and unstable growth interface, resulting in random trapping of Te-rich solution [8]. For growing high-quality grains, it is essential not only to create optimal crystal growth conditions, but also to ensure the stability of thermal and flow fields during crystal growth. However, there has been little discussion on maintaining a stable thermal environment in CdZnTe crystal growth. Another important issue for solution growth is the preparation of raw materials including seed, Te-rich solution, and feeding material. This makes the growth process complicated and also increases the possibility of contamination. The same problem occurs for the as-grown crystals for both TGSG and THM methods.

This chapter presents an overview of our efforts in solution growth of large-size $Cd_{0.9}Zn_{0.1}Te$ crystals for X-ray detectors. Firstly, a direct-mixing solution growth method is proposed to prepare high-purity CdZnTe crystals from 7 N raw material during one single step and the accelerated crucible rotation technique (ACRT) is

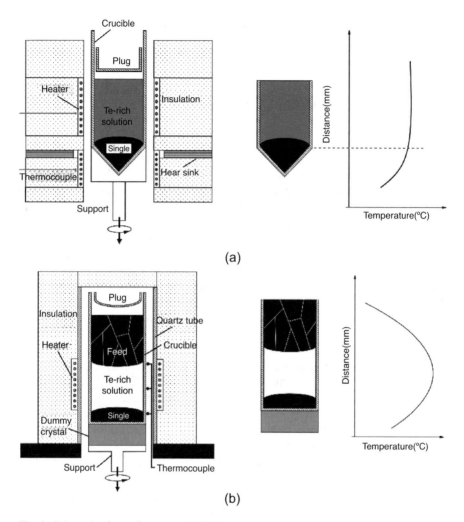

Fig. 1 Schematic of growth system, (**a**) TGSG/DMSG method and (**b**) THM method [7]

used to improve the crystal size and quality. Secondly, unseeded THM process is studied to optimize the THM process. In this part, the THM furnace is firstly designed based on numerical investigation of the relationships among the ampoule diameter, the height of Te-rich solution, and the height of heater. The thermal field stability during crystal growth is also improved by introducing a dummy crystal which has the same thermal conductivity with CdZnTe crystals. Controlling of growth interface by ampoule ration is also studied in order to get large crystals through unseeded THM growth. Finally, all the crystals are characterized by IR transmission images of Te inclusion, composition, and I–V curve of the as-made CdZnTe detectors. A comparison between the DMSG and THM method is conducted based on numerical simulation and characterization of as-grown CdZnTe crystals and some future research work to improve the crystal quality is discussed at the end of this chapter.

2 Direct Mixed Solution Growth Method

2.1 Design of Direct Mixed Solution Growth Process

The direct mixed solution growth method is a one-step method based on solution growth method which means that the preparation of CdZnTe feeding material and the growth process are combined together during the synthesis of CdZnTe single crystal [9]. Since no seed is used, the material preparation and crystal growth are achieved by enlargement of random nucleation. The growth system used for DMSG method is shown in Fig. 1a and a vertically increasing thermal profile like VB method is adopted. The furnace is a self-designed and homemade two-zone electric furnace and two heaters are separated by insulation brick and a heat sink. Modular design is used in the design of furnace so that the thermal field can be easily modified by changing parts of the furnace. The upper heater is the main heater which controls the temperature profile of the melting region. The lower heater and the heat sink are designed to modify the post-growth profile which can also act as a post-growth annealing process. The ampoule is placed onto the ampoule support which can control movement of ampoule with stepping motors. The temperature profile used in the experiment is shown in Fig. 1a and the temperature at the growth interface is 890 °C which is tested using an external thermocouple during the experiments. The temperature gradient at the growth interface can be controlled by changing the set point of the upper and lower heaters and also the position of growth interface. For example, by increasing the temperature of the upper heater and lowering the ampoule, a higher temperature gradient in the solution can be obtained.

Carbon-coated quartz ampoule of 75 mm in diameter with a 120° cone-shaped bottom is used. The ampoule is filled with the Cd, Te, and Zn raw materials and In as dopant and then vacuum sealed before the growth process. The raw materials of Cd, Te, and Zn are directly mixed to form the Te-rich solution. By controlling the reaction process of raw materials, the heat release rate is controlled, thus avoiding the explosion of sealed ampoule. The reaction inside the ampoule can be indicated by a rapid temperature increase of thermocouple at the wall of ampoule. As shown in Fig. 2, reaction occurs at 450 °C and 650 °C. Before growth, the ampoule is raised to the position with a temperature of 890 °C. At the saturation point of the source material, CdZnTe is fully mixed in the Te-rich solution by flow in the ampoule which is derived from temperature gradient and rotation of ampoule. After the source material is fully dissolved, the ampoule is pulled downward at a predetermined speed. The initial growth temperature is 890 °C, and the temperature of growth interface is lowered with the reduction of CdZnTe solute, which is controlled as a function of crystal growth rate and time.

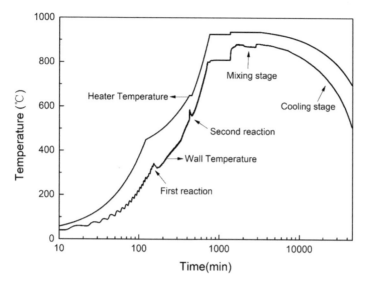

Fig. 2 Temperature evolution during DMSG process

2.2 Optimizations of Direct Mixed Solution Growth Process

Since the vertical temperature profile is monotonically increasing, which is a stable thermal structure for convection, the convection of solution is mainly caused by the lateral temperature difference across the ampoule. Figure 3a shows the numerical simulation of thermal and flow field in a typical DMSG process. It is seen that the main vortex is formed at the top of the solution and there is slight flow near the growth interface. The solution near the growth front is quite stable, for which it is easy to control the growth interface by changing the temperature field. However, the heat and mass transfer are also in a low level without solution flow, which leads to a low temperature gradient and low crystal growth rate. In order to enhance the heat and mass transfer near the growth interface, rotation of ampoule is used in the mixing stage before crystal growth.

Figure 4 shows four typical crystal growth experiments conducted in order to improve the crystal quality with DMSG method. Two different ACRT processes are introduced (see Fig. 3b), during which the Ekman flow will occur during acceleration and deceleration and cause mixing near the solid/fluid interfaces. A high rotation rate with a high frequency is necessary to destroy the stable density stratification and facilitate mixing, due to the stable thermal structure near the growth interface. Crystal sample 1 is grown at the temperature of 890 °C and the temperature gradient is measured relatively low at 20 K/cm. The growth rate is 3 mm/day for the first 2 days at the nucleation stage, after which the growth rate is kept at 5 mm/day. Longitudinal cross sections of the as-grown crystal (Fig. 4a) show that initial grains are formed at the tip of the cone, but such grains disappear soon and new grains are formed at the wall of ampoule. This leads to the formation of many small grains in

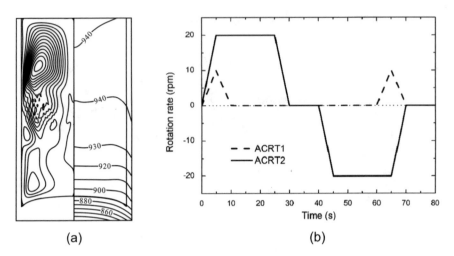

Fig. 3 (a) Numerical simulation of temperature and flow field in DMSG, (b) ACRT sequence used to enhance the heat and mass transfer in DMSG

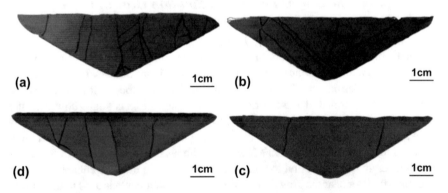

Fig. 4 Longitudinal cross section of as-grown crystals in DMSG experiments [9]

sample 1. During the growth of sample 2 (see Fig. 4b), the growth interface temperature gradient is increased to 40 K/cm while growth temperature and growth rate remain the same. However, the grain size is almost not improved. An ACRT pattern (dash line in Fig. 3b) is introduced during the growth of sample 3. From the cross sections of the as-grown sample 3, it is observed that initial grains are formed at the tip of the cone and grow continuously further without too many new grains formed at the wall of ampoule (see Fig. 4c). Another bidirectional ACRT process is used in the growth of crystal sample 4 which is shown with a solid line in Fig. 3b. The longitude cross section of the as-grown crystal in Fig. 4d shows that the initial grain is also formed at the top of the ampoule and grows continuously to the end of

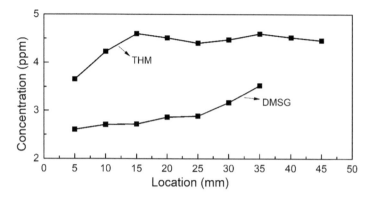

Fig. 5 Indium concentration along the growth direction during DMSG and unseeded THM method

the growth process. However, there are still unexpected new grains formed during the enlarging process. The cross section of the crystal shows that the grain size is improved than sample 3 and the number of unexpected small grains is reduced obviously. From the cross section of samples, it is seen that the introduction of ACRT process can improve the crystal size during DMSG growth of CdZnTe crystal.

Increasing temperature gradient at the growth interface makes the growth process more stable when there is a fluctuation of thermal field. In sample 3, there is a periodic stir of the solution to break the balance built by buoyancy flow. A more intense flow will increase the heat transfer in the solution. In sample 4, a bidirectional ACRT is used to increase the forced flow and this will compress the supercooling layer at the growth interface by increasing the temperature gradient. By enhancing the temperature gradient and mass transfer rate at growth interface with ACRT, large grains can be easily achieved in DMSG method. However, there is an important problem in the DMSG method, i.e., the concentration nonuniformity of dopant elements. From Fig. 5 it is shown that the axial concentration of indium in recently prepared DMSG ingot remains uniform at the beginning of the growth process and increases more and more rapidly towards the end of the ingot. This phenomenon is mainly caused by the segregation effect of indium. Another potential reason for the nonuniformity is the changing growth temperature during the growth process.

The resistivity at different locations in DMSG-prepared ingot is characterized after annealing in a Te-Cd atmosphere (see Fig. 6). It is shown that the resistivity at the tip of the ingot is quite stable near 2.0×10^{10} Ω·cm under a bias of -600 V to $+600$ V. However, the resistivity reaches 1.7×10^{12} Ω·cm under a bias of -600 V to 0 V while 3.5×10^{10} Ω·cm under a bias of 0–600 V at the tail of the ingot. As shown in Fig. 5, the dopant concentration is different between two faces of the wafers prepared by DMSG method and this is much more severe at the tail part of the ingot. Combining the resistivity analysis, the nonuniformity of dopant concentration is a potential reason for the asymmetric resistivity distribution, even though this is commonly caused by poor ohmic contacts between the contact layer and CdZnTe samples.

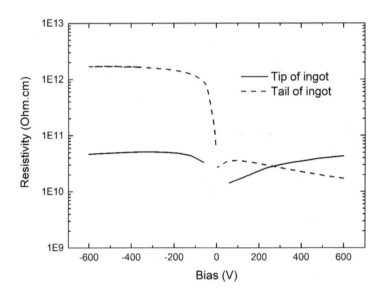

Fig. 6 Resistivity at different locations of DMSG-prepared ingot

Figure 7 shows Te inclusions in DMSG-prepared crystals before (Fig. 7a–c) and after (Fig. 7d) annealing which is captured using IR transmission microscopy. There are three types of Te inclusion formed during DMSG growth process, i.e., (1) large amount of small Te inclusions with the size smaller than 3 μm, which is shown in Fig. 7a, c; (2) small-size Te inclusions (smaller than 3 μm) along the line of dislocation and twin boundary as shown in Fig. 7b, c; and (3) occasionally formed large-size triangular or hexagon Te inclusions with the size between 15 μm and 20 μm. Since the size of most Te inclusion is smaller than 3 μm and there are huge numbers of this kind of inclusion, the density of Te inclusion is not counted. After an annealing process under Te-Cd atmosphere at 600 °C for 80 h, Te inclusions are improved as shown in Fig. 7d. After annealing, only inclusions with the size near 5 μm can be clearly seen, and the large-size inclusions disappear. Also, there will be several remaining small-size Te inclusions which are located at the dislocation before annealing. The source of the remaining 5 μm large inclusions is still unknown, and there is no evidence that can indicate that the remaining inclusion is formed by reunion of small-size inclusions or by shrinkage of large-size inclusion (the density of remaining inclusions is quite larger than large-size inclusions before annealing).

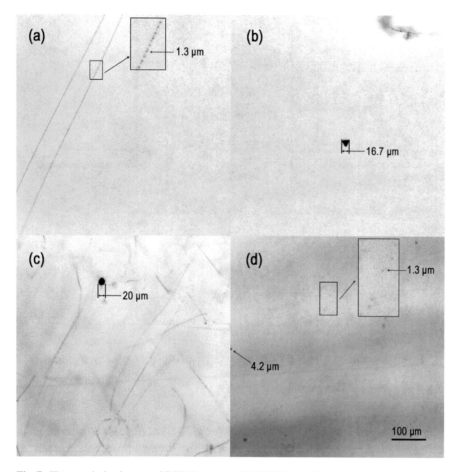

Fig. 7 IR transmission images of CdZnTe prepared in DMSG method, (**a–c**) as-grown crystals, (**d**) after annealing

3 Traveling Heater Method

As discussed in DMSG growth process, a stable and high temperature gradient near the growth interface is essential to prepare large and high-quality grains. The growth conditions could be stable in THM method when balance between melting of feeding material and solidification at the growth interface is built, and thus the longitudinal homogeneity can be hugely improved. However, as shown in Fig. 8, a vortex with an upward flow near the ampoule wall and a downward flow along the center occupies the main area of molten zone due to the buoyancy effects, while diffusion works only in thin boundary layers. Such convection, arising from a parabolic temperature profile, plays a significant role in enhancing mixing, but, on the other hand, is considered as the main cause of concave growth interfaces

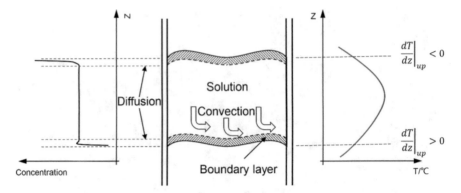

Fig. 8 Schematic of heat and mass transfer regime in the solution area

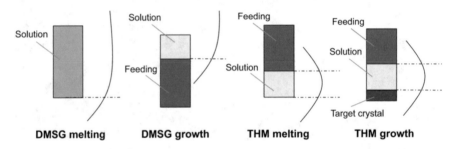

Fig. 9 Procedures of fast unseeded THM growth

[10, 11]. It is well known that a slightly convex growth interface is essential for grain enlargement and preventing polycrystalline growth through post-formed grains on the ampoule wall. In order to obtain a convex interface, many researchers start by weakening natural convection using a shorter molten zone, ampoule rotation, an additional magnetic field, and microgravity growth [12–15]. Ampoule rotation is a simple and convenient technology in experiments. Lan [16] first reported that a concave growth front due to buoyancy convection can be easily inverted into a convex one by applying a constant rotation of 3–5 RPM for THM growth of 1-in. GaAs. Dost [17] also found that a constant rotation of 5.0 RPM for a 1-in. CdTe crystal is effective in optimizing the interface. However, there has been no further experimental verifications.

Aiming for a stable and convex growth front with relatively large axial temperature gradient, fast unseeded THM growth of $Cd_{0.9}Zn_{0.1}Te$ crystals is conducted by introducing DMSG growth during feeding material preparation step. The procedures of fast unseeded THM growth are shown in Fig. 9. The schematic diagram of the THM growth system is shown in Fig. 1b. The furnace is a self-designed one-zone electric furnace with a precision of $\pm 0.1\ °C$. Before growth, polycrystalline CdZnTe feed is synthesized from 7 N metals using the DMSG method with a composition of $Cd:Zn = 9:1$. Then, the pre-synthesized CdZnTe polycrystalline with pure tellurium

Table 1 Growth process parameters used in fast unseeded THM experiments

Case#	Growth rate (mm/day)	Diameter (mm)	Length of grown crystal (mm)	Constant rotation rate (RPM)	Dummy crystal
CZT-1	3–5	27	27	/	NO
CZT-2	3–5	27	27	/	Yes
CZT-3	3–5	50	21	2.5	Yes
CZT-4	3–5	75	10	0	Yes
CZT-5	3–5	75	15	1.25	Yes

as solvent material and indium as dopant is loaded into a carbon-coated quartz ampoule and vacuum sealed at 5×10^{-4} Pa. The growth temperature is designed at 750 °C by controlling the amount of excess tellurium. The sealed ampoule and a dummy crystal are orderly placed on the support before growth. Three thermocouples are fixed at different locations on the ampoule wall to monitor the initial temperature evolution at nucleation stage.

Several CdZnTe ingots with different diameters are prepared through fast unseeded THM as listed in Table 1. A relatively low growth rate of 3 mm/day is used to control nucleation, which is subsequently increased to 5 mm/day for routine growth. In order to capture the growth interface shape, ingots are rapidly cooled to room temperature and then cut along the axial direction. Based on simulation results, the thermal structure of 2-in. and 3-in. THM furnace is redesigned to obtain a stable and high temperature gradient at the growth interface. On the basis of stable thermal profile, the control strategy for interface shape is discussed by analysis of numerical simulation and experimental results.

3.1 Thermal Optimization

3.1.1 Design of Growth System Structure

As shown in Fig. 1b, a one-zone furnace is used to build the parabolic temperature profile which is needed for the THM growth. The main conditions for stable growth of THM are a parabolic temperature profile with large axial temperature gradient, a slightly convex growth front, and a stable melt flow. Generally, these objectives are not mutually promoted, but contradictory. For instance, because of the parabolic temperature profile used in THM, the buoyancy-induced flow is stronger than DMSG method. To avoid the constitutional supercooling, introducing a large temperature gradient will inevitably increase the temperature difference across the molten zone, which further strengthens the intensity of flow. For a given growth system, the growth conditions are basically determined by the thermal structure of furnace, including the heater, insulation material of furnace, ampoule, and support of ampoule, and also there are other ways to optimize the growth conditions, such as rotation of ampoule. Normally, homogeneous insulation material is used to build the

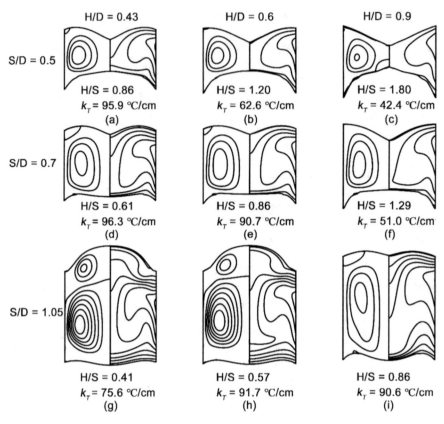

Fig. 10 The temperature field, flow field, and interface shape under the different combinations of heater and solution zone. The left of each picture is streamline and the right is isotherm. S, H, and D represent the height of the solution zone, height of heater, and ampoule diameter, respectively

furnace body, so the growth conditions are mainly determined by the ampoule diameter, height of heater, and height of solution, and also coupling between the heater and support of ampoule.

The success of the application of computational fluid dynamics to thermal design makes it capable to discuss the influence of different parameter combinations in a wider range. Numerical simulation is used to find the best relationship among the ampoule diameter, height of solution, and height of heater. Figure 10 shows the simulation results using our self-programmed software, MASTRAPP, for cases with the same growth interface temperature and different furnace structures while the ampoule diameter is fixed. As shown in Fig. 10, the growth front, axial temperature gradient, and flow pattern are significantly changed under different combinations of heater and solution zone. In terms of interface shape, it is an effective way to weaken the melt flow by using a short solution zone, but there is a risk of interface touching. For long solution cases (see Fig. 10g–i), strong upward flow near the wall and downward flow near the centerline continually scour the upper polycrystalline and lower single crystal, respectively, which results in a convex dissolution interface and more concave growth interface. For cases H/S < 0.57 and S/D > 1.05, the flow pattern has changed greatly to form two vortexes in the solution zone. The upper vortex with counterclockwise direction occupies the region adjacent to the dissolution interface, which restricts the component transport from the dissolution interface to the growth interface. Another issue of using long solution zone is the potential smaller axial temperature gradient since the interface position is far away from the edge of the heater.

For a certain height of solution zone, the height of the heater mainly affects the temperature difference across the solution zone and the axial temperature gradient. Since the growth interface temperature is expected to remain unchanged, a higher heater temperature is needed to ensure that the interface temperatures reach the set value when using a short heater. This, from another viewpoint, increases the temperature difference across the solution zone and enhances the flow, resulting in more concave interface as shown in Fig. 10d–f. When the height of the heater is larger than solution zone (see Fig. 10c, f), although the temperature difference is decreased, the axial temperature gradient is also decreased since the solution zone is placed inside the heater in that cases.

It is concluded in Fig. 10 that S/D and H/S are two main parameters in determining the flow pattern and thermal conditions for THM system. For cases with H/S > 0.86, the interface temperature gradient tends to be much smaller since the solution zone is placed inside the heater. Cases with S/D > 1 are also not recommended due to the longer transport distance between the dissolution and growth interfaces, the complex flow pattern, the strong convection flow, and the concave growth interface. However, the main difficulty of using the short solution zone is to avoid the interface touching. Based on the simulation results in Fig. 10, the recommended structures for THM are 0.5 < S/D < 0.7 and 0.61 < H/S < 0.86 while the exact limits of each parameter are not studied in this chapter.

3.1.2 Temperature Profile Stability

Figure 11 shows the surface temperature of ampoule that is measured by three thermocouples located at the ampoule wall near solution zone for ingot CZT-1. The upper thermocouple and the lower thermocouple are located at the initial interface of molten zone, while the middle thermocouple is at the middle of the molten zone (see Fig. 1b). Obviously, the ampoule wall temperature gradually increases with the pulling process of ampoule. For instance, at the position of 115 mm, the temperature difference between the upper and lower thermocouples reaches 37 °C. This is because of the deposition of low-thermal-conductivity CdZnTe on the bottom of the ampoule, which blocks the heat extraction from the crucible and makes the thermal field increased. The increasing temperature is believed to enhance the convection intensity, even change the flow structure, which further affects the concentration distribution, interface shape, and growth temperature. Figure 12 shows the evolution of growth interface with different

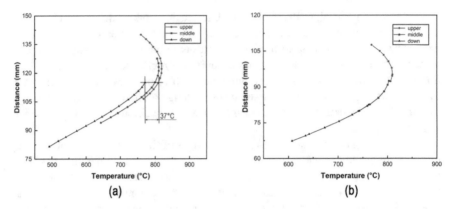

Fig. 11 The temperature profile on the wall of the ampoule in a 1-in. growth experiment without (CZT-1) and with (CZT-2) a dummy crystal [7]

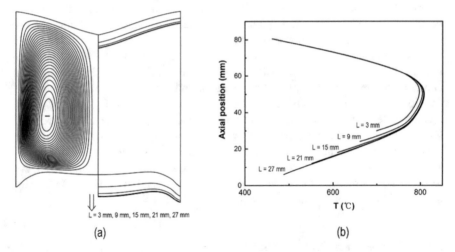

Fig. 12 (a) The left-hand side of the picture is streamlines at L = 3 mm, while the growth and dissolution interfaces under increasing thermal field are on the right-hand side. (b) The simulated wall temperature profiles [7]

lengths (L) of single crystal deposited on ampoule bottom. Because of increasing wall temperature, the average growth temperature is also increased from 748.6 °C to 766.0 °C in simulation. Since the growth interface gradually moves away from the heater and the increase of flow intensity, the growth interface shape also changes from convex to concave. The reduction of grain size and post-formed grains are considered to be related to the transition of interface shape from convex to concave.

The key factor of stabilizing the thermal field growth process is to maintain the stability of heat release near ampoule bottom at the early stage when low-thermal-conductivity CdZnTe crystals accumulate. As presented in Fig. 12b, such tempera-ture rise will disappear after a certain length of single crystal gets deposited on the ampoule bottom. Based on the analysis, a simple and easy-to-process support is designed [7] to stabilize the thermal field by choosing appropriate material support. Figure 11b is the temperature profile of the ampoule surface in the crystal growth of ingot CZT-2. It is obvious that the temperature of ampoule wall does not change during the pulling process of growth. By introducing the dummy crystal as a support of ampoule, thermal decoupling between ampoule and support is achieved, i.e., a stable thermal condition is built at the beginning of the nucleation stage, during which the change of temperature condition caused by changes in thermal structure is the main problem.

3.2 Interface Control by Rotation of Ampoule

For a growth system without rotation, buoyancy is the only driving force of melt flow. As discussed in 3.11, an upward flow near the wall and a downward flow near the centerline continuously scour the feeding and the already grown crystals, respectively, which results in a convex dissolution interface and a concave growth interface. Such interfaces are clearly observed in the 3-in. experiment (CZT-4), as shown in Fig. 14b. Figure 13 shows the effects of constant rotation on flow and thermal fields in a 2-in. THM system. Fundamental changes occur in the flow structure after applying constant rotation, even at 1.0 RPM (see Fig. 13b). The upward flow arises near the wall of the ampoule because of the temperature difference and turns inward after reaching the feed crystal. The inward melt flow is weakened after flowing through the dissolution interface in the presence of centrifugal force. After turning downward, the downward flow is further weakened because of the positive temperature gradient at the growth front and begins to flow outward under the action of centrifugal force. As a consequence, natural convection, which initially occupies the core area of the molten zone, is squeezed to the upper left part (see Fig. 13c), and a counter-rotating vortex arises in the center of the growth interface, which makes the interface convex. However, Fig. 13d shows that a further increase in rotation rate has limited positive effect on the growth interface shape, or even makes it worse. The primary large vortex breaks into many small vortices, which may weaken the transport of components between the external region near the wall and the internal region near the center. Figure 14a shows the convex growth interface obtained in a 2-in. THM growth system (CZT-3) at a constant rotation rate of 2.5 RPM while the grains are enlarging. At the same time, the simulated growth and dissolution interface shape are quite similar to the experimental results.

Such an interface transition from concave to convex is the consequence of the confrontation between thermal convection and forced convection, which has also been observed in Czochralski growth of garnets [18, 19]. With an increase in rotation rate, the intensity of thermal convection does not decrease, while the Grashof number fluctuates around 8.5×10^5. On the other hand, the ratio of the Grashof number to the square of the Reynolds number (Gr/Re^2) decreases from 0.79 at 1.0 RPM to 0.13 at 2.5 RPM and 0.06 at 3.5 RPM. This value shows that it does not

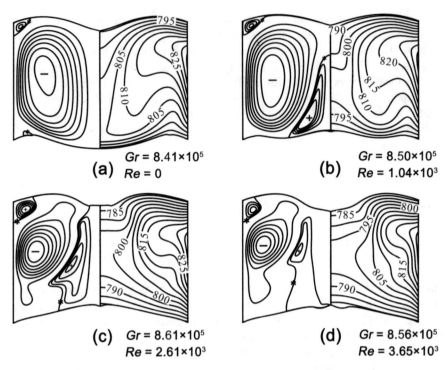

Fig. 13 The thermal and flow fields of a 2-in. crystal with different rotation rates: (**a**) 0.0 RPM, (**b**) 1. 0 RPM, (**c**) 2.5 RPM, and (**d**) 3.5 RPM. The streamlines are spaced at 1.2×10^{-6} for positive values and 1.2×10^{-7} for negative values, and the asterisk represents the zero streamline [7]

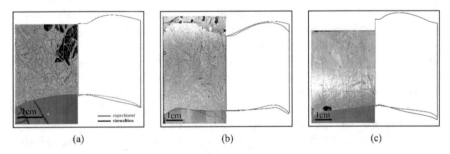

Fig. 14 Experimental and simulated interfaces of crystals: (**a**) 2-in. with 2.5 RPM, (**b**) 3-in. without rotation, (**c**) 3-in. with 1.25 RPM (end of growth) [7]

require a higher rotation rate to completely suppress thermal convection ($Gr/Re^2 < 0.1$) with the purpose of controlling the growth interface shape.

A 3-in. ingot (CZT-5) is grown by THM to repeat these results in larger crystals. Since the Gr/Re^2 is proportional to $1/w^2R$, a smaller rotation rate than 2.5 RPM is assumed to be valid in a 3-in. THM growth system. From Fig. 14b and c, it is

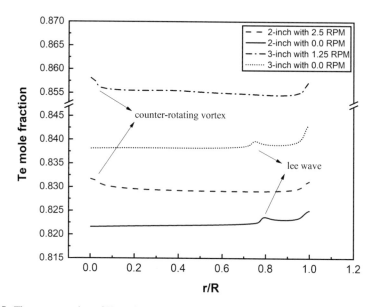

Fig. 15 The concentration of Te at the growth interface with different diameters and rotation rates [7]

obvious that the growth interface changes from concave to convex, even though only 1.25 RPM is used. In contrast to the elliptical convex growth interface in a 2-in. ingot, there is an obvious uplift in the center of the growth interface, while a depression is formed on the uplift in the simulation (see Fig. 14c). The formation of uplift shows that the constant rotation of 1.25 RPM is still too large for a 3-in. THM growth system. Apart from making growth interface convex, on the other hand, a counter-rotating vortex leads to the accumulation of excess Te in the center. From the CdTe binary phase diagram, the liquidus temperature of crystals decreases with an increase in the Te amount, which results in a depression at the center of the growth interface. However, uncontrollable factors in the experiments, such as geometric deviation, fluctuation of furnace temperature, eccentricity of rotation, and mechanical shaking, destroy the symmetry of two-dimensional model and, therefore, avoid the accumulation of Te to a certain extent.

Another concern in crystal growth is the uniformity of concentration in both axial and radial directions. Due to strong natural convection in THM, the radial distribution of Te is relatively uniform over most of the growth interface, while a counter-rotating vortex adjacent to the growth interface leads to the accumulation of Te in the local region, as shown in Fig. 15. Without rotation, the formation of lee wave [14] causes Te accumulation around 4/5R in 2-in. and 3-in. crystals. After introducing rotation, the original lee-wave structure disappears, and a counter-rotating vortex arises in the center, which moves Te-rich area to the center.

3.3   Characterization of THM Samples

As a comparison, the ingot prepared by THM (CZT-3 in Table 1) with a largest single-crystal volume of 3.26 cm^3 is characterized by concentration of indium, resistivity, and Te inclusion. The distribution of indium concentration for ingot CZT-3 is shown in Fig. 5. Different from DMSG, the concentration of indium increases at the early stage of the growth (the tip of the ingot) and then keeps stable at 4.5 ppm when the stable status is built between melting and solidification. This is a great advantage of THM compared with DMSG method in controlling the uniformity of dopant, since indium is dopant twice in both the feeding preparation process and the THM growth process. The concentration of dopant is higher than that of DMSG process. The reason why there is an increase of dopant concentration is not fully understood since the concentration in the solution is quite high at the beginning of fast-unseeded THM process.

In order to evaluate the THM-grown crystals, I-V curve of detector made by wafer at the stable part is measured and shown in Fig. 16. It is seen that the resistivity of the crystal is symmetrical due to the uniformity of dopant concentration. However, the I-V curve is not smooth which indicates that the resistivity is not stable under different biases. The resistivity of the as-grown crystal is 8×10^9 Ω·cm under a bias of -600 V and drops to 1.6×10^9 Ω·cm quickly. Under the bias of $+60$ V– $+600$ V, the resistivity of the as-grown crystal keeps at 1.0×10^{10} Ω·cm. The wafer is annealed under the same conditions as the case of DMSG in Fig. 6, i.e., an annealing process under a Te-Cd atmosphere at 600 °C for 80 h. From the I-V

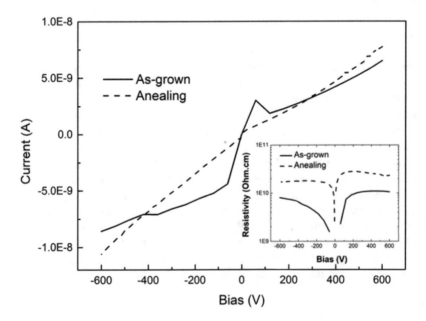

Fig. 16 The I-V curve of THM-made crystals before and after annealing

curve, it can be seen that the quality of the sample is improved on both the stability and the absolute value of the resistivity. The measured resistivity varies from 1.7×10^{10} $\Omega \cdot$cm to 2.7×10^{10} $\Omega \cdot$cm. By comparing the difference between resistivity of samples prepared by two methods, it can be seen that the resistivity is more stable under different biases in THM-grown crystals, but the absolute resistivity value of DMSG method is much higher than that of THM method. This phenomenon is thought to be caused by the difference in dopant concentration of the samples. Since the indium concentration of the THM sample is higher than the DMSG sample, the concentration of THM should be reduced. However, the optimization of doping process has not been studied. How to adjust the amount of indium in two doping processes is to be studied in the future fast-unseeded THM process.

Additionally, Te inclusions in the THM samples before and after annealing are characterized by IR transmission images which are shown in Fig. 17. Figure 17a shows a typical image of Te inclusions of the as-grown crystals which contains Te inclusions with the size of 10 μm and 3–5 μm. The statistical data of Te inclusion counts in Fig. 17c shows that the inclusions are mainly distributed near 3 μm and 10 μm and there are also inclusions with a large size of 20 μm. Figure 17b shows the results of Te inclusions after annealing. It is shown that there are still inclusions with the size near 10 μm, but most inclusions have quite small size of about 2 μm. The statistical data in Fig. 17d also show that the inclusions with the size near 5 μm and 10 μm are reduced and no inclusions larger than 15 μm are captured. The density of inclusions with the size of 2μm is dramatically increased, which is also shown in Fig. 17b. However as mentioned by Chen [6], inclusions smaller than 3 μm with a concentration of $<10^{6}$ cm^{-3} can be tolerated in CdZnTe detectors with thickness up to 15 mm. Generally, annealing is an effective way to reduce the inclusions and improve the performance of THM-grown crystals.

4 Conclusion

Solution growth is an effective method for preparation of detector-grade $Cd_{0.9}Zn_{0.1}Te$ crystals. By controlling the reaction rate during synthesis process, the DMSG method is achieved which can prepare high-purity CdZnTe crystals from 7 N raw material during one single step. Like the VB method, a stable temperature profile makes the convection quite weak in the melt which leads to easy control of the growth interface by changing the temperature field. However, the heat and mass transfer are also at quite low level without solution flow, which results in a low temperature gradient and low crystal growth rate. Thus, ACRT with high rotation rate is used to get large size and high-quality CdZnTe crystals. The resistivity of annealed crystals reached higher than $10^{10}\Omega \cdot$cm and even reached $10^{12}\Omega \cdot$cm; however, the segregation effect leads to a nonuniform dopant concentration. And this restricts the application of DMSG in mass production of CdZnTe crystals, but this method can be used in preparing high-quality seeds and feedings for

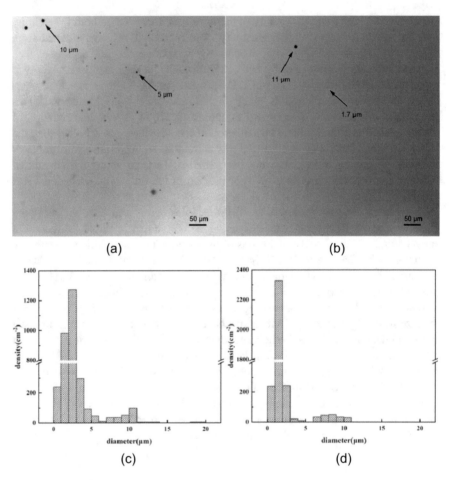

Fig. 17 Te inclusions in the THM samples, IR transmission images of Te inclusions before (**a**) and after (**b**) annealing; statistical data of Te inclusions before (**c**) and after (**d**) annealing

the THM method. Based on the DMSG method, fast-unseeded THM process is conducted to improve the THM growth system and optimize the process. The THM furnace is designed based on numerical investigation of the relationships among the ampoule diameter, the height of Te-rich solution, and the height of the heater. The thermal field stability during crystal growth is also improved by introducing a dummy crystal which has the same thermal conductivity with CdZnTe crystals. Controlling of growth interface by ampoule rotation is also introduced in order to get large-size crystals. The resistivity of THM-prepared ingot can also reach higher than $10^{10}\Omega\cdot$cm after annealing under a Te-Cd atmosphere, which is more stable than that of DMSG method. Also, the annealing process can dramatically reduce the large-size Te inclusions of the ingot and is an effective way to improve the quality of the crystal. However, there are still problems to be solved to improve the crystal quality:

1. Even though the resistivity is improved after annealing, there are still inclusions with the size near 10 μm left. How to furtherly reduce the Te inclusions through annealing has not been studied.
2. Based on the improved growth system, high-quality seeds can be used in the THM method to eliminate the random nucleation and achieve large-size crystal growth.
3. The concentration of indium dopant plays a determining role in the performance of detectors, since the doping is conducted twice during THM process and a higher indium concentration is probably responsible for the low resistivity of THM than DMSG. The control strategy of indium concentration is not fully understood and needs a further investigation in the future work.

References

1. Li, L.X., et al.: Studies of Cd-vacancies, indium dopant and impurities in CdZnTe crystals (Zn = 10%). In: 2003 IEEE Nuclear Science Symposium. Conference Record (IEEE Cat. No. 03CH37515) (2003)
2. Xu, Y., et al.: Characterization of CdZnTe crystals grown using a seeded modified vertical Bridgman method. IEEE Trans. Nucl. Sci. **56**(5), 2808–2813 (2009)
3. Shiraki, H., et al.: THM growth and characterization of 100 mm diameter CdTe single crystals. IEEE Trans. Nucl. Sci. **56**(4), 1717–1723 (2009)
4. MacKenzie, J., et al.: Recent advances in THM CZT for nuclear radiation detection. Nucl. Rad. Detect. Mater. 2009. **1164**, 155 (2010)
5. Yin, L.Y., et al.: The effects of ACRT on the growth of ZnTe crystal by the temperature gradient solution growth technique. Crystals. **7**(3), 1–12 (2017)
6. Chen, H., et al.: Characterization of traveling heater method (THM) grown cd(0.9)Zn(0.1)Te crystals. IEEE Trans. Nucl. Sci. **54**(4), 811–816 (2007)
7. Hong, B., et al.: Studies on thermal and interface optimization for CdZnTe crystals by unseeded traveling heater method. J. Cryst. Growth. **546**, 125776 (2020)
8. Roy, U.N.: Macro- and microscopic growth interface study of CdZnTe ingots by THM technique.**7805**(1):780502–780508 (2010)
9. Zhang, S., et al.: Controlling Te inclusion during direct mixed solution growth of large size CdZnTe crystal. In: Payne, S.A., et al. (eds.) Hard X-Ray, Gamma-Ray, and Neutron Detector Physics Xx. Spie-Int Soc Optical Engineering, Bellingham (2018)
10. Roy, U.N., Burger, A., James, R.B.: Growth of CdZnTe crystals by the traveling heater method. J. Cryst. Growth. **379**(10), 57–62 (2013)
11. Liu, Y., et al.: A three-dimensional numerical simulation model for the growth of CdTe single crystals by the travelling heater method under magnetic field. J. Cryst. Growth. **254**(3–4), 285–297 (2003)
12. Wang, Y., et al.: Growth interface of CdZnTe grown from Te solution with THM technique under static magnetic field. J. Cryst. Growth. **284**(3–4), 406–411 (2005)
13. Roy, U.N.: Macro- and microscopic growth interface study of CdZnTe ingots by THM technique (2010)
14. Peterson, J.H., Fiederle, M., Derby, J.J.: Analysis of the traveling heater method for the growth of cadmium telluride. J. Cryst. Growth. **454**, 45–58 (2016)

15. Zhou, B.R., et al.: Modification of growth interface of CdZnTe crystals in THM process by ACRT. J. Cryst. Growth. **483**, 281–284 (2018)
16. Lan, C.W., Chian, J.H.: Effects of ampoule rotation on vertical zone-melting crystal growth: steady rotation versus accelerated crucible rotation technique (ACRT). J. Cryst. Growth. **203**(1), 286–296 (1999)
17. Dost, S., Liu, Y.C.: Controlling the growth interface shape in the growth of CdTe single crystals by the traveling heater method. C. R. Mec. **335**(5–6), 323–329 (2007)
18. Zydzik, G.: Interface transitions in Czochralski growth of garnets. Mater. Res. Bull. **10**(7), 701–708 (1975)
19. Carruthers, J.R.: Flow transitions and interface shapes in Czochralski growth of oxide crystals. J. Cryst. Growth. **36**(2), 212–214 (1976)

Laser-Induced Transient Currents in Radiation Detector Materials

Kazuhiko Suzuki

Abstract In this chapter, the utilization of the transient charge technique for the development of semiconductor radiation detector materials is discussed. In "Basic Formalism of Transient Current Waveforms," a basic formalism for the theoretical reconstruction of experimentally observed current waveforms is described. The formalism combines the one-dimensional drift-diffusion equation with Poisson's equation. A kind of numerical finite difference technique developed in the field of computational fluid dynamics referred to as the constrained interpolation profile method (CIP) is employed to solve the drift-diffusion equation. In "Monte Carlo (MC) Simulation of Transient Current," another theoretical approach for the simulation of the drift-diffusion processes using the Monte Carlo (MC) simulation technique is presented. In "Experimental Methods," some experimental techniques for obtaining high-fidelity current waveforms are described. Then, as an example of waveform analysis, experimental results for voltage-induced polarization under space charge-free transport in ohmic-type cadmium telluride (CdTe) and cadmium zinc telluride (CZT) detectors are discussed. In addition, the temperature dependence of the mobilities for some detector materials is also reviewed in detail. In the last section, the effect of space charge perturbation on the transient current waveforms is discussed based on the temporal evolution of the charge density and internal electric field derived from the solution of a one-dimensional drift-diffusion equation in combination with Poisson's equation by the CIP and MC methods.

1 Introduction

The mobility-lifetime ($\mu\tau$) product is the most important material property for the development of semiconductor radiation detectors. A variety of experimental methods, such as thermally stimulated current (TSC) [1] and photo-induced transient

K. Suzuki (✉)
Faculty of Engineering, Department of Electrical and Electronic Engineering, Hokkaido University of Science, Sapporo, Japan
e-mail: suzukik@hus.ac.jp

© The Author(s), under exclusive license to Springer Nature Switzerland AG 2022   307
K. Iniewski (ed.), *Advanced Materials for Radiation Detection*,
https://doi.org/10.1007/978-3-030-76461-6_14

current spectroscopy (PICTS) [2], have been applied to investigate deep-level characteristics related to the improvement of the carrier lifetime in detector materials. Understanding the charge transport processes under an applied electric field charac- terized by the mobility is another key to the development of high-energy resolution semiconductor radiation detectors. An experimental method known as the transient charge technique (TCT) is especially useful for understanding the carrier dynamics after an excitation. The technique monitors the induced current in an external circuit for the observation of the charge carrier drift generated by an instantaneous excita- tion under an applied electric field. The advantage of TCT over the previously mentioned characterization techniques is that not only the mobility, but also the trapping and detrapping time of both charge carriers in the detector materials, can be obtained. Because the operation of radiation detectors is based on the same concepts that underlie this technique, the method has long been used in the development of semiconductor radiation detectors. Historically, this method has been rediscovered many times in research on a variety of materials, not only in solids but also in gases and liquids [3]. Therefore, various names have been given. For instance, the method was referred to as the "drift mobility technique" in [4], as the "time of flight" (TOF) in [5–7], and as the "transient charge technique" in [8, 9]. In this chapter, we use all of these names according to the context.

The transport properties of cadmium telluride (CdTe) and the compensation mechanism necessary to achieve the high resistivity required for radiation detection have been extensively studied in the 1960s to the 1970s, mainly by a collaboration between Modena University and some American groups using the temperature dependence of TCT [6, 7]. From these works, a compensation model was constructed in which a doped shallow donor (Cl, Br) induces the formation of complex defects, each of which comprises the donor and a Cd vacancy. Furthermore, the mobilities, trapping times, activation energies, and trap concentrations were measured for both holes and electrons. Two electron traps at 25 meV and 50 meV below the conduction band and two hole traps at 140 meV and 350 meV above the valence band have been reported [6]. Hot electron transport [7] and Poole-Frenkel effect [5] of hole transport in CdTe have also been studied by the group.

With the development of new detector materials, this method has regained the attention of many researchers. The temperature dependence of the electron and hole mobilities in CdTe:Cl grown by the vertical gradient freeze method was reported in [8]. The evolution of the transport properties along a CdTe ingot was reported in [9]. The electron and hole mobilities of high-pressure Bridgman-grown cadmium zinc telluride (CZT) have been reported by several authors [10, 11]. A change in the transport properties of solution-grown CZT by thermal annealing was reported in [12]. Rafiei et al. reported the electron mobility and mobility-lifetime product of CdMnTe: In [13]. The shallow trap-controlled transport of electrons and holes in commercially available CdTe:Cl and CZT was investigated by measuring the tem- perature dependence of the electron and hole mobilities [14]. By using a tunable laser, Pousset et al. measured the TOF currents in CdTe:Cl excited at different wavelengths from 800 to 1800 nm to investigate the effect of deep levels on the photoinduced current transients [15]. In addition, the transient current waveforms obtained by this method contain rich information about the internal electric field. Fink et al. evaluated the internal electric field distribution by measuring the transient current signals under DC bias conditions in CdTe and CZT using [241]Am as an excitation source [16]. We will discuss the change in the internal electric field distribution by DC and pulsed bias application elucidated from the waveform analysis of the TCT results [17] in detail in "Waveform Analysis Under Space Charge-Free Conditions." A novel method to characterize the fast polarization of n-CdTe detectors was proposed by Musiienko et al. by utilizing transient current measurements in combination with alpha pulse height spectrum measurements [18]. The temporal evolution of the internal electric fields due to the voltage-induced polarization of Schottky CdTe:Cl detectors has also been analyzed using this method [19].

In relation to radiation-induced polarization, we will discuss the space charge perturbation of the internal electric field revealed by measuring the excitation intensity dependence of the TOF current waveforms [20] in "Excitation Intensity Dependence." To extract this information from the experimental waveforms, the theoretical reconstruction of the waveforms is required, which in turn requires the evolution of the charge distribution and the internal electric field. In the following sections, we will show two different approaches to theoretically obtain the TCT current waveforms. In addition, some of the experimental techniques to ensure the high-fidelity observation of the current waveforms will be discussed in "Experimental Methods".

2 Basic Formalism of Transient Current Waveforms

2.1 One-Dimensional Drift-Diffusion Equation

The shape of the transient current waveforms in CdTe detector materials was investigated theoretically in the 1960s mainly by two groups [21, 22]. Only the transient current response for times less than the transit time (T_R) with trapping and detrapping from a single level can be expressed in a closed form [21]. The general expressions for times longer than the transit time are quite complicated. Therefore, transport, including the effect of several trapping and detrapping centers, is usually analyzed by the numerical solution of the drift-diffusion equation in combination with Poisson's equation. Unlike typical semiconductor devices, the carriers in radiation detector materials drift for very large distances of as long as several millimeters up to approximately a centimeter under a relatively high electric field. Therefore, the drift dominates over the diffusion, and the situation is very close to what is referred to as the advection of particles in different disciplines such as fluid dynamics. The Scharfetter-Gummel discretization scheme [23] is often used for the solution of drift-diffusion equations. However, when the drift dominates over the diffusion, the method is essentially the same as upwind discretization, which is known to suffer from numerical diffusion in the case of advection [24].

Here, we briefly describe the theory of the small-signal transient response of a plane-parallel semiconductor detector that contains some trapping centers. The transient current arises from the excess free carriers generated by a short-pulsed laser excitation as they are trapped and detrapped while drifting to the counter electrode. The following analysis is confined to one-dimensional charge transport in a crystal with two trapping levels, one located shallow in the bandgap with trapping and detrapping times much shorter than the transit time of the carriers, and the other located deep in the bandgap with a detrapping time much longer than the transit time of the carriers. Such trap-controlled transport has been reported in detector materials such as CdTe and CZT [6, 14]. Further, the analysis will make the basic approximations that the material is homogeneous, and that there is no direct transition of carriers between the shallow and deep traps. In an actual experiment, both electrons and holes are created at the incident surface of the crystal; however, only one of the two carrier types will be displaced through the crystal by an electric field. Except for the surface recombination process, the density functions within the material will, therefore, describe only a single-carrier system, chosen here to be the electrons. The one-dimensional drift-diffusion equation for the free charge $n(x, t)$ is given by

$$\frac{\partial n}{\partial t} = -\frac{\partial}{\partial x}(\mu n E) + D_n \frac{\partial^2 n}{\partial x^2} - \frac{\partial n_s}{\partial t} - \frac{\partial n_d}{\partial t} \tag{1}$$

where μ is the electron mobility, D_n is the electron diffusion constant, $n_s(x, t)$ and $n_d(x, t)$ are the shallow and deep-trapped charge densities, and E is the electric field. Charge conservation in the shallow and deep traps is given as

$$\frac{\partial n_s}{\partial t} = c_s n(N_s - n_s) - e_s n_s \tag{2}$$

$$\frac{\partial n_d}{\partial t} = c_d n(N_d - n_d) - e_d n_d \tag{3}$$

where $c_{s(d)}$ is the shallow (deep) capture coefficient and $e_{s(d)}$ is the shallow (deep) emission rate. In addition to these trapping-detrapping events, the surface recombination of generated electrons and holes at the incident surface should be considered. Because the process takes place at the surface region at $t = 0$, we include this effect by considering the initial charge density n_0 as [25]

$$n_0 = \frac{\mu E_0}{s + \mu E_0} n_{00} \tag{4}$$

where s is the surface recombination velocity, E_0 is the electric field at the surface, n_{00} is the number of electron-hole pairs generated by the laser excitation at $t = 0$, and n_0 is the number of electrons escaping from the surface recombination and drifting into the bulk of the sample. The initial conditions for each state are therefore given by Eqs. (5–7):

$$n(x,0) = n_0 \exp(-\alpha x), \tag{5}$$

$$n_s(x,0) = 0 \quad \text{for } x > 0, \tag{6}$$

$$n_d(x,0) = 0 \quad \text{for } x > 0. \tag{7}$$

Here, α is the absorption coefficient of the sample at the incident laser wavelength.

2.2 Constrained Interpolation Profile Method

Equation (1) can be rewritten as

$$\frac{\partial n}{\partial t} + \frac{\partial}{\partial x}(nv) = D_n \frac{\partial^2 n}{\partial x^2} - \frac{\partial n_s}{\partial t} - \frac{\partial n_d}{\partial t} = h \tag{8}$$

where $v = \mu E$ is the drift velocity of the electrons. Transposing the spatial derivative of velocity to the right side gives

$$\frac{\partial n}{\partial t} + v \frac{\partial n}{\partial x} = h - n \frac{\partial v}{\partial x} \equiv H. \tag{9}$$

Various numerical methods have been proposed to solve this type of nonlinear advection equation. The constrained interpolation profile (CIP) method [24] is a useful scheme for the numerical solution of the advection equation as shown in Eq. (9), and is widely used in the field of computational fluid dynamics. The CIP method is less diffusive and more stable compared to other finite difference methods. In the CIP method, the spatial profile of the density function n within the discretized segment is approximated by a cubic polynomial [24]. The coefficients of this polynomial are determined by the condition that the density and its spatial derivative are continuous on the mesh boundaries. The spatial derivative of Eq. (9) is given as

$$\frac{\partial(\partial_x n)}{\partial t} + v \frac{\partial(\partial_x n)}{\partial x} = \partial_x H - \frac{\partial n}{\partial x} \frac{\partial v}{\partial x}. \tag{10}$$

In the CIP method, Eqs. (9) and (10) are divided into two parts called the advection and non-advection phases, as shown below.

The advection phase is

$$\frac{\partial n}{\partial t} + v\frac{\partial n}{\partial x} = 0,$$ (11a)

$$\frac{\partial(\partial_x n)}{\partial t} + v\frac{\partial(\partial_x n)}{\partial x} = 0.$$ (11b)

The non-advection phase is

$$\frac{\partial n}{\partial t} = H,$$ (12a)

$$\frac{\partial(\partial_x n)}{\partial t} = \partial_x H - \frac{\partial n}{\partial x}\frac{\partial v}{\partial x}.$$ (12b)

The details of this procedure can be found in [24].

The relation between the free charges and the electric field strength is given by Poisson's equation:

$$\frac{\partial E}{\partial x} = \frac{\rho}{\epsilon} = -\frac{q}{\epsilon}(n - p + N)$$ (13)

where ρ is the charge density, and $N \equiv N_A^- - N_D^+$ is the net ionized impurity concentration including the abovementioned traps. The product of the free carrier density $n(x, t)$ and the electric field distribution $E(x,t)$ thus obtained are integrated over the sample thickness to obtain the transient current density $I(t)$:

$$I(t) = \frac{1}{d}\int_0^d qn\mu E dx$$ (14)

where d is the sample thickness and $1/d$ is the weighting field of the plane-parallel detector [26].

3   Monte Carlo (MC) Simulation of Transient Current

The numerical diffusion problem encountered in the solution of the drift-diffusion equation by the finite difference technique can alternatively be avoided by using the MC technique to simulate the drift-diffusion process [17]. Figure 1 schematically illustrates the basic concept of our one-dimensional MC simulation for the motion of photo-generated carriers under an applied electric field perpendicular to the planar electrodes. We assume that the sample contains N particle sites. At the beginning of the simulation ($t = 0$), all the sites are occupied by photo-generated carriers. We divide the sample into a thin surface layer and the rest, which is treated as the bulk. At each calculation time step t_k, the status of each carrier is determined by the MC technique. In the surface region, there are only two possibilities for each carrier. One possibility is

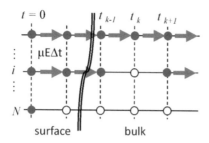

Fig. 1 Basic concept of Monte Carlo simulation of carrier drift. Filled circles indicate carriers in the free state. At each time step Δt, they can drift toward the counter electrode by $\mu E \Delta t$. Open circles indicate carriers that are trapped or surface recombined (after [17])

that the carrier drifts and diffuses into the bulk at $\mu E(x)\Delta t + \left(\sqrt{6D\Delta t}\right)_x$, where $E(x)$ is the applied electric field at position x from the surface, D is the diffusion constant, subscript x indicates the x-direction component of the three-dimensional diffusion length [27], and Δt is the calculation time step size. The other possibility is that the carrier is annihilated by surface recombination. The probability of surface recombination p_s is derived from Eq. (4) as

$$p_S = \frac{s}{s + \mu E_0}. \tag{15}$$

This procedure is repeated until the carriers arrive at the bulk region.

In the bulk region, the carriers have three possible motions, namely, drift and diffusion in the conduction (valence) band, being trapped at the i-th trap level, and being detrapped from the j-th level. The probability of trapping at the i-th trap level with the trap concentration of N_{Ti} is given by the inverse of the trapping time τ_i+:

$$1/\tau_i^+ = N_{Ti}\sigma_i v_{th}. \tag{16}$$

Here, σ_i is the capture cross section of the i-th trap and v_{th} is the thermal velocity of the electrons.

The detrapping probability for carriers trapped in the previous step is given by the inverse of the detrapping time τ_{Di}, which is given by

$$1/\tau_{Di} = \nu_0 \exp\left(-\frac{E_{Ti}}{kT}\right). \tag{17}$$

Here, ν_0 is the attempt to escape frequency, and E_{Ti} is the energy depth of the i-th trap level. We do not assume a direct transition of the carriers between the i-th and j-th traps. This procedure continues until the arrival of each carrier at the counter electrode.

At each time step, the internal electric field distribution is calculated by Eq. (13), and the number of mobile carriers is summed to calculate the induced charge on the electrode based on Ramo's theorem [28]. The contribution of each carrier to the total current induced at the electrode is given by $qv(x)$ irrespective of its position x, where q is the electronic charge, and $v(x)$ is the drift velocity of the carrier at position x.

4   Experimental Methods

4.1   Samples

Usually, planar-type detectors with dimensions of $4 \times 4 - 5 \times 5$ mm^2 and a thickness of 1 mm are used for our TCT measurements. This simple structure allows for one-dimensional analysis of the current transport. In addition, this size corresponds to a geometrical capacitance of approximately 2 pF, which is the upper limit of our system to maintain the necessary response time for high-fidelity waveform measurements of the electron transport in CdTe and CZT. In the case of TlBr, the capacitance of a sample with the above dimensions exceeds 6 pF because of the high dielectric constant of the material. However, the electron mobility is at most 1/30 that of CdTe or CZT, and as a result, the effect of the capacitance on the response time can be neglected.

In principle, a pair of semitransparent metal electrodes with a blocking nature are required to generate carriers just beneath the contact and avoid electrical injection of the carriers. Empirically, however, it is possible to detect signals from samples with approximately 1 μm thick Au or Pt electroless plating, which are usually used for CdTe detectors. This means that the transport properties of a detector as is can be evaluated by TCT measurements.

4.2   Experimental Setup

The basic principle of signal generation and the various excitation methods for TCT measurements are discussed in detail in [3]. Some of the experimental setups suitable for the development of semiconductor radiation detector materials such as CdTe, CZT, and TlBr will be described here. Figure 2 shows our recent experimental setup. UV light pulses from a diode-pumped passively Q-switched solid-state laser (355 nm) running at a repetition rate of 10 Hz and a nominal pulse duration of 1 ns are employed as the excitation source. To ensure complete recovery (depolarization) of the sample to its dark state, the repetition rate of the laser is selected to be much lower than the inverse of the dielectric relaxation time ($\rho\varepsilon$) of the sample. The UV light source is chosen in consideration of its applicability to various new detector materials with relatively large energy gaps such as TlBr. However, as long as the energy of the laser is higher than the gap of the material, longer laser wavelengths are more desirable to avoid surface recombination of the generated carriers. In fact, several authors have used visible light sources for CZT and CdTe measurements

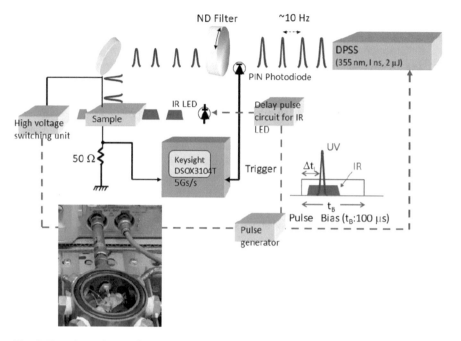

Fig. 2 Experimental setup for TCT measurements. The inset shows a direct connection between the sample and the oscilloscope

[18]. The duration of the light pulses should be much shorter than the drifting time of the carriers to determine the transit time (T_R) accurately. To avoid the space charge effect, the intensity of the incident light pulses is reduced by neutral density filters so that the photo-generated charge is small compared to CV, where C is the geometrical capacitance of the sample and V is the applied voltage. In space charge-perturbed transport, a transit time cusp (of approximately 0.7 T_R) appears before the actual transit time. We will discuss the intensity dependence of the transient current waveforms in a later section.

Our system can control the duration of the bias pulse (t_B) within the range of 100 µs to several tens of milliseconds. Further, to investigate the effect of bias-induced polarization on the current waveforms, it is possible to vary the suspension time (Δt_L), defined as the bias duration prior to the laser excitation, from approximately 25 µs to 37 ms. The biggest advantage of using artificial excitation sources over natural sources like ^{241}Am is that it is possible to synchronize the bias application with carrier generation. The timing between the laser pulse and the application of the bias pulse is schematically shown in Fig. 2. A minimum duration of 25 µs is required for stabilizing the applied pulse bias in our pulse bias supply system. In addition, it is possible to apply a DC bias. As we will discuss in the next section, depending on the contact material used, bias-induced polarization begins less than 1 ms after the bias application. Precise control of Δt_L is hence required to maintain a constant electric field condition. Additional below-gap pulse excitation at the side face of the sample can be applied by using the IR LED.

Because the excitation intensity of the laser pulse (2 μJ) is high enough for the direct measurement of transient current signals, we can feed the current signals from the sample directly into a digital oscilloscope (5 GS/s) through a 50 Ω load resistor. For this, the length of the cable from the sample surface electrode to the input terminal of the storage scope should be less than 10 cm to reduce the effect of any stray capacitance on the measured current waveforms. As can be seen from the inset of Fig. 2, literally, a direct connection between the sample and the oscilloscope is established. The advantage of this direct connection is that it avoids the complicated deconvolution procedures required to retrieve the original waveform shape from the signal convolved with the transfer function of the preamplifier electronic setup [29]. However, a charge of approximately 10 pC is required for this direct feed, which is near the upper limit for space charge-free transport. Further, the charge is approximately three orders of magnitude larger than the charge in the actual radiation detection event. The measurements can be carried out at temperatures ranging from approximately 100 K to 330 K using liquid nitrogen.

5   Waveform Analysis Under Space Charge-Free Conditions

In this section, we will discuss the measurements under space charge-free conditions where the generated charge Q is less than the CV of the samples. The most straightforward quantity that can be extracted from the TCT method is the transit time (T_R) of the charge carriers and hence the drift mobility (μ_d) of the electrons and holes in the material, which is given by

$$\mu_d = \frac{d^2}{T_R V} \tag{18}$$

where d is the thickness of the plane-parallel detector and V is the applied bias voltage, provided that a homogeneous electric field E given by V/L is established inside the detector. Figure 3 shows an example of the bias dependence of the electron transient current waveforms for an ohmic-type CdTe detector measured at 280 K for different pulse bias voltages with the duration of 100 μs and $\Delta t_L = 25$ μs. As can be seen from the figure, the generated charge is estimated to be approximately 10 pC (0.1 mA × 100 ns at 100 V), which is small enough to keep Q << CV (2 pF × 100 V = 200 pC). Except for an avoidable high-frequency oscillation due to the incomplete impedance matching of the 50 Ω load resistors [17, 29], the induced currents show a nearly flat plateau, indicating the good quality of the crystal and that the carrier lifetime is much longer than the transit time. As can be seen from the inset of Fig. 3, the drift velocity increases linearly with the electric field strength.

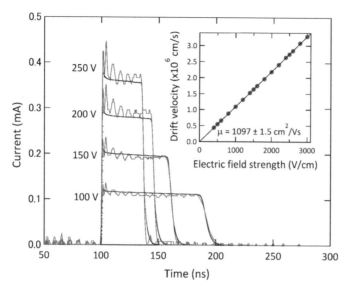

Fig. 3 TOF current waveforms (solid lines) for an ohmic-type CdTe detector measured at 280 K under different pulse biases from −100 to −250 V. The dotted lines are the corresponding MC results assuming the transport parameters described in the text. The inset shows the drift velocity versus the applied electric field strength (after [30])

This indicates that the carriers are in thermal equilibrium with the lattice within the range of the applied electric field strength. The electron drift mobility is thus determined from the slope as 1097 cm^2/Vs for this sample. Such an ohmic behavior does not always hold, especially in the very high electric field regime where the electron temperature is higher than that of the lattice [7]. The dotted lines in Fig. 3 are the result of the MC simulation of the transient current at each bias voltage. The following transport parameters were assumed for the simulation: a shallow trapping time τ^+ of 30 ns, a corresponding detrapping time τ_d of 2 ns, a deep trapping time τ_D^+ of 2 μs, and a scattering mobility of 1150 cm^2/Vs. As observed from the figure, the simulation results are in good agreement with the experimental results. The assumed scattering mobility (μ_0) at 280 K [14] is in reasonable agreement with the observed drift mobility (μ_d) if we consider the trap-controlled transport described by

$$\mu_d = \frac{\tau^+}{\tau^+ + \tau_d}\mu_0. \tag{19}$$

Further, because the electrons and the lattice are in thermal equilibrium (ohmic behavior), the above equation can be written as

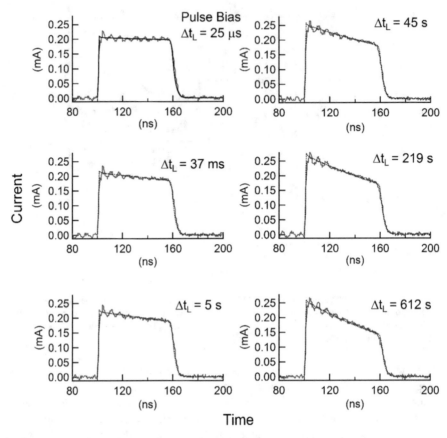

Fig. 4 TOF current waveforms (solid lines) measured at the bias voltage of 150 V with different suspension times, Δt_L. The dashed lines are the results of the least square fit of Eq. (21). The dotted lines are the results of the MC simulations assuming the space charge density estimated by the least square fit (after [30])

$$\mu_d = \mu_0 \left[1 + \frac{N_T}{N_C} \exp \left(\frac{E_T}{kT} \right) \right]^{-1}. \qquad (20)$$

Here, N_C is the effective density of states for the conduction band.

It should be noted that the simulations assumed a constant electric field for each of the bias voltages given by the applied voltage divided by the sample thickness. Therefore, at least under the pulse bias condition with a Δt_L of 25 μs, the internal electric field of the actual sample can be well approximated by a constant throughout the entire sample thickness.

The homogeneity of the internal field, however, does not hold when a DC bias is applied. Inhomogeneous electric field distributions in CZT and CdTe detectors have been reported and analyzed based on the MC simulation [17]. Figure 4 shows the current transients measured at 277 K and a constant bias voltage of 150 V applied on a 1 mm thick ohmic-type CdTe detector at different suspension times Δt_L ranging from 25 μs to 612 s. The constant transient current (current plateau) observed under a pulse bias turns into a decaying current with the increase of Δt_L. Generally, such decay in the transient current waveform occurs because of a short trapping time compared to the carrier transit time, provided that the internal electric field is constant. However, in the present case, as only the suspension time is changed between the current waveforms shown in Fig. 4, a change in the transport properties, such as a decrease in the trapping time, can be ruled out. The observed change in the current waveforms can be attributed to the change in the internal electric field with the increase in the suspension time Δt_L. The dotted lines shown in each of the current waveforms are the results of the MC simulations assuming each a constant positive space charge inside the sample. The good correspondence between the results and the experimental observations suggests the emergence of a positive space charge inside the sample and its evolution with Δt_L. It should be noted, however, that unlike the catastrophic distortion of the internal field reported in Schottky CdTe detectors [19], the change observed in ohmic-type CdTe was limited and no reduction of the collected charge with time was observed [30]. Quite stationary distributions of the internal field over time have also been observed via Pockels effect measurements [31]. A simple evaluation method for the space charge density from the current waveform was first proposed by Fink et al. in [16] in the absence of trapping. This method was further developed by Uxa et al. for the case of charge loss by deep trapping [32]. Following the numerical treatment in [19], the induced current $i(t)$ is given by

$$i = \frac{Q\mu E_0 \exp(-a\mu t)}{d} \tag{21}$$

where Q is the total charge traveling through the sample, and a is related to the space charge density N_{sc} through the electronic charge q and the dielectric constant of the material ε by

$$a = \frac{qN_{sc}}{\varepsilon}. \tag{22}$$

By measuring the evolution of the TCT current waveforms at various temperatures from 277 K to 315 K, it was concluded that the main cause of the polarization in the ohmic p-type CdTe detector is ascribed to the ionization of the deep donor state located at about 0.64 eV from the conduction band with a capture cross section of 1.6×10^{-17} cm^2 [30]. Further, it was revealed from the measurements that a very fast component of the polarization due to hole injection from the anode exists. It is interesting to note that the injection of holes from the anode was suggested to be the cause of the stabilizing polarization in ohmic p-type CdTe [31]. It is also interesting to note that for n-type CdTe with Au (blocking) contacts, a deep level with an energy of 0.74 eV from the conduction band with a capture cross section of 9×10^{-13} cm^2 was reported by TCT measurements [18].

The voltage-induced polarization of Schottky-type CdTe detectors is one of the most serious challenges for the development of a high-performance room-temperature radiation detector with high energy resolution. The nature of the deep defect responsible for the space charge buildup by this voltage-induced polarization has been extensively studied by direct measurements of the internal field using the Pockels electro-optic effect [31] or by the lowering of the Schottky barrier estimated using current-voltage characteristic measurements [33]. The temperature dependence of the TCT in Schottky-type CdTe detectors allows the very early time stage of the polarization process at less than 1 ms after the application of the bias voltage to be assessed. Such an early stage of the polarization cannot be assessed in principle by the techniques mentioned above. Three relaxation processes with time constants of 22 ms, 30 s, and 600 s were identified, and the two longer processes were attributed to the ionization of two deep acceptor defects at 0.54 eV and 0.6 eV [19].

5.1 Temperature Dependence Measurements

Even if the detectors are used only at room temperature, understanding the scattering processes of electrons and holes is very important for possible improvement of the mobility. Here, we briefly summarize the scattering processes and trap-controlled transport commonly observed in typical detector materials. Electron mobilities in II-VI semiconductors have been extensively studied by Road [34]. It was reported that longitudinal polar optical (LO) mode scattering and acoustic deformation potential (DP) scattering dominate the lattice-limited mobility of CdTe. In addition, ionized impurity scattering limits the overall mobility at low temperatures, depending on the purity of the material.

Figure 5 shows the temperature dependence of the (a) electron and (b) hole mobility for a CdTe:Cl detector from Acrorad Co., Ltd. [14]. The electron mobility shows a monotonous increase with decreasing temperature until approximately 100 K, and then saturates as the temperature decreases further. The theoretical temperature dependence of the mobility (μ_0), including contributions from the polar optical phonon scattering (μ_{po}), acoustic deformation potential scattering (μ_{ac}), and ionized impurity scattering (μ_{io}), was obtained by solving the Boltzmann equation using an iteration method by Rode [35]. As can be seen from the figure, it is not possible to reproduce the observed saturation behavior only by theoretical mobility as determined by the scattering mechanisms mentioned above. To reproduce the observed saturation behavior, a trap-controlled mobility μ_d defined in Eq. (20) is assumed to dominate the temperature dependence, provided that thermal equilibrium between the lattice and electrons has been established. The solid line in Fig. 5a denoted as μ_d was obtained by assuming $N_T = 1.0 \times 10^{16}$ cm^{-3} and $E_T = 28$ meV [14].

The hole mobility shows a slight increase with decreasing temperature until approximately 260 K, and then decreases gradually and shows a rapid decrease at

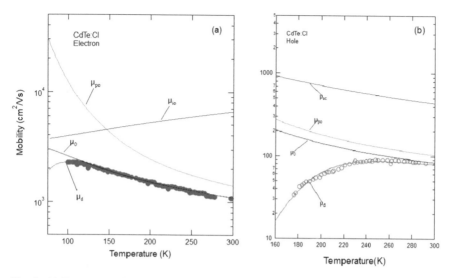

Fig. 5 (**a**) Temperature dependence of electron mobility (solid circles). (**b**) Temperature dependence of hole mobility (open circles). The contributions of ionized impurity scattering (μ_{io}) and polar optical phonon scattering (μ_{po}) to the overall mobility μ_0 are also depicted. The acoustic deformation potential is also included for the calculation (not shown here). The trap-controlled mobility is denoted as μ_d (after [14])

temperatures lower than 220 K. The theoretical mobility determined only by scattering cannot account for the observed experimental temperature dependence of hole transport as well. Trap-controlled mobility with a trap at 0.14 eV above the valence band reproduces the experimental temperature dependence very well, as shown in Fig. 5b. It is very interesting to note that such trap-controlled transport is commonly observed in several radiation detector materials such as CZT [14] and TlBr [36]. In addition to the above scattering mechanisms, alloy scattering should be considered for the mobility of CZT. Chattopadhyay [35] used a very small alloy scattering potential ΔU of 0.3 eV (almost half of the usually predicted value based on the bandgap difference or the electron affinity difference) to explain the experimental temperature dependence reported in [10]. On the other hand, a widely accepted value of $\Delta U = 0.78$ eV, corresponding to the electron affinity difference, was used in [11] to successfully explain the overall scattering mobility of high-pressure-grown CZT. This means that although the dominant scattering mechanism for CdTe and CZT is LO-phonon scattering, because of alloy scattering, the lattice mobility of CZT will not be higher than that of CdTe.

The drift mobility of TlBr was extensively studied in the 1960s to the 1970s from the perspective of polaron physics. The electron and hole drift mobilities of TlBr and their temperature dependence have been determined experimentally by Kawai et al. [36] using the TOF technique. They reported an electron mobility of 30 cm^2/Vs and a hole mobility of 2.5 cm^2/Vs at room temperature. Furthermore, the temperature dependence of the mobility of their samples was dominated by strong

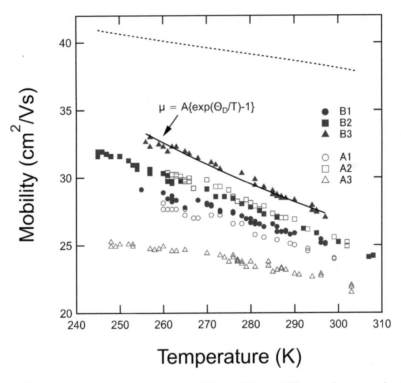

Fig. 6 Temperature dependence of electron drift mobilities in TlBr samples grown by the Bridgman method. The dashed line is an extrapolation of the theoretical polaron mobility proposed by Okamoto and Takeda [38]. The solid line represents $\mu = A[\exp(\Theta_D/T)\text{-}1]$ with $A = 37$ and $\Theta_D = 162$ K (after [39])

electron-optical phonon coupling (polaron transport). In addition to this strong interaction, some of the samples showed a reduction in mobility due to shallow trapping and detrapping. Two shallow trap levels were identified: one with a trap energy of 0.04 eV and the other with a trap energy of 0.12 eV [36]. Several theoretical models have been proposed [37, 38] during this period to explain the polaron mobility. Figure 6 shows the temperature dependence of the electron mobility measured in recent Bridgman-grown TlBr crystals for detector applications [39]. As can be seen from the figure, the temperature dependence of most of the samples is affected by strong LO-phonon interactions. In addition, some of the samples denoted as A1 and A3 in Fig. 6 show a reduction in mobility due to shallow trapping and detrapping.

6 Excitation Intensity Dependence

Attempts to use CZT pulse-mode detectors for fast-scanning high-flux X-ray applications such as computed tomography have been reported [40]. However, polarization at high X-ray fluxes, which degrades the spectroscopic performance and detection efficiency, has been recognized and has become a focus of interest. Several authors have studied this phenomenon by using the Pockels effect to reveal the defects responsible for the trapping of carriers. Owing to the limited acquisition time of the Pockels effect, these investigations [41, 42] focused on the deformation of the internal electric field due to the trapped and, hence, immobile charges. Before such catastrophic degradation starts, however, the perturbation of the internal field by the drifting charges themselves should occur even at much lower excitation intensities. We will discuss such space charge perturbation in detector materials in this section. Figure 7 shows the TOF current waveforms of electron drift in the CZT detector under different optical excitation intensities (solid lines) at a bias voltage of -200 V applied on the incident surface [43]. The charge Q at each measurement was estimated by integrating each transient current waveform. The transient current at $Q = 9.8$ pC with a transit time T_R of approximately 50 ns is a conventional TOF current under the SCF condition. The drift mobility of the present sample estimated by the least square fit of v_d versus the applied electric field at room temperature is 960 cm^2/Vs.

The dash-dotted line at the $Q = 9.8$ pC signal is the result of the MC simulation, and the dotted line is the result of the CIP calculation assuming a shallow trapping time τ_s^+ of 20 ns, a shallow detrapping time τ_{Ds} of 1.8 ns, and a deep trapping time τ_d^+ of 3.0 μs. Further, we assume a scattering mobility of 1050 cm^2/Vs and a constant electric field strength calculated as the applied voltage divided by the sample thickness. As can be seen from the figure, the two simulation models agree

Fig. 7 Excitation intensity dependence of electron drift signals for the CZT sample (solid lines) at a bias voltage of -200 V. The corresponding CIP (dotted lines) and MC (dashed-dotted lines) results are also shown (after [43])

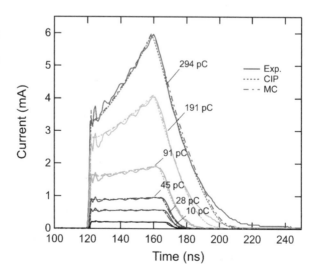

completely. As discussed in [14], the drift mobility of high-resistivity CZT is controlled by the trapping and detrapping at a shallow level located 0.023 eV from the conduction band. The observed drift mobility is in line with the assumed scattering mobility if we consider the reduction of mobility by trapping-detrapping at the shallow trap. Because the material is fully compensated and a high reverse bias field is applied, the recombination of electrons with holes is unlikely during TOF measurements. Instead, deep trapping determines the charge collection efficiency. Possible candidates for the electron traps located at 0.7 and 0.9 eV from the conduction band have been reported in [41] for a CZT detector with the electron mobility-lifetime product of 1.5×10^{-3} cm^2/Vs.

As shown in Fig. 7, although the applied bias voltage is kept constant at -200 V, the transient time increases with increasing excitation intensity owing to a long tail. Further, the plateau observed at lower excitation intensities, which indicates a constant drift velocity, shows a rapid increase with increasing excitation intensity. Finally, a cusp appears at approximately $t \approx 0.7T_R$ for the higher excitation-intensity waveforms. Here, T_R is the carrier transit time for the $Q = 9.8$ pC signal. This deformation of the current waveforms is obviously due to the perturbation of the internal electric field by the drifting carriers themselves. Similar changes in transient currents have been reported in insulating materials under intense optical excitation [44, 45]. The cusp indicates the arrival of the front edge of the charge distribution at the counter electrode. The corresponding timescale is denoted as T_1 ($\approx 0.7T_R$) [46]. Namely, under high excitation, the internal electric field is no longer a constant. It increases with increasing distance from the cathode. The SCP state is characterized by β, which is defined as [46]

$$\beta = \frac{CV}{Q_0} \tag{23}$$

where Q_0 is the charge generated at the surface. Experimentally, however, it is not possible to measure the generated charge at $t = 0$; instead, we estimate β from the collected charge Q, provided that $\tau_d^+ >> T_R$. Hereafter, we designate this value as β'.

The theoretical analysis of carrier transport under the SCP condition has been performed only for very limited cases such as trapping from a deep state without detrapping [46]. However, the carriers in these detector materials suffer from frequent trapping and detrapping as they drift to the counter electrode. We adopted both the MC and CIP methods to reconstruct the observed current waveforms. The dash-dotted lines at the higher excitation intensity curves in Fig. 7 are the results of the MC calculations and the dotted lines are the CIP calculation results in which the transport parameters are the same as those in the SCF case, but only the initial charge density is increased. Although a slight discrepancy is evident in the tail regions of the transient currents at 191 and 294 pC, which is probably due to the slight changes in the trapping parameters, the overall agreement between the experimental and both the MC and CIP results indicates the appropriateness of our calculation model. Therefore, it is worth discussing the charge distribution and electric field evolution derived from these simulation results.

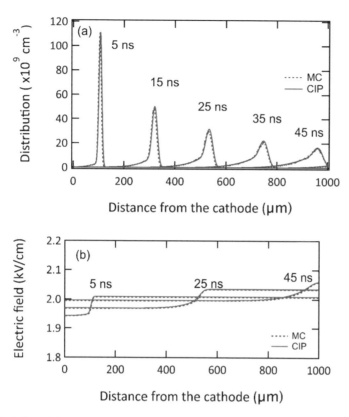

Fig. 8 Evolution of the (**a**) charge distribution and (**b**) electric field calculated using the CIP (solid line) and the MC (dotted line) methods under the SCF condition (10 pC) (after [43])

Figure 8a shows the calculated charge distribution under the SCF condition every 5 ns after charge generation at the cathode ($t = 0$) at a bias voltage of -200 V. In principle, the transport is described as nondispersive. However, the long tail at the trailing edge, mainly due to shallow trapping and detrapping, is remarkable. The corresponding electric field profiles at $t = 5$, 25, and 45 ns are shown in Fig. 8b. A step corresponding to the movement of the front edge of the charge distribution from the cathode to the anode is observed. Therefore, strictly speaking, the electric field is perturbed by carrier drift even under the so-called SCF condition. This difference in the electric field strength on each side of the distribution is an additional reason for the long tail at the trailing edge of the charge distribution shown in Fig. 8a.

Figure 9a shows the charge distribution under the SCP condition at a bias voltage of -200 V, which corresponds to $\beta' = 1.4$. Because the high-density electron distribution screens the externally applied field, the internal electric fields on each side of the charge distribution show significant differences (see Fig. 9b). This further broadens the charge distribution with time evolution owing to the dispersion of the

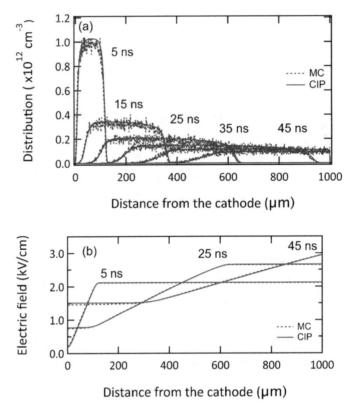

Fig. 9 Evolution of the (**a**) charge distribution and (**b**) electric field calculated using the CIP (solid line) and the MC (dotted line) methods under the SCP condition (294 pC) (after [43])

drift velocity. As a result, the charge distribution extends to nearly the entire sample length at $t \geq 45$ ns, as shown in Fig. 9a.

In this section, the effect of intense optical excitation on the TOF current waveforms of CZT detectors was investigated by using the MC simulation technique and the semi-Lagrangian-type advection scheme CIP method to solve the drift-diffusion equation. These two methods successfully reproduced the charge distribution and evolution of the internal field as well as the transient current for the CZT detector under a wide range of excitation intensities from SCF to SCP conditions and can be used interchangeably. It should be noted, however, that the advantage of the CIP method compared to the MC technique can be seen in Fig. 9a. The charge distribution obtained by the MC technique suffers from the contribution of unavoidable noise (statistical uncertainty) because of the limited number of simulated particles, whereas this problem does not exist in principle for the CIP method.

7 Summary

The basic formalism of laser-induced transient currents in radiation detector materials as well as some of the experimental techniques for high-fidelity measurements was presented. The applicability of the CIP method to the finite-difference drift-diffusion equation is demonstrated by the good correspondence with the experimental results and the results from the MC technique. It was shown that these two numerical approaches may be used both at the low excitation intensity regime (SCF) and at the enhanced intensity regime (SCP), at which the transient charge distribution is significantly affected by the Coulomb broadening of the charge cloud. Because of the simple sample geometry used in TCT, only single-carrier (electron) transport was treated in this chapter. In future outlook, however, the inclusion of hole transport and expansion of the dimensionality will open the possibility of accessing the carrier dynamics in actual radiation detector devices.

The TCT technique was originally developed to study carrier transport in high-resistivity materials. Recently, however, the technique has shown more power for the investigation of the internal electric field and its dynamics and contributed to the understanding of polarization phenomena. In particular, the very early stages of the dynamics that are inaccessible to conventional methods such as the Pockels effect have been revealed by the TCT analysis.

Acknowledgements I would like to thank Dr. A. Shiraki of Acrorad Co., Ltd. for providing the CdTe samples. Also, thanks to Prof. S. Seto of the National Institute of Technology, Ishikawa College, for his stimulating discussions. I am grateful to my graduate students for assistance in the experiments.

References

1. Scharager, C., Muller, J.C., Stuck, R., Siffert, P.: Determination of deep levels in semi-insulating cadmium telluride by thermally stimulated current measurements. Phys. Stat. Sol (a). **31**, 247–253 (1975)
2. Samimi, M., Biglari, B., Hage-Ali, M., Koebel, J.M., Siffert, P.: About the origin of the 0.15 to 0.20 eV defect level in cadmium telluride. Phy. Stat. Sol(a). **100**, 251–258 (1987)
3. Martini, M., Mayer, J.W., Zanio, K.R.: Drift velocity and trapping in semiconductors-transient charge technique. In: Wolfe, R. (ed.) Materials and Device Research, vol. 3, p. 181. Academic Press, New York (1972)
4. Spear, W.E.: Drift mobility techniques for the study of electrical transport properties in insulating solids. J. Non-Cryst. Sol. **1**, 197–214 (1969)
5. Ottaviani, G., Canali, C., Jacoboni, C., Alberigi Quaranta, A., Zanio, K.: Hole mobility and Pool-Frenlkel effect in CdTe. J. Appl. Phys. **44**, 360–371 (1973)
6. Canali, C., Ottaviani, G., Bell, R.O., Wald, F.V.: Self-compensation in CdTe. J. Phys. Chem. Sol. **35**, 1405–1413 (1974)
7. Canali, C., Martini, M., Ottaviani, G., Zanio, K.R.: Transport properties of CdTe. Phys. Rev. B. **4**, 422–431 (1971)
8. Suzuki, K., Seto, S., Tanaka, A., Kawashima, K.: Carrier drift mobilities and PL spectra of high resistivity cadmium telluride. J. Cryst. Growth. **101**, 859–863 (1990)

9. Suzuki, K., Tanaka, A.: Evolution of transport properties along a semi-insulating CdTe crystal grown by vertical gradient freeze method. Jpn. J. Appl. Phys. **31**, 2479–2482 (1992)
10. Burshtein, Z., Jayatirtha, H.N., Burger, A., Butler, J.F., Apotovsky, B., Doty, F.P.: Charge carrier mobilities in $Cd_{0.8}Zn_{0.2}Te$ single crystals used as nuclear radiation detectors. Appl. Phys. Lett. **63**, 102–104 (1993)
11. Suzuki, K., Seto, S., Iwata, A., Bingo, M., Sawada, T., Imai, K.: Transport properties of undoped $Cd_{0.9}Zn_{0.1}Te$ grown by high-pressure Bridgman technique. J. Electron. Mater. **29**, 704–707 (2000)
12. Suzuki, K., Seto, S., Dairaku, S., Takojima, N., Sawada, T., Imai, K.: Drift mobility and photoluminescence measurements on high resistivity $Cd_{1-x}Zn_xTe$ crystals grown from Te-rich solution. J. Electron. Mater. **25**, 1241–1246 (1996)
13. Rafiei, R., Reinhard, M.I., Sarbutt, A., Uxa, S., Boardman, D., Watt, G.C., Belas, E., Kim, K., Bolotnikov, A.E., James, R.B.: Characterization of CdMnTe radiation detectors using current and charge transients. J. Semicond. **34**, 073001 (2013)
14. Suzuki, K., Seto, S., Sawada, T., Imai, K.: Carrier transport properties of HPB CdZnTe and THM CdTe:Cl. IEEE Trans. Nucl. Sci. **49**, 1287–1291 (2002)
15. Pousset, J., Farella, I., Gambino, S., Cola, A.: Subgap time of flight: a spectroscopic study of deep levels in semi-insulating CdTe:Cl. J. Appl. Phys. **119**, 105701 (2016)
16. Fink, J., Krüger, H., Lodomez, P., Wermes, N.: Characterization of charge collection in CdTe and CZT using the transient current technique. Nucl. Instrum. Methods Res. Sect. A. **560**, 435–443 (2006)
17. Suzuki, K., Sawada, T., Imai, K.: Effect of DC bias field on the time-of-flight current waveforms of CdTe and CdZnTe detectors. IEEE Trans. Nucl. Sci. **58**, 1958–1963 (2011)
18. Musiienko, A., Grill, R., Pekárek, J., Belas, E., Praus, P., Pipek, J., Dědič, V., Elhadidy, H.: Characterization of polarizing semiconductor radiation detectors by laser-induced transient currents. Appl. Phys. Lett. **111**, 082103 (2017)
19. Suzuki, K., Sawada, T., Seto, S.: Temperature-dependent measurements of time-of-flight current waveforms in Schottky CdTe detectors. IEEE Trans. Nucl. Sci. **60**, 2840–2844 (2013)
20. Suzuki, K., Ichinohe, Y., Seto, S.: Effect of intense optical excitation on internal electric field evolution in CdTe gamma-ray detectors. J. Electron. Mater. **47**, 4322–4327 (2018)
21. Zanio, K.R., Akutagawa, W.M., Kikuchi, R.: Transient currents in semi-insulating CdTe characteristic of deep traps. J. Appl. Phys. **39**, 2818–2828 (1968)
22. Tefft, W.E.: Trapping effects in drift mobility experiments. J. Appl. Phys. **38**, 5265–5272 (1967)
23. Kulikovsky, A.A.: A more accurate Scharfetter–Gummel algorithm of electron transport for semiconductor and gas discharge simulation. J. Comput. Phys. **119**, 149–155 (1995)
24. Takewaki, H., Yabe, Y.: Cubic-interpolated pseudo particle (CIP) method. J. Comput. Phys. **70**, 355–372 (1987)
25. Levi, A., Schieber, M.M., Burshtein, Z.: Carrier surface recombination in HgI_2 photon detectors. J. Appl. Phys. **54**, 2472–2476 (1983)
26. He, Z.: Review of Shockley-Ramo theorem and its application in semiconductor gamma-ray detectors. Nucl. Instrum. Methods Res. Sect. A. **463**, 250–267 (2001)
27. Gould, H., Tobochnik, J., Christian, W.: An introduction to computer simulation methods: applications to physical systems, 2nd Ed, p. 250. Addison-Wesley (2007). Appendix 10, "Random Walks and the diffusion Equation"
28. Cavalleri, G., Fabri, G., Gatti, E., Svelto, V.: On the induced charge in semiconductor detectors. Nucl. Instrum. Methods. **21**, 173–178 (1963)
29. Praus, P., Belas, E., Bok, J., Grill, R., Pekárek, J.: Laser induced transient current pulse shape formation in (CdZn)Te detectors. IEEE TNS. **63**, 246–251 (2016)
30. Suzuki, K., Sawada, T., Seto, S.: Electric field inhomogenity in ohmic-type CdTe detectors measured by time-of-flight technique. Phys. Stat. Sol(c). **13**, 656–660 (2016)
31. Farella, I., Montagna, G., Mancini, A.M., Cola, A.: Study on instability phenomena in CdTe diode-like detectors. IEEE Trans. Nucl. Sci. **56**, 1736–1742 (2009)

32. Uxa, Š., Belas, E., Grill, R., Praus, P., James, R.B.: Determination of electric-field profile in CdTe and CdZnTe detectors using transient-current technique. IEEE Trans. Nucl. Sci. **59**, 2402–2408 (2012)
33. Toyama, H., Higa, A., Yamazato, M., Maehama, T., Ohno, R., Toguchi, M.: Quantitative analysis of polarization phenomena in CdTe radiation detectors. Jpn. J. Appl. Phys. **45**, 8842–8847 (2006)
34. Rode, D.L.: Electron mobility in direct-gap polar semiconductors. Phys. Rev. B. **2**, 1012–1024 (1970)
35. Chattopadhyay, D.: Electron mobility in $Cd_{0.8}Zn_{0.2}Te$. Solid State Commun. **91**, 149–151 (1994)
36. Kawai, T., Kobayashi, K., Kurita, M., Makita, Y.: Drift mobilities of electrons and holes in Thallous bromide. J. Phys. Soc. Jpn. **30**, 1101–1105 (1971)
37. Osaka, Y.: Polaron mobility at finite temperature (weak coupling limit). J. Phys. Soc. Jpn. **21**, 423–433 (1966)
38. Okamoto, K., Takeda, S.: Polaron mobility at finite temperature in the case of finite coupling. J. Phys. Soc. Jpn. **37**, 333–339 (1974)
39. Suzuki, K., Shorohov, M., Sawada, T., Seto, S.: Time of flight measurements on TlBr detectors. IEEE Trans. Nucl. Sci. **62**, 433–436 (2015)
40. Prokesch, M., Bale, D.S., Szeles, C.: Fast high-flux response of CdZnTe X-ray detectors by optical manipulation of deep level defect occupations. IEEE Trans. Nucl. Sci. **57**, 2397 (2010)
41. Franc, J., Dedic, V., Zazvorka, J., Hakl, M., Grill, R., Sellin, P.J.: Flux-dependent electric field changes in semi-insulating CdZnTe. J. Phys. D. Appl. Phys. **46**, 235306 (2013)
42. Cola, A., Farella, I.: CdTe X-ray detectors under strong optical irradiation. Appl. Phys. Lett. **105**, 203501 (2014)
43. Suzuki, K., Mishima, Y., Masuda, T., Seto, S.: Simulation of the transient current of radiation detector materials using the constrained profile interpolation method. Nucl. Instrum. Methods Phys. Res. Sect. A. **971**, 164128 (2020)
44. Gibbons, D.J., Papadakis, A.C.: Transient space-charge perturbed currents in orthorhombic sulphur. J. Phys. Chem. Solids. **29**, 115 (1968)
45. Many, A., Simhony, M., Weisz, S.Z., Levinson, J.: Studies of photoconductivity in iodine single crystals. J. Phys. Chem. Solids. **22**, 285 (1961)
46. Papadakis, A.C.: Theory of transient space-charge perturbed currents in insulators. J. Phys. Chem. Solids. **28**, 641 (1967)

Cadmium Zinc Telluride Detectors for Safeguards Applications

Peter Schillebeeckx, Alessandro Borella, Michel Bruggeman, and
Riccardo Rossa

Abstract Cadmium zinc telluride (CZT) semiconductor detectors can be used as
medium-resolution gamma-ray spectrometers, with the main advantage of a high
portability and operation at room temperature. Their characteristics in terms of
detection efficiency and resolution fall between the performance of sodium iodide
and germanium detectors. One of the disadvantages of a CZT detector is the
presence of a pronounced low-energy tail in the peak shape that complicates the
analysis of spectra with overlapping peaks. The range of applications of CZT
detectors includes verification measurements of fresh and irradiated nuclear material
for safeguarding nuclear materials.

In this chapter nuclear safeguards applications using CZT detectors are reviewed.
We discuss the verification of uranium isotopic abundances by both the enrichment
meter principle and methods relying on a multi-peak spectrum analysis and also
discuss applications at the back end of the fuel cycle. The latter include the
characterization of spent nuclear fuel assemblies and spent fuel pin counting by
means of passive gamma-ray emission tomography.

1 Introduction

Cadmium zinc telluride (CZT) detectors are semiconductor detectors for gamma-ray
spectrometry applications that can be operated at room temperature. CZT detectors
have an energy resolution at 662 keV in the order of 1–3%, when expressed in full
width at half maximum (FWHM). The resolution strongly depends on the detector
characteristics. CZT detectors are considered medium-resolution detectors with a

P. Schillebeeckx
Joint Research Centre—JRC Geel, Geel, Belgium
e-mail: peter.schillebeeckx@ec.europa.eu

A. Borella (✉) · M. Bruggeman · R. Rossa
SCK CEN Belgian Nuclear Research Centre, Mol, Belgium
e-mail: alessandro.borella@sckcen.be; michel.bruggeman@sckcen.be;
riccardo.rossa@sckcen.be

© The Author(s), under exclusive license to Springer Nature Switzerland AG 2022
K. Iniewski (ed.), *Advanced Materials for Radiation Detection*,
https://doi.org/10.1007/978-3-030-76461-6_15

resolution between high-purity germanium (HPGe) detectors (high resolution, 0.2%) and NaI(Tl) detectors (low resolution, 7%) [1]. The main drawbacks of CZT for gamma-ray spectroscopic applications are the limited detection efficiency and the low-energy tail in the full-energy peak shapes. The latter is due to mechanisms related to the charge carrier transport and trapping in the detector crystal [2, 3].

In the past, the detection efficiency of CZT detectors was strongly limited to difficulties in growing crystals with a volume larger than 1 cm^3. In addition, charge trapping effects increase with the crystal size. Recent technological improvements allow to grow crystals with volumes up to 20 cm^3, keeping the FWHM of the full-energy peak at 662 keV below 1% [4].

In this chapter, the use of CZT detectors for nuclear safeguards applications is discussed. The main objective of nuclear safeguards is to ensure the peaceful use of nuclear energy, preventing the misuse of nuclear facilities and deterring from the diversion of nuclear materials. The implementation of nuclear safeguards implies that nuclear materials can be independently verified by measurements. Nondestructive assay (NDA) is a main component in verifying nuclear material accountancy by nuclear safeguards authorities. NDA techniques are used to identify missing items (gross defects) and/or to measure specific characteristics or attributes of items to detect whether a fraction (partial defect) of a declared item is missing. The most widely used NDA instruments rely on the detection of gamma rays and/or neutrons emitted by the materials under investigation.

CZT detectors are used in NDA systems for nuclear safeguards to identify gross and partial defects in both irradiated and unirradiated nuclear materials [5–15]. They offer a more practical and cost-effective alternative to HPGe detectors, as they operate at room temperature and are more compact and easy to operate and maintain. Due to their performance and small size, they are suitable detectors for handheld instruments. They are preferred for applications where it is impossible or inconvenient to use HPGe detectors, such as underwater verification of fuel assemblies, in situ measurements in high-gamma-ray fields, or space restrictions. We discuss NDA of unirradiated material containing ^{235}U based on gamma-ray spectra obtained with CZT detectors and describe the use of CZT detectors for the verification of spent nuclear fuel assemblies, either as attribute tester or as part of a tomographic device to detect removed or replaced fuel pins.

2 Verification of Unirradiated Material

The IAEA (International Atomic Energy Agency) and other international organizations such as ESARDA (European Safeguards Research and Development Association) are working since several years to use medium-resolution gamma-ray spectroscopy (MRGS) detectors such as the CZT detector for an isotopic analysis and mass assay of U and Pu samples [16, 17]. The status of MRGS, including room-temperature CZT detectors, for nuclear safeguards applications was reported at the International Workshop on Uranium and Plutonium Isotopic Analysis by NDA

Techniques for Nuclear Safeguards [18]. While progress has been made into developing codes for the analysis of both U and Pu, further work needs to be done to improve the performances and uncertainty evaluation.

A typical application is the verification of fresh nuclear fuel in order to measure in a nondestructive way the ^{235}U isotopic abundance, frequently called ^{235}U enrichment.

2.1   Determination of ^{235}U Enrichment

Nondestructive determination of the ^{235}U enrichment, of nonirradiated nuclear material, relies primarily on gamma-ray spectroscopy [19]. Figure 1 compares spectra of a low-enriched U_3O_8 CBNM standard sample of the EC NRM 171 series [20] recorded with a HPGe, a LaBr$_3$, and CZT detector. Specific characteristics of spectra obtained measuring low-enriched uranium samples are shown in Fig. 1. The low-energy region of the spectrum between 80 keV and 110 keV shows a group of peaks due to the emission of both gamma- and X-rays following the decay of ^{235}U and ^{238}U and their daughters. The most pronounced peaks at 98.43 keV and 94.6 keV result from self-induced fluorescence. The peak with the highest intensity in the whole-energy region is observed at 185.7 keV. This peak together with those at 143.8, 163.3, and 205.3 keV is due to the decay of ^{235}U and those at 258.2, 766.4, and 1001.1 keV due to the decay of the ^{234m}Pa, which is a daughter product of ^{238}U. All signatures of ^{238}U are the result of the decay of daughter products, i.e., ^{234}Th, ^{234m}Pa, and ^{234}Pa. They are in secular equilibrium with ^{238}U after about 5 months. Two main methods to determine the ^{235}U enrichment can be distinguished: the enrichment meter principle requiring additional calibration measurements [21] and an absolute method based on a spectrum analysis of multiple peaks

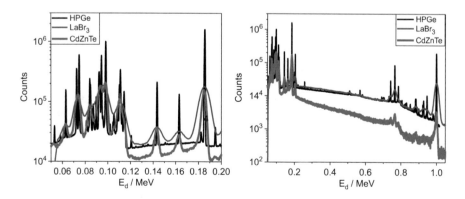

Fig. 1 Spectra obtained with a HPGe, LaBr$_3$, CZT detector for a U_3O_8 sample enriched to 4.5%$_w$ in ^{235}U. The sample is one of the CBNM EC NRM 171 series

[21]. Both principles are explained in the following paragraphs where CZT detectors are used to implement the methods.

2.1.1   Enrichment Meter Principle

The enrichment meter principle supposes a proportionality between the [235]U enrichment and the observed emission rate of a gamma ray following the decay of [235]U [21, 23, 24]. In most cases the full-energy peak of the 186 keV gamma ray is used. This proportionality is achieved when the infinite thickness condition for the detected gamma rays is fulfilled. The sample should be thick enough such that a further increase in thickness does not result in more gamma rays reaching the detector. This requires relatively thick samples and proper collimation conditions, as illustrated in Fig. 2. The proportionality constant is derived from measurements in the same conditions with representative calibration samples. The first measurements applying this principle relied on the use of NaI detectors and simple counting using two regions of interest [23]. This approach has its limitations in particular when there is a need for a correction for gamma-ray attenuation in the wall of the container holding the uranium sample. The method may also be influenced by gamma rays of other radionuclides in the peak region of interest. [25]. The performance of the enrichment measurement principle using low- and medium-resolution detectors, such as NaI and CZT detectors, is improved by performing a shape analysis of the observed full-energy peak to account for overlapping contributions of other gamma rays and for background components due to, e.g., Compton scattering. This approach, which is implemented in the NaIGEM code [26], allows to correct for differences in gamma-ray attenuation in the container walls of the unknown sample and the calibration standard [27–29].

The NaIGEM code was used by Gunnink and Arlt [30] to demonstrate the use of a 1500 mm^3 CZT detector to determine the [235]U enrichment by measurements of the CBNM EC NRM 171 standards [20]. The possibility of using a CZT detector to determine the [235]U enrichment following a similar concept as in NaIGEM is demonstrated in [31, 32]. In the work of Mortreau results from measurements of

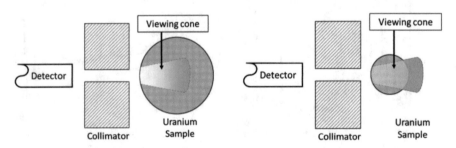

Fig. 2 Geometry illustrating the infinite thickness condition with the viewing cone of the detector completely inside the sample (left) and a geometry not fulfilling the infinite thickness condition (right)

uranium oxide powder samples with enrichments ranging from 0.31% to 92.42% obtained with a 15 mm^3 × 15 mm^3 × 7.5 mm^3 CZT detector are compared with those obtained with a NaI, LaBr, and HPGe detectors. Evidently, the best results are obtained with a HPGe detector. The comparison reveals that the LaBr detector does not present any clear advantage compared to a relatively large-volume CZT detector. The added value of CZT detectors for the verification of UF$_6$ cylinders is discussed in [33]. This study reveals that with a cluster of four 1500 mm^3 CZT detectors similar results as with a HPGe detector can be obtained. However, with the CZT detectors also stacked cylinders can be verified.

2.1.2 Multiple Peak Analysis

The basic principles to apply the enrichment meter principles are not always fulfilled, e.g., in case of samples with a complex matrix or thin samples which do not fulfil the infinite thickness condition. In case of samples with an arbitrary shape and composition and/or a not well-defined counting geometry, a method proposed by Parker and Reilly [22] is to be preferred. This method was originally developed for the assay of plutonium samples. It relies on an analysis of multiple full-energy gamma-ray peaks in the spectrum of the unknown sample to determine the energy dependence of the overall detection efficiency, i.e., including gamma-ray attenuation in the sample, solid angle, and intrinsic gamma-ray detection efficiency. An overview of different analytical models to parameterize the energy dependence of the relative overall detection efficiency is given in [34]. The parameters of the model describing the detection efficiency together with the isotopic abundances are derived from a least square fit to the experimental data. The concept is illustrated in Fig. 3, where the ratio between the full-energy net peak areas and the decay constant multiplied by the gamma-ray emission probability for different gamma rays of

Fig. 3 Illustration of the determination of the ^{235}U enrichment based on a multiple peak analysis. The full-energy net peak areas (A) divided by decay constant (λ) and gamma-ray emission probability (P$_\gamma$) are plotted as a function of gamma-ray energy. The full lines represent the energy dependence of the overall detection efficiency

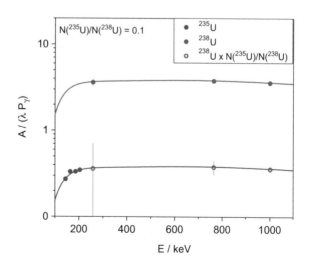

^{235}U and ^{238}U is plotted as a function of gamma-ray energy. The full lines represent the energy dependence of the overall detection efficiency. Accounting for the correct ratio of the two abundances brings the data to the same level [35]. This method avoids the need of calibration measurements with representative reference samples.

Various authors proposed such a self-calibrating method to determine the ^{235}U abundance in uranium-bearing samples. Two main approaches can be distinguished: an analysis based on the region below 210 keV [36–40] and an analysis relying on the energy region above 120 keV including the higher energy gamma ray from ^{234m}Pa [35, 41, 42]. The former is the basis of the MGAU code [37, 38] and the latter of the FRAM code [33]. In the latest version of MGAU the energy region was extended to 210 keV to include a correct treatment of the K-shell absorption edge and reduce bias effects on the determination of the ^{234}U [43]. A performance comparison of different approaches is reported in, e.g., [44, 45].

The use of a CZT detector to determine the ^{235}U enrichment with the MGAU code was demonstrated in [46] using uranium standards with a ^{235}U enrichment between 3% and 75% and by Alrt et al. [47] for samples with an enrichment between 2% and 90%. Borella et al. [42] performed measurements with a 10 mm × 10 mm × 5 mm quasi-hemispherical CZT detector [48] using the CBNM EC NRM 171 standards [20]. The spectra were analyzed in the energy region above 120 keV using response functions that were determined separately from measurements with radionuclide sources. The results of a parameterization of the spectrum in the region of the full-energy peak of 143.8, 185.7, 766.4, and 1001 keV are shown in Fig. 4. Monte Carlo simulations were used to determine the energy dependence of the intrinsic gamma-ray detection efficiency and the gamma-ray attenuation in the sample. This study shows the importance of the 258 keV full-energy peak in this procedure, as also mentioned in [33, 34].

3   Verification of Spent Nuclear Fuel

Verification of fuel assemblies at reactor sites is a major nuclear safeguards activity given the number of items to be measured and the variety of assembly types that are currently used [49, 50]. Both fresh fuel, i.e., fuel that has not been inserted in a reactor core, and spent fuel, i.e., fuel that has been irradiated in a reactor core, have to be verified for their fissile material content. After removal from the reactor core a spent nuclear fuel assembly is stored in a fuel storage pool to shield against the radiation and ensure appropriate cooling. After a cooling period it may be transferred to a wet or dry spent fuel storage facility until it will go for reprocessing or for final geological disposal [51]. Hence, routine verification of fresh and spent fuel assemblies by NDA requires detectors that can operate in various conditions.

Spent nuclear fuel contains a large number of radionuclides formed as a result of neutron-induced reactions and radioactive decay [52, 53]. The decay of fission products is the main contributor to the gamma-ray emission of spent nuclear fuel originating from irradiated fresh UO_2. For cooling times between 20 and 200 years,

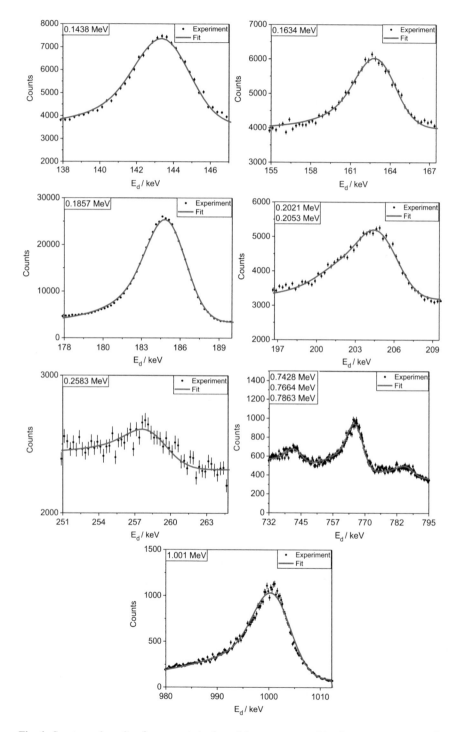

Fig. 4 Spectra and results of a parameterization of the response resulting from measurements of a low-enriched uranium sample with a CZT detector [42]

Table 1 Radionuclides present in spent nuclear fuel with their half-life and most intense gamma rays with energies above 600 keV

Nuclide	Half-life	Gamma-ray energies, keV
^{60}Co	5.27 a	1173, 1332
^{95}Zr	64.0 d	724, 757
^{95}Nb	35.0 d	766
^{106}Ru/^{106}Rh	1.02 a/30.1 s	622, 1050
^{134}Cs	2.06 a	605, 796
^{137}Cs	30.1 a	662
^{140}Ba/^{140}La	12.8 d/1.7 d	816, 1596
^{144}Ce/^{144}Pr	284.9 d/17.3 min	696, 1489, 2186
^{154}Eu	8.59 a	723, 1274

the 661 keV gamma ray due to the decay of ^{137}Cs dominates the spectrum. For cooling times between 10 and 20 years, there is a substantial additional contribution from the decay of ^{134}Cs and ^{154}Eu. For cooling times between 1 year and 10 years the emission spectrum is also influenced by short-lived fission products such as ^{106}Ru/^{106}Rh and ^{134}Cs. At shorter cooling times gamma rays following the decay of ^{95}Zr, ^{95}Nb, ^{140}Ba/^{140}La, and ^{144}Ce/ ^{144}Pr become important. Evidently, also gamma rays following the decay of activation products such as ^{60}Co due to the presence of cladding and structural materials will be observed. Table 1 lists radionuclides and the corresponding decay data of interest for NDA of spent nuclear fuel. The intensity of gamma rays emitted by fission products can be used to determine the initial enrichment, burnup, and cooling time (or time after the end of irradiation) [54–56] and to differentiate between spent fuel originating from a MOX (mixed oxides) or UO$_2$ assembly [57]. The intensity of the 1596 keV gamma ray due to the decay of ^{140}Ba/^{140}La is a good measure of the power level during a few weeks prior to shutdown [54, 58].

3.1 Attribute Testing of Spent Fuel Assemblies

Attribute testers for spent fuel assemblies are safeguards instruments designed for gross and/or partial defect verification of the assemblies. They are used to confirm specific characteristics of the assemblies (e.g., initial enrichment, burnup, cooling time) and to differentiate between fuel and structural materials, to distinct between spent fuel and fresh fuel and between spent fuel originating from fresh UO$_2$ and MOX fuel [59].

Several authors [14, 30, 47, 56, 60–65] demonstrated that CZT detectors operating at ambient temperature can resolve gamma-ray emission from, e.g., ^{60}Co, ^{134}Cs, ^{137}Cs, and ^{154}Eu and even from ^{95}Zr and ^{95}Nb in case of short cooling time in high-radiation fields of gamma rays and neutrons. This triggered the use of CZT detectors for verifications of spent fuel assemblies by an analysis of peaks due to the detection of gamma rays emitted by fission products or by activation products contained in irradiated structural materials. Various spent fuel attribute testers using CZT

Fig. 5 Schematic illustration of the deployment of a SFAT device in a spent fuel pool

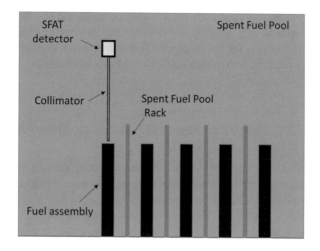

detectors have been developed for in situ verification under industrial conditions. To verify assemblies in storage ponds, the devices are housed in a watertight box with appropriate shielding and collimation. A multi-wire cable connects the underwater measurement head with a data acquisition system that is operated above water. Often a CZT detector is combined with a neutron detector, mostly a fission chamber, to record the total neutron emission rate [63, 64, 66, 67].

The irradiated fuel attribute tester (IRAT) and spent fuel attribute tester (SFAT) are equipped with a collimated CZT detector [59]. A schematic drawing of the SFAT device is shown in Fig. 5. The detection of gamma rays from fission products and from activation products such as ^{60}Co provides attributes for the presence of spent fuel and structural materials, respectively. The CANDU Bundle Verifier (CBVB) is an attribute tester that was especially designed to verify spent fuel from a CANDU reactor by detection of gamma rays with a highly collimated and shielded CZT detector [59]. It includes an automatic scanning system to provide gamma-ray spectral data as a function of position. The Fork Detector (FDET) system, originally developed at the Los Alamos National Laboratory [68], is one of the main NDA safeguards instruments to verify spent nuclear fuel from light water reactors (LWR). The detector consists of two arms that are positioned around a LWR assembly. Each arm contains three detectors: two fission chambers and one ionization chamber to record the total neutron and gamma-ray emission rate, respectively. The version proposed by [63] includes a 20 mm^3 CZT detector to record gamma-ray spectra from which the signatures of ^{134}Cs, ^{137}Cs, and ^{106}Ru/^{106}Rh can be obtained [63, 64]. The Safeguards MOX Phyton (SMOPY) device is another system combining the results of gross neutron counting and low-resolution gamma-ray spectroscopic data from a CZT detector to characterize irradiated LEU and MOX fuels [66, 67]. The Cask Radiation Profiling System (CRPS) is a general-purpose attribute tester to measure the presence of spent fuel in closed dry storage casks [69]. The measurement head of the CRPS is composed of a small fission chamber and a CZT detector with tungsten collimator and shielding. Using a speed-controlled motor the neutron and gamma-ray emissions from the cask are recorded as the detector head moves along the cask axis. Consistency between the measured and fingerprint profiles is used as an attribute for the presence of spent fuel in the measured cask [69]. More information on the abovementioned systems can be found in [59].

3.2 Passive Gamma-Ray Emission Tomography

Since the late 1980s the IAEA investigated through various Member States Safeguards Support Programmes the use of passive gamma-ray emission tomography for the verification of spent nuclear fuel from LWR [70–75]. The studies resulted in the development and testing of a prototype instrument for passive gamma emission tomography (PGET) [76–79] performing three simultaneous measurements: gross neutron counting, medium-resolution gamma-ray spectrometry, and two-dimensional gamma-ray emission tomography. Two objectives were identified. The system should allow (1) an independent determination of the number of active pins present in the spent fuel assembly and (2) a quantitative assessment of individual pin properties, e.g., activity of key radionuclides [74, 75].

The latest version consists of a torus-shaped watertight detection head that contains neutron and gamma-ray detectors which are arranged in two opposite banks on a rotating horizontal plate. Each detector bank contains a tungsten-shielded ^{10}B-lined proportional counter as neutron detector and can host up to 104 hemispherical CZT detectors (3.5 mm × 3.5 mm × 1.75 mm) [80] for gamma-ray spectroscopic and tomographic measurements. The detectors view the assembly through a parallel-hole tungsten collimator that is 1.5 m wide, 100 mm thick, and tapered in the axial direction of the assembly from 70 mm close to the assembly to 10 mm at the detector face. A schematic horizontal cross section of a PGET system is shown in Fig. 6. The number of detectors that is used varies with the dimensions of the spent fuel assembly to assay. Each CZT crystal is connected to an amplification circuit and a four-channel discriminator enabling data recording in four different energy

Fig. 6 Horizontal cross section of a PGET system consisting of two detector banks that each can host up to 104 CZT detectors. The actual number of detectors used varies between 83 and 104, depending on the dimensions of the spent fuel assembly. The figure is a courtesy of IAEA and STUK

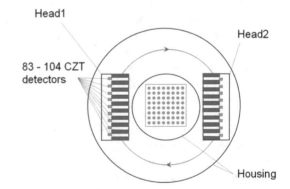

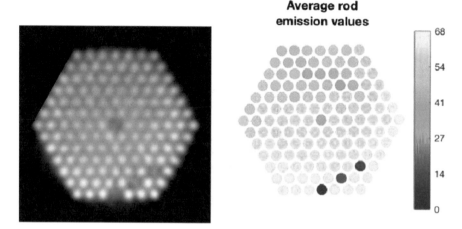

Fig. 7 Example of a contrast tomogram (left) and the two-dimensional activity distribution (right) of the gamma-ray emitters (or fuel pins) reconstructed from the measured data. From this tomogram, three missing pins are observed from the low emission rates (shown in blue). The figure is a courtesy of IAEA and STUK

windows. Measurements are performed underwater, with the spent fuel assembly placed in the center of a toroidal shaped detector platform.

The performance of the PGET system was tested by measurements of research reactor spent fuel rods at the JRC Ispra [81] and measurements of PWR, BWR, and VVER spent fuel assemblies at spent fuel ponds of nuclear power plants in Finland and Sweden [76–78, 82, 83]. At JRC Ispra three items were measured using fuel rods with a burnup ranging between 4 GWd/t and 30 GWd/t with cooling times between 40 and 50 years. The PWR, BWR, and VVER spent fuel assemblies covered burnups between 5 GWd/t and 58 GWd/t and cooling times between 2 years and 27 years. Figure 7 shows an example of a two-dimensional activity distribution of fuel pins that was reconstructed from the measured data on a VVER spent fuel assembly. These results demonstrate the capability of the PGET system to identify missing fuel pins. However, to determine in addition the activity level of individual pins improved reconstruction processing techniques are required.

Based on the positive outcome of the measurements at the power plants in Finland verification of spent fuel assemblies prior to final geological disposal will rely on an integrated NDA system combining a PGET system to check the integrity of the assembly and a passive neutron albedo reactivity (PNAR) system to assess the fissile content [84, 85].

4 Conclusion

Portability, compactness, and the fact of being operational at room temperature make the CZT detector a very interesting device for the verification of nuclear materials that are of nuclear safeguards concern. The limited sensitivity of the CZT that might be a problem in other applications is not a restriction for safeguards applications in which the gamma emissions of the objects to verify result in clear signals in the detector. The medium energy resolution of the CZT still restricts the possibilities of the analysis of very complex spectra compared to high-resolution detectors, but some typical spectral applications such as the determination of uranium enrichment and determination of characteristics of fresh and spent nuclear fuel are successfully obtained from CZT spectrometers. Due to their compactness they are in particular suited to construct an array of collimated detectors for tomographic applications in nuclear safeguards, e.g., for the verification of spent fuel assemblies.

Acknowledgement This work was partly funded by the European Union's Horizon 2020 Research and Innovation Programme under the grant agreement No. 847593 (EURAD).

References

1. Russo, P.A., Vo, D.T Gamma-Ray Detectors For Nondestructive Analysis, Los Alamos Report, LA-UR-05-3813 (2005)
2. Kalemci, E., Matteson, J.L., Skelton, R.T., Hink, P.L., Slavis, K.R.: Model calculations of the response of CZT strip detectors. In: Proc. SPIE 3768, Hard X-Ray, Gamma-Ray, and Neutron Detector Physics, p. 360 (1999). https://doi.org/10.1117/12.366601
3. Dardenne, Y.X., Wang, T.F., Lavietes, A.D., Mauger, G.J., Ruhter, W.D., Kreek, S.A.: Nucl. Instr. Meth. Phys. Res. Sect. A. **A422**, 159 (1999). https://doi.org/10.1016/S0168-9002(98)00947-4
4. Bolotnikov, A.E., Brands, H., Camarda, G.S., Cui, Y., Gul, R., De Geronimo, G., Fried, J., Hossain, A., Hoy, L., Liang, F., Preston, J., Vernon, E., Yang, G., James, R.B.: CdZnTe position-sensitive drift detectors for spectroscopy and imaging of gamma-ray sources. In: Proceedings of the INMM 57th Annual Meeting Conference (2016)
5. Arlt, R., Czock, K.-H., Rundquist, D.E.: Overview of the use of CdTe detectors for the verification of nuclear material in nuclear safeguards. Nucl. Instrum. Methods Phys. Res. Sect. A. **322**, 575–582 (1992)
6. Arlt, R., Rundquist, D.E., Bot, D., Siffert, P., Richter, M., Khusainov, A., Ivanov, V., Chrunov, A., Petuchov, Y., Levai, F., Desi, S., Tarvainen, M., Ahmed, I.: Use of room temperature semiconductor detectors for the verification of nuclear material in international safeguards - recent advances. In: Material Research Society Symposium, April 1993, San Francisco, U.S.A., Proceedings, vol. 302, pp. 19–29 (1993)
7. Arlt, R., Rundquist, D.E.: Room temperature semiconductor detectors for safeguards measurements. Nucl. Instrum. Methods Phys. Res. Sect. A. **380**(1–2), 455–461 (1996)
8. Ivanov, V., Dorogov, P., Arlt, R.: Development of large volume hemispheric CdZnTe detectors for use in safeguards applications. In: ESARDA European Safeguards Research and Development Association, 19th Annual Symposium on Safeguards and Nuclear Material Management, Proceedings, pp. 447. Le Corum, Montpellier, France 13–15 May, 1997 (1997)

9. Khusainov, A.K., Zhukov, M.P., Krylov, A.P., Lysenko, V.V., Morozov, V.F., Ruhter, W.D., Clark, D., Lavietes, T., Arlt, R.: High resolution pin type CdTe detectors for the verification of nuclear material. In: ESARDA European Safeguards Research and Development Association, 19th Annual Symposium on Safeguards and Nuclear Material Management, Proceedings, pp. 411. Le Corum, Montpellier, France 13–15 May, 1997 (1997)
10. Aparo, M., Arlt, R.: Development and safeguards use of advanced CdTe and CdZnTe detectors. In: Proceedings of INMM 39th Annual Meeting, Naples, Florida, 26–30 July, 1988 (1998)
11. Ruhter, W.D.: Application of CZT detectors in nuclear materials safeguards. In: Proc. SPIE 3446, Hard X-Ray and Gamma-Ray Detector Physics and Applications, p. 204 July 1, 1988 (1998). https://doi.org/10.1117/12.312892
12. Aparo, M., Arenas Carrasco, J., Arlt, R., Bytchkov, V., Esmailpour, K., Heinonen, O., Hiermann, A.: Development and implementation of compact gamma spectrometers for spent fuel measurements. In: ESARDA Symposium on Safeguards and Nuclear Material Management, Sevilla, Spain, 4–6 May, 1999 (1999)
13. Arlt, R., Ivanov, V., Parnham, K.: Advantages and use of CdZnTe detectors in safeguards measurements. In: Presented at MPA&C Conference, Obninsk, Russia, May 2000; to Be Published in the Conference Proceedings in November, 2001 (2001)
14. Czock, K.H., Arlt, R.: Use of CZT detectors to analyze gamma emission of safeguards samples in the field. Nucl. Instrum. Methods Phys. Res. Sect. A. **458**(1–2), 175–182 (2001)
15. Carchon, R., et al.: Gamma radiation detectors for safeguards applications. NIMA. **579**, 380 (2007)
16. Berlizov, A., Koskelo, M., McGinnis, B., Carbonaro, J: Report on the Results from Phase I of the Intercomparison Exercise on U and Pu Isotopic Analysis with Medium Resolution Gamma-Ray Spectrometers (MRGS) (2018)
17. Berlizov, A., Koskelo, M., McGinnis, B., Carbonaro, J.: Report on the Results from Phase II of the Inter-Comparison Exercise on U and Pu Isotopic Analysis with Medium Resolution Gamma-Ray Spectrometers (MRGS) (2019)
18. IAEA. Proceedings of the International Workshop on Uranium and Plutonium Isotopic Analysis by Nondestructive Assay Techniques for Nuclear Safeguards (2021). https://www.iaea.org/events/evt1906701, in preparation
19. Smith: Passive Nondestructive Assay of Nuclear Material, Reilly, Ensslin, Smith, and Kreiner, was published in 1991 (Chapter 7, smith) LA-UR-90-0732 (1991)
20. De Bievre, P., et al.: EC Nuclear Reference Material 171 Certification Report, CEC-COM 4153 (1985)
21. Reilly, T.D., Walton, R.B., Parker, J.L.: Nuclear Safeguards Research and Development LA-4605-MS, Los Alamos Scientific Laboratory, pp. 19–21 (1970)
22. Parker, J.L., Reilly, T.D.: Plutonium isotopic determination by gamma-ray spectroscopy. In: Robert Keepin, G., (eds.) Nuclear Analysis Research and Development Program Status Report, January–April 1974, Los Alamos Scientific Laboratory report LA-5675-PR (1974)
23. Kull, L.A., Ginaven, R.O.: Guidelines for Gamma-Ray Spectroscopy Measurements of 235U Enrichment, BNL 50414 (1974)
24. Matussek, P.: Accurate determination of the 235U isotope abundance by gamma spectrometry a User's manual for the certified reference material EC-NRM-171 I NBS-SRM-969, KfK 3752 Mai 1985
25. Nagel, W., Quick, N.: A New Approach for the High-Precision Determination of the Elemental Uranium Concentration in Uranium Ore by Gamma-Ray Spectrometry, CEC Report EUR 14659 ISSN 1018–5593 (1993)
26. Gunnink, R., Arlt, R., Berndt, R.: New Ge and NaI analysis methods for measuring 235U enrichment. In: Proceedings of the 19th Annual ESARDA Symposium on European Safeguards Research and Development Association, Luxembourg, ESARDA Joint Research Centre, Ispra, Italy, May 13–17, 1997 (1997)
27. Bracken, D.S. et al.: Peak Fitting Applied to Low-Resolution Enrichment Measurements, LA-UR-98-2436 (1998)

28. Mortreau, P., Berndt, R.: Determination of the uranium enrichment with the NaIGEM code. NIMA. **530**, 559–567 (2004)
29. Solodov, A.: NaIGEM analysis for Monte Carlo code generated uranium spectra July 2015. Conference: Institute of Nuclear Materials Management Annual Meeting (2015)
30. Gunnink, R., Arlt, R.: Methods for evaluating and analyzing CdTe and CdZnTe spectra. NIMA. **458**, 196–205 (2001)
31. Mortreau, P.: Reinhard Berndt, "determination of 235U enrichment with a large volume CZT detector". NIMA. **556**, 219–227 (2006)
32. Mortreau, P.: R.Berndt "measurement of 235U enrichment with a LaBr3 scintillation detector". NIMA. **620**, 324–331 (2010)
33. Berndt, R., Mortreau, P.: 235U enrichment determination on UF6 cylinders with CZT detectors. NIMA. **886**, 40–47 (2018)
34. Sampson, T.E., Kelley, Vo: Application Guide to Gam-Ray Isotopic Analysis Using the FRAM Software, LA-14018 (2003)
35. Harry, R.J.S., et al.: Gamma-Spectrometric Determination of Isotopic Composition Without Use of Standards. International Atomic Energy Agency (IAEA): IAEA, Vienna (1976)
36. Hagenauer, R.C.: Nondestructive determination of uranium enrichment using low energy X and gamma rays. J. INMM XL:216–220 (1983)
37. Gunnink, R. et al.: MGAU: a new analysis code for measuring U-235 enrichments in arbitrary samples. In: IAEA Symposium, January, 1994, UCRL-JC-114713 (1994)
38. Abousahl, S.: Applicability and limits of the MGAU code for the determination of the enrichment of uranium samples. NIMA. **368**, 443–448 (1996)
39. Morel, J., et al.: Uranium enrichment measurements by X- and g-ray spectrometry with the "URADOS" process. Applied Radiation andIsotopes. **49**, 1251–1257 (1998)
40. Yucel, H., et al.: 235U isotopic characterization of natural and enriched uranium materials by using multigroup analysis (MGA) method at a defined geometry using different absorbers and collimators. Nukleonika. **60**, 615–620 (2015)
41. Korob, R.O., Nuno: A simple method for the absolute determination of uranium enrichment by high-resolution γ spectrometry. Appl. Radiat. Isot. **64**, 525–531 (2006)
42. Borella, A., et al.: Peak shape calibration of a cadmium zinc telluride detector and its application for the determination of uranium enrichment. NIMA. **986**, 164718 (2021)
43. Berlizov, A., et al.: Performance testing of the upgraded uranium isotopics multi-group analysis code MGAU. NIMA. **575**, 498–506 (2007)
44. Morel, J., et al.: Results from the international evaluation exercise for uranium enrichment measurements. Appl. Radiat. Isot. **52**, 509–522 (2000)
45. Darweesh, M., Shawky, S.: Study on the performance of different uranium isotopic codes used in nuclear safeguards. Heliyon. **5**, e01470 (2019)
46. Ruhter, W., Gunnink: Application of cadmium-zinc-telluride detectors in 235U enrichment measurements. NIMA. **353**, 716 (1994)
47. Arlt, et al.: Gamma spectrometric characterization of various CdTe and CdZnTe detectors. NIMA. **428**, 127–137 (1999)
48. Ritec: Spectrometric Detection Probe with Large Volume CdZnTe Detector SDP 500 (S), Operator's Manual (2013)
49. Vaccaro, S., et al.: Advancing the Fork Detector for Quantitative Spent Nuclear Fuel Verification. United States: N. p., 2018. Web. https://doi.org/10.1016/j.nima.2018.01.066
50. Park W. S., et al., 2014. "Safeguards by design at the encapsulation plant in Finland". Proceedings of the 2014 IAEA safeguards symposium
51. International Atomic Energy Agency: Storage of Spent Nuclear Fuel, IAEA Safety Standards Series No. SSG-15, IAEA, Vienna (2012)
52. Phillips, J.R.: Passive Nondestructive Assay of Nuclear Material, Reilly, Ensslin, Smith, and Kreiner, was published in 1991 (Chapter 18, Smith) LA-UR-90-0732 (1991)

53. Žerovnik, G., Schillebeeckx, P., Govers, K., Borella, A., Ćalić, D., Fiorito, L., Kos, B., Stankovskiy, A., Van den Eynde, G., Verwerft, M.: Observables of Interest for the Characterisation of Spent Nuclear Fuel. JRC Technical Reports, EUR 29301 EN (2018)
54. Hsue, S.T., Crane, T.W., Talbert, W.L., Lee, J.C.: Nondestructive Assay Methods for Irradiated Nuclear Fuels. Los Alamos Report, LA-6923 (1978)
55. Favalli, A.: Determining initial enrichment, burnup, and cooling time of pressurized-water-reactor spent fuel assemblies by analyzing passive gamma spectra measured at the Clab interim-fuel storage facility in Sweden. NIMA. **820**, 102–111 (2016)
56. Lebrun, A., et al.: Gamma spectrometric characterization of short cooling time nuclear spent fuels using hemispheric CdZnTe detectors. Nucl. Instrum. Methods Phys. Res. Sect. A. **448**, 598–603 (2000). https://doi.org/10.1016/S0168-9002(00)00296-5
57. Willman, C., et al.: A nondestructive method for discriminating MOX fuel from LEU fuel for safeguards purposes. Ann. Nucl. Energy. **33**, 766 (2006)
58. Matsson, I.: LOKET - a gamma-ray spectroscopy system for in-pool measurements of thermal power distribution in irradiated nuclear fuel. NIMA. **569**, 872 (2006)
59. International Atomic Energy Agency. Safeguards Techniques and Equipment, ISBN 978-92-0-118910-3, 2011 Edition
60. Carrasco, J.A.: International Atomic Energy Agency, Department of Safeguards, Vienna (Austria); 1990 p (1999); [8 p.]; IAEA Symposium on International Safeguards; Vienna (Austria); 13–17 Oct 1997; IAEA-SM--351/149
61. Ahmed, I. et al.: Safeguards Verification of Short Cooling Time KANUPP Irradiated Fuel Bundles Using Room Temperature Semiconductor Detectors. IAEA-SM-367/7/03/P (2001)
62. Mortreau, P., Berndt, R.: Characterisation of cadmium zinc telluride detector spectra – application to the analysis of spent fuel spectra. NIMA. **458**, 183 (2001)
63. Tiitta, A. et al.: Spent BWR Fuel Characterisation Combining a Fork Detector with Gamma Spectrometry. STUK-YTO-TR 175 (2001)
64. Tiitta, A., Hautamäki, J.: Spent VVER fuel characterisation combining a Fork detector with gamma spectrometry, STUK-YTO-TR 181 (2001)
65. Berndt, R., Mortreau, P.: Spent-fuel characterization with small CZT detectors. NIMA. **564**, 290 (2006)
66. Lebrun, A. et al.: SMOPY a new NDA tool for safeguards application on LEU and MOX spent fuel. IAEA-SM-367/14/03, Vienna, Austria, 29 October-2 November 2001 (2001)
67. Gallozzi-Ulmann, A., et al.: Underwater characterization of control rods for waste disposal using SMOPY. In: Conference: Advancements in Nuclear Instrumentation Measurement Methods and their Applications (ANIMMA) at: Lisbon (Portugal), IEEE Xplore Publications (2015)
68. Philips, J.R., Bosler, G.E., Halbig, J.K., Klosterbuer, S.F., Menlove, H.O., Rinard, P.M.: Experience using a spent fuel measurement system. Nucl. Mater. Manage. XII, 175–181 (1983)
69. Kane, S.: Reference Manual, Cask Radiation Profiling System, (CRPS). SG-EQ-CRPS-RM-0001, IAEA (2008)
70. Lévai, F., et al.: Use of High Energy Gamma Emission Tomography for Partial Defect Verification of Spent Fuel Assemblies. STUK-YTO-TR 56 (1993)
71. Lévai, F., et al.: Feasibility of Gamma Emission Tomography for Partial Defect Verification of Spent LWR Fuel Assemblies. STUK-YTO-TR 189 (2002)
72. Jansson, P., et al.: Monte Carlo simulations of a universal gamma-ray emission tomography device. Proceedings of the 37th ESARDA Annual meeting, pp. 937–948, Manchester, UK (2015)
73. Jacobsson Svärd, S.: Tomographic determination of spent fuel assembly pin-wise burnup and cooling time for detection of anomalies. Proceedings of the 37th ESARDA Annual meeting, pp. 961–972, Manchester, UK (2015)
74. Smith, E.: A Viability Study of Gamma Emission Tomography for Spent Fuel Verification: JNT 1955 Phase I Technical Report. PNNL – 25995 (2016)

75. Jacobsson Svärd, S.: Outcomes of the JNT 1955 phase I viability study of gamma emission tomography for spent fuel verification. ESARDA Bull. **55**, 10–28 (2017)
76. Honkama, T.: A prototype for passive gamma emission tomography. In Symposium, IAEA-CN-220, p. 281, 20–24 October 2014 (2014)
77. White, T.: Application of passive gamma emission tomography (PGET) for the verification of spent nuclear fuel. INMM (2016)
78. Mayorov, M.: Gamma emission tomography for the inspection of spent nuclear fuel. IEEE **2017** (2017)
79. White, T.: Performance evaluation framework for the passive gamma emission tomography (PGET) system. In: International Workshop on Numerical Modelling of NDA Instrumentation and Methods for Nuclear Safeguards, 16–17 May, 2018, pp. 9–15 (2018a)
80. Ivanov, V., Aleksejeva, L., Dorogov, P., Luchanskii, A.: CZT quasi-hemispherical detectors with improved spectrometric characteristics. 2009 IEEE Nuclear Science Symposium Conference Record (NSS/MIC), pp. 1696–1699, Orlando, FL (2009), https://doi.org/10.1109/NSSMIC.2009.5402238
81. Berndt, R., et al.: JRC scientific and polity reports, EUR 25672 EN 2012 (2012)
82. White, T.: Verification of spent nuclear fuel using passive gamma emission tomography. IAEA Symposium, 5–8 November 2018 (2018b)
83. Virta, R., Backholm, R., Bubba, T.A., Helin, T, Moring, M., Siltanen, S., Dendooven, P., Honkamaa, T.: Fuel Rod Classification from Passive Gamma Emission Tomography (PGET) of Spent Nuclear Fuel Assemblies. arXiv:2009.11617v1 [physics.ins-det] (61st ESARDA bulletin) (2020)
84. Tobin, S.J., Peura, P., Bélanger-Champagne, C., Moring, M., Dendooven, P., Honkamaa, T.: Utility of including passive neutron albedo reactivity in an integrated NDA system for encapsulation safeguards. ESARDA Bull. **56**, 12–18 (2018)
85. Tupasela, T., et al.: Passive neutron albedo reactivity measurements of spent nuclear fuel. NIMA. **986**, 164707 (2021)

Index

Printed in the United States
by Baker & Taylor Publisher Services